Tharwat F. Tadros
Formulations
De Gruyter Graduate

Also of Interest

Emulsions
Formation, Stability, Industrial Applications
Tadros, 2016
ISBN 978-3-11-045217-4, e-ISBN 978-3-11-045224-2

Nanodispersions
Tadros, 2015
ISBN 978-3-11-029033-2, e-ISBN 978-3-11-029034-9

Interfacial Phenomena and Colloid Stability:
Volume 1 Basic Principles
Tadros, 2015
ISBN 978-3-11-028340-2, e-ISBN 978-3-11-028343-3

Interfacial Phenomena and Colloid Stability:
Volume 2 Industrial Applications
Tadros, 2015
ISBN 978-3-11-037107-9, e-ISBN 978-3-11-036647-1

An Introduction to Surfactants
Tadros, 2014
ISBN 978-3-11-031212-6, e-ISBN 978-3-11-031213-3

Electrospinning
A Practical Guide to Nanofibers
Agarwal, Burgard, Greiner, Wendorff, 2016
ISBN 978-3-11-033180-6, e-ISBN 978-3-11-033351-0

Tharwat F. Tadros

Formulations

In Cosmetic and Personal Care

DE GRUYTER

Author
Prof. Tharwat F. Tadros
89 Nash Grove Lane
Workingham RG40 4HE
Berkshire, UK
tharwat@tadros.fsnet.co.uk

ISBN 978-3-11-045236-5
e-ISBN (PDF) 978-3-11-045238-9
e-ISBN (EPUB) 978-3-11-045240-2

Library of Congress Cataloging-in-Publication Data
A CIP catalog record for this book has been applied for at the Library of Congress.

Bibliographic information published by the Deutsche Nationalbibliothek
The Deutsche Nationalbibliothek lists this publication in the Deutsche Nationalbibliografie; detailed bibliographic data are available on the Internet at http://dnb.dnb.de.

© 2016 Walter de Gruyter GmbH, Berlin/Boston
Cover image: studio22comua/iStock/thinkstock
Typesetting: PTP-Berlin, Protago-T$_E$X-Production GmbH, Berlin
Printing and binding: CPI books GmbH, Leck
♾ Printed on acid-free paper
Printed in Germany

www.degruyter.com

Preface

Several cosmetic formulations can be identified, namely skincare products, e.g. lotions and hand creams, nanoemulsions, multiple emulsions, liposomes, shampoos and hair conditioners, sunscreens and colour cosmetics. The ingredients used must be safe and should not cause any damage to the organs that they come in contact with. Cosmetic and toiletry products are generally designed to deliver a functional benefit and to enhance the psychological well-being of consumers by increasing their aesthetic appeal. In order to have consumer appeal, cosmetic formulations must meet stringent aesthetic standards such as texture, consistency, pleasing colour and fragrance, convenience of application, etc. In most cases this results in complex systems consisting of several components of oil, water, surfactants, colouring agents, fragrants, preservatives, vitamins, etc. The formulation of these complex multiphase systems requires understanding the interfacial phenomena and colloid forces responsible for their preparation, stabilization and application. These disperse systems contain "self-assembly" structures, e.g. micelles (spherical, rod-shaped, lamellar), liquid crystalline phases (hexagonal, cubic or lamellar), liposomes (multilamellar bilayers) or vesicles (single bilayers). They also contain "thickeners" (polymers or particulate dispersions) to control their rheology. In addition, several techniques must be designed to assess their quality, application and assessment of the long-term physical stability of the resulting formulation.

This book will deal with the basic principles of formulating cosmetic and personal care products and their applications. Chapter 1 highlights the complexity of cosmetic formulations and the necessity of using safe ingredients for their production. The various classes of cosmetic products are briefly described. Chapter 2 describes the various surfactant classes that are used in cosmetic and personal care products. A section is devoted to the properties of surfactant solutions and the process of micellization, with definition of the critical micelle concentration (cmc). The ideal and non-ideal mixing of surfactants is analysed to indicate the importance of using surfactant mixtures to reduce skin irritation. The interaction between surfactants and polymers is described at a fundamental level. Chapter 3 deals with the use of polymeric surfactants in cosmetic formulations. It starts with the description of the structure of polymeric surfactants, namely homopolymers, block copolymers and graft copolymers. This is followed by a section on the adsorption and conformation of the polymeric surfactants at the interface. The advantages of using polymeric surfactants in cosmetic formulations are highlighted by giving some practical examples. Chapter 4 describes the self-assembly structures produced by surfactants present in cosmetic formulations, with particular reference to the various liquid crystalline structures, namely hexagonal, cubic and lamellar phases. The driving force responsible for the production of each type is described at a fundamental level. Chapter 5 describes the various interaction forces between particles or droplets in a dispersion.

Three main types are distinguished, namely van der Waals attraction, electrostatic (double layer) repulsion and steric repulsion produced in the presence of adsorbed layers of non-ionic surfactants or polymers. Combining van der Waals attraction with double layer repulsion results in the general theory of colloid stability due to Deryaguin–Landau–Verwey–Overbeek (DLVO theory) which shows the presence of an energy barrier that prevents flocculation. The flocculation of dispersions that are electrostatically stabilized is described in terms of the reduction of the energy barrier by addition of electrolytes. Combining van der Waals attraction with steric repulsion forms the basis of the theory of steric stabilization. The factors responsible for effective steric stabilization are described. This is followed by sections on flocculation of sterically stabilized dispersions. Four types of flocculation can be distinguished: weak (reversible) flocculation in the presence of a shallow minimum in the energy distance curve; incipient flocculation produced when the solvency of the stabilizing chain is reduced; depletion flocculation caused by the presence of "free" (non-adsorbing) polymer; and bridging flocculation whereby the polymer chain becomes attached to two or more particles or droplets. Chapter 6 describes the formulation of cosmetic emulsions. The benefits of using cosmetic emulsions for skincare are highlighted. The various methods that can be applied for selection of emulsifiers for the formulation of oil/water and water/oil emulsions are described. The control of the emulsification process for producing the optimum droplet size distribution is described. The various methods that can be used for preparation of emulsions are described. The use of rheology modifiers (thickeners) for control of the physical stability and the consistency of the system is also described. Three different rheological techniques are applied, namely steady state (shear stress-shear rate measurements), dynamic (oscillatory techniques) and constant stress (creep) measurements. Chapter 7 describes the formulation of nanoemulsions in cosmetics. It starts with a section highlighting the main advantages of nanoemulsions. The origin of stability of nanoemulsions in terms of steric stabilization and the high ratio of adsorbed layer thickness to droplet radius is described at a fundamental level. A section is devoted to the problem of Ostwald ripening in nanoemulsions and how the rate can be measured. The reduction of Oswald ripening by incorporation of a small amount of highly insoluble oil and/or modification of the interfacial film is described. Chapter 8 deals with the formulation of multiple emulsions in cosmetics. Two types are described, namely Water/Oil/Water (W/O/W) and Oil/Water/Oil (O/W/O) multiple emulsions. The formulation of multiple emulsions using a two-stage process is described. The various possible breakdown processes in multiple emulsions are described and the factors affecting the long-term stability of the formulation are analysed. The characterization of multiple emulsions using optical microscopy and rheological techniques is described. Chapter 9 describes liposomes and vesicles in cosmetic formulations. The procedures for preparation of liposomes and vesicles are described together with the methods that can be applied for assessment of their stability. The enhancement of liposome stability by incorporation of block copolymers is described. Chapter 10 deals with the formulation of shampoos.

The different components in a shampoo formulation are described together with the necessity of addition of an amphoteric surfactant to the anionic surfactant in a shampoo to reduce skin irritation. Surfactants added to enhance the foaming characteristics of the shampoo are described. This is followed by a section on the mechanism of dirt and oil removal by the shampoo. The enhancement of the viscosity of the shampoo formulation by addition of electrolytes that produce rod-shaped micelles is described. Chapter 11 deals with the formulation of hair conditioners in the shampoo. The surface properties of hair and the role of adding cationically modified polymer to neutralize the negative charge on hair are described. The importance of hair conditioners in management of hair during combing is analysed. Chapter 12 describes the formulation of sunscreens for UV protection. The importance of control against UVA (wavelength 320–400 nm) and UVB (wavelength 290–320 nm) radiation is emphasized. The use of organic and semiconductor (titania) sunscreen agents in the formulation enables one to protect against UVA and UVB radiation. Chapter 13 describes the formulation of coloured cosmetic products. Several colour pigments are used in cosmetic formulations ranging from inorganic pigments (such as red iron oxide) to organic pigments of various types. The formulation of these pigments in colour cosmetics requires a great deal of skill since the pigment particles are dispersed in an emulsion (oil-in-water or water-in-oil). A section describing the fundamental principles of preparation of pigment dispersion is given. Chapter 14 gives some examples of industrial cosmetic and personal care formulations: (i) shaving formulations; (ii) bar soaps; (iii) liquid hand soaps; (iv) bath oils; (v) bubble baths; (vi) after-bath formulations; (vii) skincare products; (viii) haircare formulations; (ix) sunscreens; (x) make-up products.

This book gives a comprehensive overview of the various applications of colloid and interface science principles in cosmetic and personal care formulations. It provides the reader with a systematic approach to the formulation of various cosmetic and personal care products. It also provides the reader with an understanding of the complex interactions in various cosmetic disperse systems. The book will be valuable to research workers engaged in formulation of cosmetic and personal care products. It will also provide the industrial chemist with a text that can enable him/her to formulate the product using a more rational approach. Therefore, this book is valuable for chemists and chemical engineers both in academic and industrial institutions.

August 2016   Tharwat Tadros

Contents

Preface —— V

1	General introduction —— 1	
2	Surfactants used in cosmetic and personal care formulations, their properties and surfactant–polymer interaction —— 11	
2.1	Surfactant classes —— 11	
2.1.1	Anionic surfactants —— 11	
2.1.2	Cationic surfactants —— 14	
2.1.3	Amphoteric (zwitterionic) surfactants —— 16	
2.1.4	Nonionic surfactants —— 17	
2.1.5	Surfactants derived from mono- and polysaccharides —— 20	
2.1.6	Naturally occurring surfactants —— 21	
2.1.7	Polymeric (macromolecular) surfactants —— 21	
2.1.8	Silicone surfactants —— 22	
2.2	Physical properties of surfactant solutions and the process of micellization —— 23	
2.2.1	Thermodynamics of micellization —— 29	
2.3	Micellization in surfactant mixtures (mixed micelles) —— 35	
2.4	Surfactant–polymer interaction —— 39	
2.4.1	Factors influencing the association between surfactant and polymer —— 41	
2.4.2	Interaction models —— 42	
2.4.3	Driving force for surfactant/polymer interaction —— 44	
2.4.4	Structure of surfactant/polymer complexes —— 45	
2.4.5	Surfactant/hydrophobically modified polymer interaction —— 46	
2.4.6	Interaction between surfactants and polymers with opposite charge (surfactant/polyelectrolyte interaction) —— 46	
3	Polymeric surfactants in cosmetic formulations —— 51	
3.1	Introduction —— 51	
3.2	General classification of polymeric surfactants —— 51	
3.3	Polymeric surfactant adsorption and conformation —— 53	
3.3.1	Measurement of the adsorption isotherm —— 57	
3.3.2	Measurement of the fraction of segments p —— 58	
3.3.3	Determination of the segment density distribution $\rho(z)$ and adsorbed layer thickness δ_h —— 58	

3.4	Examples of the adsorption results of nonionic polymeric surfactant —— 60	
3.4.1	Adsorption isotherms —— 60	
3.4.2	Adsorbed layer thickness results —— 64	
3.5	Kinetics of polymer adsorption —— 66	
3.6	Emulsions stabilized by polymeric surfactants —— 67	
4	**Self-assembly structures in cosmetic formulations —— 73**	
4.1	Introduction —— 73	
4.2	Self-assembly structures —— 73	
4.3	Structure of liquid crystalline phases —— 74	
4.3.1	Hexagonal phase —— 74	
4.3.2	Micellar cubic phase —— 74	
4.3.3	Lamellar phase —— 75	
4.3.4	Discontinuous cubic phases —— 76	
4.3.5	Reversed structures —— 76	
4.4	Driving force for liquid crystalline phase formation —— 76	
4.5	Identification of the liquid crystalline phases and investigation of their structure —— 79	
4.6	Formulation of liquid crystalline phases —— 80	
4.6.1	Oleosomes —— 81	
4.6.2	Hydrosomes —— 81	
5	**Interaction forces between particles or droplets in cosmetic formulations and their combination —— 83**	
5.1	Van der Waals attraction —— 83	
5.2	Electrostatic repulsion —— 85	
5.3	Flocculation of electrostatically stabilized dispersions —— 89	
5.4	Criteria for stabilization of dispersions with double layer interaction —— 91	
5.5	Steric repulsion —— 92	
5.5.1	Mixing interaction G_{mix} —— 93	
5.5.2	Elastic interaction G_{el} —— 95	
5.5.3	Total energy of interaction —— 95	
5.5.4	Criteria for effective steric stabilization —— 96	
5.5.5	Flocculation of sterically stabilized dispersions —— 97	
6	**Formulation of cosmetic emulsions —— 105**	
6.1	Introduction —— 105	
6.2	Thermodynamics of emulsion formation —— 105	
6.3	Emulsion breakdown processes and their prevention —— 107	
6.3.1	Creaming and sedimentation —— 108	

6.3.2	Flocculation —— 109	
6.3.3	Ostwald ripening (disproportionation) —— 110	
6.3.4	Coalescence —— 111	
6.3.5	Phase Inversion —— 111	
6.4	Selection of emulsifiers —— 111	
6.4.1	The Hydrophilic-Lipophilic Balance (HLB) concept —— 111	
6.4.2	The Phase Inversion Temperature (PIT) concept —— 117	
6.4.3	The Cohesive Energy Ratio (CER) concept —— 120	
6.4.4	The Critical Packing Parameter (CPP) for emulsion selection —— 122	
6.5	Manufacture of cosmetic emulsions —— 124	
6.5.1	Mechanism of emulsification —— 125	
6.5.2	Methods of emulsification —— 130	
6.6	Rheological properties of cosmetic emulsions —— 139	
7	**Formulation of nanoemulsions in cosmetics —— 147**	
7.1	Introduction —— 147	
7.2	Preparation of nanoemulsion by the use of high pressure homogenizers —— 149	
7.3	Low-energy methods for preparation of nanoemulsions —— 158	
7.3.1	Phase Inversion Composition (PIC) principle —— 159	
7.3.2	Phase Inversion Temperature (PIT) principle —— 160	
7.3.3	Preparation of nanoemulsions by dilution of microemulsions —— 162	
7.4	Practical examples of nanoemulsions —— 163	
7.5	Nanoemulsions based on polymeric surfactants —— 172	
8	**Formulation of multiple emulsions in cosmetics —— 179**	
8.1	Introduction —— 179	
8.2	Types of multiple emulsions —— 180	
8.3	Breakdown processes of multiple emulsions —— 180	
8.4	Preparation of multiple emulsions —— 181	
8.5	Characterization of multiple emulsions —— 186	
8.5.1	Droplet size analysis —— 186	
8.5.2	Dialysis —— 186	
8.5.3	Rheological techniques —— 186	
8.6	Summary of the factors affecting stability of multiple emulsions and criteria for their stabilization —— 190	
9	**Liposomes and vesicles in cosmetic formulations —— 193**	
9.1	Introduction —— 193	
9.2	Nomenclature of liposomes and their classification —— 194	
9.3	Driving force for formation of vesicles —— 195	

10	**Formulation of shampoos** —— 201	
10.1	Introduction —— 201	
10.2	Surfactants for use in shampoo formulations —— 201	
10.2.1	Anionic surfactants —— 201	
10.2.2	Amphoteric surfactants —— 202	
10.2.3	Nonionic surfactants —— 203	
10.3	Properties of a shampoo —— 203	
10.4	Components of a shampoo —— 204	
10.4.1	Cleansing agents —— 204	
10.4.2	Foam boosters —— 205	
10.4.3	Thickening agents —— 205	
10.4.4	Preservatives —— 206	
10.4.5	Miscellaneous additives —— 206	
10.5	Role of the components —— 206	
10.5.1	Behaviour of mixed surfactant systems —— 206	
10.5.2	Cleansing function —— 207	
10.5.3	Foam boosters —— 208	
10.5.4	Thickeners and rheology modifiers —— 209	
10.5.5	Silicone oil emulsions in shampoos —— 211	
10.6	Use of associative thickeners as rheology modifiers in shampoos —— 211	
11	**Formulation of hair conditioners in shampoos** —— 217	
11.1	Introduction —— 217	
11.2	Morphology of hair —— 217	
11.3	Surface properties of hair —— 220	
11.3.1	Wettability investigations —— 220	
11.3.2	Electrokinetic studies —— 223	
11.4	Role of surfactants and polymers in hair conditioners —— 224	
12	**Formulation of sunscreens for UV protection** —— 231	
12.1	Introduction —— 231	
12.2	Mechanism of absorbance and scattering by TiO_2 and ZnO —— 232	
12.3	Preparation of well-dispersed particles —— 233	
12.4	Experimental results for sterically stabilized TiO_2 dispersions in nonaqueous media —— 237	
12.5	Competitive interactions in sunscreen formulations —— 245	
13	**Formulation of colour cosmetics** —— 249	
13.1	Introduction —— 249	
13.2	Fundamental principles for preparation of a stable colour cosmetic dispersion —— 250	

13.2.1	Powder wetting —— 250	
13.2.2	Powder dispersion and milling (comminution) —— 254	
13.2.3	Stabilization of the dispersion against aggregation —— 255	
13.3	Classes of dispersing agents —— 260	
13.4	Assessment of dispersants —— 262	
13.4.1	Adsorption isotherms —— 262	
13.4.2	Measurement of dispersion and particle size distribution —— 263	
13.4.3	Rheological measurements —— 263	
13.5	Application of the above fundamental principles to colour cosmetics —— 265	
13.6	Principles of preparation of colour cosmetics —— 267	
13.7	Competitive interactions in colour cosmetic formulations —— 269	
14	**Industrial examples of cosmetic and personal care formulations —— 271**	
14.1	Shaving formulations —— 271	
14.2	Bar soaps —— 273	
14.3	Liquid hand soaps —— 274	
14.4	Bath oils —— 275	
14.5	Foam (or bubble) baths —— 275	
14.6	After bath preparations —— 275	
14.7	Skincare products —— 276	
14.8	Haircare formulations —— 277	
14.9	Sunscreens —— 280	
14.10	Make-up products —— 282	

Index —— 287

1 General introduction

Several cosmetic formulations can be identified: Lotions, hand creams (cosmetic emulsions), nanoemulsions, multiple emulsions, liposomes, shampoos and hair conditioners, sunscreens and colour cosmetics. The formulation of these complex multiphase systems requires understanding the interfacial phenomena and colloid forces responsible for their preparation, stabilization and application. The ingredients used must be safe and should not cause any damage to the organs that they come in contact with. The fundamental principles of interface and colloid science that are responsible for the formulation of cosmetic formulations must be considered.

Cosmetic and toiletry products are generally designed to deliver a functional benefit and to enhance the psychological well-being of consumers by increasing their aesthetic appeal. Thus, many cosmetic formulations are used to clean hair, skin, etc. and impart a pleasant odour, make the skin feel smooth and provide moisturizing agents, provide protection against sunburn etc. In many cases, cosmetic formulations are designed to provide a protective, occlusive surface layer, which either prevents the penetration of unwanted foreign matter or moderates the loss of water from the skin [1–3]. In order to have consumer appeal, cosmetic formulations must meet stringent aesthetic standards such as texture, consistency, pleasing colour and fragrance, convenience of application, etc. In most cases this results in complex systems consisting of several components of oil, water, surfactants, colouring agents, fragrants, preservatives, vitamins, etc. In recent years, there has been considerable effort in introducing novel cosmetic formulations that provide great beneficial effects to the customer, such as sunscreens, liposomes and other ingredients that may keep skin healthy and provide protection against drying, irritation, etc. All these systems require the application of several interfacial phenomena such as charge separation and formation of electrical double layers, the adsorption and conformation of surfactants and polymers at the various interfaces involved and the main factors that affect the physical stability/instability of these systems. In addition several techniques must be designed to assess their quality, application and prediction of the long-term physical stability of the resulting formulation.

Since cosmetic products come in thorough contact with various organs and tissues of the human body, a most important consideration for choosing ingredients to be used in these formulations is their medical safety. Many of the cosmetic preparations are left on the skin after application for indefinite periods of time and, therefore, the ingredients used must not cause any allergy, sensitization or irritation. The ingredients used must be free of any impurities that have toxic effects.

One of the main areas of interest in cosmetic formulations is their interaction with the skin [3]. A cross section through the skin is shown in Fig. 1.1 [4].

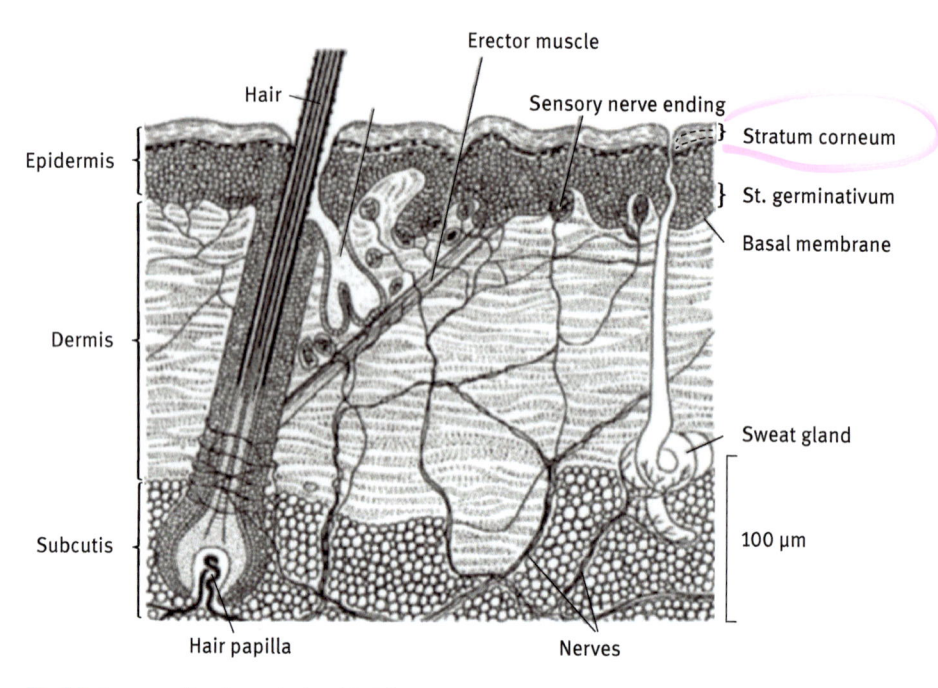

Fig. 1.1: Cross section through the skin [4].

The top layer of the skin, which is the man barrier to water loss, is the stratum corneum which protects the body from chemical and biological attack [5]. This layer is very thin, approximately 30 µm, and it consists of ≈ 10 % by weight of lipids that are organized in bilayer structures (liquid crystalline) which at high water content is soft and transparent. A schematic representation of the layered structure of the stratum corneum, suggested by Elias et al. [6], is given in Fig. 1.2. In this picture, ceramides were considered as the structure-forming elements, but later work by Friberg and Osborne [7] showed the fatty acids to be the essential compounds for the layered structure and that a considerable part of the lipids are located in the space between the methyl groups. When a cosmetic formulation is applied to the skin, it will interact with the stratum corneum and it is essential to maintain the "liquid-like" nature of the bilayers and pre-

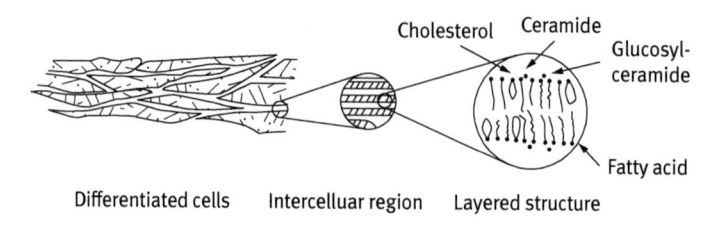

Fig. 1.2: Schematic representation of the stratum corneum structure.

vent any crystallization of the lipids. This happens when the water content is reduced below a certain level. This crystallization has a drastic effect on the appearance and smoothness of the skin ("dry" skin feeling).

To achieve the above criteria "complex" multiphase systems are formulated [8, 9]: (i) Oil-in-Water (O/W) emulsions; (ii) Water-in-Oil (W/O) emulsions; (iii) Solid/Liquid dispersions (suspensions); (iv) Emulsions-Suspension mixtures (suspoemulsions); (v) Nanoemulsions; (vi) Nanosuspensions; (vii) Multiple emulsions. As mentioned above, all these disperse systems require fundamental understanding of the interfacial phenomena involved such as the adsorption and conformation of the various surfactants and polymers used for their preparation. This will determine the physical stability/instability of these systems, their application and shelf life.

All the above disperse systems contain "self-assembly" structures: (i) Micelles (spherical, rod-shaped, lamellar); (ii) Liquid crystalline phases (hexagonal, cubic or lamellar); (iii) Liposomes (multilamellar bilayers) or vesicles (single bilayers). They also contain "thickeners" (polymers or particulate dispersions) to control their rheology. All these self-assembly systems involve an interface whose property determines the structures produced and their properties.

The above complex multiphase systems require fundamental understanding of the colloidal interactions between the various components. Understanding these interactions enables the formulation scientist to arrive at the optimum composition for a particular application. One of the most important aspects is to consider the property of the interface, in particular the interactions between the surfactants and/or polymers that are used for formulating the product and the interface in question. In most cases such mixtures produce synergy for the interfacial region which is essential for ease of preparation of the disperse system. The fundamental principles involved also help in predicting the long-term physical stability of the formulations.

A summary of some of the most commonly used formulations in cosmetics is given below:

(i) Lotions (moisturizing emulsions): These can be oil-in-water (O/W) or water-in-oil (W/O) emulsions (cold cream, emollient cream, day cream, night cream, vanishing cream, etc.). Moisturizing lotions with high water content and body milks are also used. Their bases may contain the components listed in Tab. 1.1.

Lotions are formulated in such a way (see the chapter on cosmetic emulsions) to give a shear thinning system. The emulsion will have a high viscosity at low shear rates ($0.1\,s^{-1}$) in the region of few hundred Pas, but the viscosity decreases very rapidly with increase in shear rate reaching values of few Pas at shear rates greater than $1\,s^{-1}$. These lotions are mostly more viscous than elastic and this provides a convenient system for ease of application.

(ii) Hand creams: These are formulated as O/W or W/O emulsions with special surfactant systems and/or thickeners to give a viscosity profile similar to that of lotions, but

Tab. 1.1: Base for moisturizing emulsions.

Ingredient	%
Water	60–80
Various oil components, consistency regulators and fats	20–40
Emulsifiers	2–5
Humectants	0–5
Preservatives	As required
Fragrances	As required

with orders of magnitude greater viscosities. The viscosity at low shear rates ($< 0.1\,s^{-1}$) can reach thousands of Pas and they retain a relatively high viscosity at high shear rates (of the order of few hundred Pas at a shear rate $> 1\,s^{-1}$). These systems are sometimes described as having a "body", mostly in the form of a gel-network structure that may be achieved by the use of surfactant mixtures to form liquid crystalline structures. In some case, thickeners (hydrocolloids) are added to enhance the gel network structure. In general, hand creams are more elastic than viscous and they are beneficial to form an occlusive layer on the skin thus preventing loss of water from the stratum corneum.

Both hand creams and lotions are required to serve many functions: (i) primary functions as moisturizing and blocking harmful UV radiation; (ii) secondary functions such as smoothness, pleasant smell and appearance; (iii) physical, chemical and microbiological stability; (iv) non-sensitizing and non-irritating. Formulating creams as O/W emulsions promotes easy spreading and absorption, while allowing both water and oil-soluble components to be contained in a single product. The aqueous phase (typically 60–80 vol.%) contains several components such as humectants (e.g. glycerol, sorbitol) to prevent water loss, cosolvents to solubilize fragrances or preservatives, water soluble surfactants to stabilize the emulsion, rheology modifiers such as gums, hydroxyethyl cellulose, xanthan gum and proteins, water soluble vitamins and minerals.

Lotions and creams usually contain emollients, oils that provide a smooth lubricating effect when applied to the skin. The oleic phase (typically 20–40 vol.% of the overall emulsion) contains oils and waxes (e.g. silicone oil, mineral oil, petrolatum or lipids such as triacylglycerols or wax esters), dyes and perfumes, oil soluble surfactants to stabilize the emulsion. They may also be formulated with ingredients designed to penetrate the outer layer of the skin (the stratum corneum) such as liposomes which form lamellar liquid crystalline structures on the surface of the skin, thus preventing skin irritation.

Tab. 1.2 shows a typical composition of O/W night cream, whereas Tab. 1.3 shows a W/O baby cream.

Tab. 1.2: O/W night cream.

Ingredient	Concentration (%)	Function
Orange roughy oil	8	Conditioner
Dimethicone silicon oil (200–300 cS)	1	Emollient
Cetylacetate/acetylated lanolin alcohol	1	Emollient
Meristly myristate	3	Emollient
PEG-24 stearate	3	Emulsifier
Cetearyl alcohol	2	Emulsifier
Glyceryl stearate	7	Emulsifier
Propylene glycol	3	Moisturizer
Soluble collagen in water (0.3 %)	11	Moisturizer
Water	to 100	

Tab. 1.3: W/O baby cream.

Ingredient	Concentration (%)	Function
Lanolin alcohols	2	Emulsifier
Lanolin	4.5	Emollient/moisturizer
Mineral oil (70 cS)	17	Emollient
White petroleum jelly	13.3	Emollient
Butylated hydroxytoluene (BHT)	0.01	Antioxidant
Glycerine	5	Emollient
Water	to 100	
Zinc oxide	7	Sunscreen

Stability is important for cosmetic skin products (lotions and hand creams) from the points of view of function and also shelf life. This will be discussed in detail in the chapter on cosmetic emulsions. The rheological properties of cosmetic lotions and creams are an important aspect of both product appearance and consumer acceptance. This subject will be dealt with in detail in the chapter on cosmetic emulsions.

(iii) Lipsticks and lip balms: These are suspensions of solid oils in a liquid oil or a mixture of liquid oils. They contain a variety of waxes (such as beeswax, carnauba wax, etc.) which give the lipstick its shape and ease of application. Solid oils such as lanolin, palm oil, butter are also incorporated to give the lipstick its tough, shiny film when it dries after application. Liquid oils such as castor oil, olive oil, sunflower oil provide the continuous phase to ensure ease of application. Other ingredients such as moisturizers, vitamin E, collagen, amino acids and sunscreens are sometimes added to help keep lips soft, moist and protected from UV. The pigments give the lipstick its colour, e.g. soluble dyes such as D&C Red No. 21, and insoluble dyes (lakes) such as D&C Red No. 34. Pink shades are made by mixing titanium dioxide with various red dyes. Surfactants are also used in the formulation of lipsticks. The product should show good thermal stability during storage and rheologically it behaves as a viscoelastic solid. In other words, the lipstick should show small deformation at low stresses and this deformation should recover on removal of the stress. Such information could be obtained using creep measurements [10].

(iv) Nail polish: These are pigment suspensions in a volatile non-aqueous solvent. The system should be thixotropic (showing decrease of viscosity with time at a given shear rate and its recovery on removal of the shear). On application by the brush it should show proper flow for even coating but should have enough viscosity to avoid "dripping". After application, "gelling" should occur in a controlled time scale. If "gelling" is too fast, the coating may leave "brush marks" (uneven coating). If gelling is too slow, the nail polish may drip. The relaxation time of the thixotropic system should be accurately controlled to ensure good levelling and this requires the use of surfactants.

(v) Shampoos: Formulating a shampoo generally meet the following criteria: mild detergency, good foaming, conditioning, adequately preserving and aesthetically appealing. Synthetic surfactants such as ether sulphates are commonly used in shampoos. By addition of electrolyte they produce "gelled" surfactant solution of well-defined associated structures, e.g. rod-shaped micelles. A thickener such as a polysaccharide may be added to increase the relaxation time of the system. In addition, some surfactants such as amine oxides are added to enhance foaming of the shampoo on application. The interaction between the surfactants and polymers at the interface is of great importance in arriving at the right formulation. To reduce skin and eye irritation, these anionic surfactants are mixed with amphoteric or non-ionic surfactants which produce non-ideal mixing (see chapter on surfactants in cosmetics) thus reducing the critical micelle concentration (cmc) and hence the monomer concentration in the shampoo. In many formulations, silicone surfactants are added to function as emulsifiers (for silicone oils) but also to improve feel, gloss, sheen, emolliency, conditioning and foam stabilization.

(vi) Antiperspirants and deodorants: Antiperspirants (commonly used in the US) act both to inhibit sweating and to deodorize. In contrast, deodorants (commonly used in Europe) only inhibit odour. Human skin is almost odourless, but when decomposed by bacteria on the skin, an unpleasant odour develops. There are several possible methods of combating this smell; masking with perfume oils, oxidation of the odoriferous compounds with peroxides, adsorption by finely dispersed ion-exchange resins, inhibition of the skin's bacterial flora (the basis of most deodorants), or the action of surfactants, especially appropriate ammonium compounds. Antiperspirants contain astringent substances that precipitate proteins irreversibly and these prohibit perspiration. The general composition of antiperspirant or deodorant is 60–80 % water, 5 % polyol, 5–15 % lipid (stearic acid, mineral oil, beeswax), 2–5 % emulsifiers (polysorbate 40, sorbitan oleate), antiperspirant (aluminium chlorohydrate), 0.1 % antimicrobial, 0.5 % perfume oil. These antiperspirants are thus suspensions of solid actives in a surfactant vehicle. Other ingredients such as polymers that provide good skin feel are added. The rheology of the system should be controlled to avoid particle sedimentation. This is achieved by addition of thickeners. Shear thinning of the final product is essential to ensure good spreadability. In stick application, a "semi-solid" system is produced.

(vii) Foundations: These are complex systems consisting of a suspension-emulsion system (sometimes referred to as suspoemulsions). Pigment particles are usually dispersed in the continuous phase of an O/W or W/O emulsion. Volatile oils such as cyclomethicone are usually used. The system should be thixotropic to ensure uniformity of the film and good levelling.

(viii) Aerosol products: A number of personal care products are produced as aerosols, which contain gas mixed with a liquid under very high pressure. These include cosmetic foams like hair-styling mousse, shaving foam and even shampoos. Originally these tended to use chlorofluorocarbons (CFCs) as the pressurized propellant phase, but due to regulations limiting the use of volatile organic compounds (VOCs) the propellant has been replaced by propane-butane blend, and now volatile methylsiloxanes are being substituted for hydrocarbon-based solvents. These products are formulated as emulsions in the pressurized containers and apart from the emulsifier used to stabilize the emulsion, other surfactants are used to stabilize the foam that is produced during the use of the aerosol product.

This book will deal with the basic principles of formulating cosmetic and personal care products. Chapter 2 will describe the various surfactant classes that are used in cosmetic and personal care products. A section is devoted to the properties of surfactant solutions and the process of micellization, with definition of the critical micelle concentration (cmc). The dependence of the cmc on the alkyl chain length of the surfactant molecule and the nature of the head group is also described. The thermodynamics of micelle formation, both kinetic and equilibrium aspects, is described with a section on the driving force for micelle formation. The ideal and non-ideal mixing of surfactants is analysed at a fundamental level. The interaction between surfactants and polymers is described at a fundamental level.

    Chapter 3 deals with the use of polymeric surfactants in cosmetic formulations. It starts with the description of the structure of polymeric surfactants, namely homopolymers, block copolymers and graft copolymers. Examples of the various polymeric surfactants that are used in cosmetic formulations are given. This is followed by a section on the adsorption and conformation of the polymeric surfactants at the interface. Examples are given for the adsorption of polymers on model particles, namely polystyrene latex. The advantages of using polymeric surfactants in cosmetic formulations are highlighted by giving some practical examples. Chapter 4 describes the self-assembly structures produced by surfactants present in cosmetic formulations, with particular reference to the various liquid crystalline structures, namely hexagonal, cubic and lamellar phases. The driving force responsible for the production of each type is described at a fundamental level.

Chapter 5 describes the various interaction forces between particles or droplets in a dispersion. Three main types are distinguished, namely van der Waals attraction, electrostatic (double layer) repulsion and steric repulsion produced in the presence of adsorbed layers of non-ionic surfactants or polymers. Combining van der Waals attraction with double layer repulsion results in the general theory of colloid stability due to Deryaguin–Landau–Verwey–Overbeek (DLVO theory) [11, 12] which shows the presence of an energy barrier that prevents flocculation. The factors that affect the height of the energy barrier, namely surface or zeta potential and electrolyte concentration and valency are analysed in terms of the DLVO theory. The flocculation of dispersions that are electrostatically stabilized is described in terms of the reduction of the energy barrier by addition of electrolytes. A distinction can be made between fast flocculation (in the absence of an energy barrier) and slow flocculation (in the presence of an energy barrier). This allows one to define the stability ratio W which is the ratio between the rate of fast flocculation to that of slow flocculation. A plot of log W versus electrolyte concentration C allows one to define the critical coagulation concentration (CCC) and its dependence on electrolyte valency. Combining van der Waals attraction with steric repulsion forms the basis of the theory of steric stabilization [13]. The factors responsible for effective steric stabilization are described. This is followed by sections on flocculation of sterically stabilized dispersions. Four types of flocculation can be distinguished: weak (reversible) flocculation in the presence of a shallow minimum in the energy distance curve; incipient flocculation produced when the solvency of the stabilizing chain is reduced; depletion flocculation caused by the presence of "free" (non-adsorbing) polymer; and bridging flocculation whereby the polymer chain becomes attached to two or more particles or droplets.

Chapter 6 describes the formulation of cosmetic emulsions. The benefits of using cosmetic emulsions for skincare are highlighted. The main factors that need to be controlled in formulation of cosmetic emulsions are described at a fundamental level. The various methods that can be applied for selection of emulsifiers for the formulation of oil/water and water/oil emulsions are described. These include the hydrophilic-lipophilic balance (HLB), the phase inversion temperature (PIT), the cohesive energy ratio (CER) and the critical packing parameter (CPP) methods. The control of the emulsification process for producing the optimum droplet size distribution is described. The various methods that can be used for preparation of emulsions are described. These include the use of high speed mixers (rotor-stator mixers), high pressure homogenization and membrane emulsification. The use of rheology modifiers (thickeners) for control of the physical stability and the consistency of the system is also described. Three different rheological techniques are applied, namely steady state (shear stress-shear rate measurements), dynamic (oscillatory techniques) and constant stress (creep) measurements.

Chapter 7 describes the formulation of nanoemulsions in cosmetics. It starts with a section highlighting the main advantages of nanoemulsions: transparency, lack of creaming, flocculation and coalescence as well as enhancement of deposition on the rough texture of the skin and increased skin penetration for actives in the formulation. The origin of stability of nanoemulsions in terms of steric stabilization and the high ratio of adsorbed layer thickness to droplet radius is described at a fundamental level. A section is devoted for the problem of Ostwald ripening in nanoemulsions and how the rate can be measured. The reduction of Oswald ripening by incorporation of a small amount of highly insoluble oil and/or modification of the interfacial film is described. Practical examples of nanoemulsion systems are given highlighting the most important variables that affect the stability of nanoemulsions.

Chapter 8 deals with the formulation of multiple emulsions in cosmetics. Two types are described, namely Water/Oil/Water (W/O/W) and Oil/Water/Oil (O/W/O) multiple emulsions. The formulation of multiple emulsions using a two-stage process is described. The various possible breakdown processes in multiple emulsions are described and the factors affecting the long-term stability of the formulation are analysed. The characterization of multiple emulsions using optical microscopy and rheology techniques is described.

Chapter 9 describes liposomes and vesicles in cosmetic formulations. The procedures for preparation of liposomes and vesicles are described together with the methods that can be applied for assessment of their stability. The enhancement of liposome stability by incorporation of block copolymers is described. Finally the main advantages of incorporation of liposomes and vesicles in skincare formulations are briefly discussed.

Chapter 10 deals with the formulation of shampoos. The different components in a shampoo formulation are described together with the necessity of adding an amphoteric surfactant to the anionic surfactant in a shampoo to reduce skin irritation. Surfactants added to enhance the foaming characteristics of the shampoo are described. This is followed by a section on the mechanism of dirt and oil removal by the shampoo. The enhancement of the viscosity of the shampoo formulation by addition of electrolytes that produce rod-shaped micelles is described at a fundamental level.

Chapter 11 deals with the formulation of hair conditioners in the shampoo. The surface properties of hair and role of adding cationically modified polymer to neutralize the negative charge on hair are described. The importance of hair conditioners in management of hair during combing is analysed.

Chapter 12 describes the formulation of sunscreens for UV protection. The importance of control against UVA (wavelength 320–400 nm) and UVB (wavelength 290–320 nm) radiation is emphasized. The use of organic and semiconductor (titania) sunscreen agents in the formulation enables one to protect against UVA and UVB radiation. The importance of particle size reduction of the titania particles is described at a fundamental level. The stabilization of titania dispersions against flocculation in a sunscreen formulation is essential as well as prevention of emulsion coalescence.

Chapter 13 describes the formulation of coloured cosmetic products. Several colour pigments are used in cosmetic formulations ranging from inorganic pigments (such as red iron oxide) to organic pigments of various types. The formulation of these pigments in colour cosmetics requires a great deal of skill since the pigment particles are dispersed in an emulsion (oil-in-water or water-in-oil). A section describing the fundamental principles of preparation of pigment dispersion is given. These consist of three main topics, namely wetting of the powder, its dispersion, wet milling (comminution) and stabilization against aggregation.

Chapter 14 gives some examples of industrial cosmetic and personal care formulations: (i) shaving formulations; (ii) bar soaps; (iii) liquid hand soaps; (iv) bath oils; (v) bubble baths; (vi) after bath formulations; (vii) skincare products; (viii) haircare formulations; (ix) sunscreens; (x) make-up products.

References

[1] Breuer, M. M., in "Encyclopedia of Emulsion Technology", P. Becher (ed.), Marcel Dekker, NY (1985), Vol. 2, Chapter 7.
[2] Harry, S., in "Cosmeticology", J. B. Wilkinson and R. J. Moore (eds.), Chemical Publishing, NY (1981).
[3] Friberg, S. E., J. Soc. Cosmet. Chem., **41**, 155 (1990).
[4] Czihak, G., Langer, H., and Ziegler, H., "Biologie", Springer-Verlag, Berlin, Heidelberg, New York (1981).
[5] Kligman, A. M., in "Biology of the Stratum Corneum in Epidermis", W. Montagna (ed.), Academic Press, NY, pp. 421–46 (1964).
[6] Elias, P. M., Brown, B. E., Fritsch, P. T., Gorke, R. J., Goay, G. M., and White, R. J., J. Invest. Dermatol., **73**, 339 (1979).
[7] Friberg S. E. and Osborne, D. W. J. Disp. Sci. Technol., **6**, 485 (1985).
[8] Tadros, Th. F., "Applied Surfactants", Wiley-VCH, Germany (2005).
[9] Tadros, Th. F., "Formulation of Disperse Systems", Wiley-VCH, Germany (2014).
[10] Tadros, Th. F., "Rheology of Dispersions", Wiley-VCH, Germany (2010).
[11] Deryaguin, B. V. and Landau, L. Acta Physicochem. USSR, 14, 633 (1941).
[12] Verwey, E. J. W. and Overbeek, J. Th. G., "Theory of Stability of Lyophobic Colloids", Elsevier, Amsterdam (1948).
[13] Napper, D. H., "Polymeric Stabilisation of Colloidal Dispersions", Academic Press, London (1983).

2 Surfactants used in cosmetic and personal care formulations, their properties and surfactant–polymer interaction

2.1 Surfactant classes

As mentioned in Chapter 1, surfactants used in cosmetic formulations must be completely free of allergens, sensitizers and irritants. To minimize medical risks, cosmetic formulators tend to use polymeric surfactants which are less likely to penetrate beyond the stratum corneum and hence they are less likely to cause any damage.

Conventional surfactants of the anionic, cationic, amphoteric and nonionic types are used in cosmetic systems [1–3]. Besides the synthetic surfactants that are used in preparation of cosmetic systems such as emulsions, creams, suspensions, etc., several other naturally occurring materials have been introduced and there is a trend in recent years to use such natural products more widely, in the belief that they are safer for application. As mentioned below, polymeric surfactants of the A–B, A–B–A, and BA$_n$ are also used in many cosmetic formulations.

Several synthetic surfactants that are applied in cosmetics may be listed as shown below.

2.1.1 Anionic surfactants

These are widely used in many cosmetic formulations. The hydrophobic chain is a linear alkyl group with a chain length in the region of 12–16 C atoms and the polar head group should be at the end of the chain. Linear chains are preferred since they are more effective and more degradable than the branched chains. The most commonly used hydrophilic groups are carboxylates, sulphates, sulphonates and phosphates. A general formula may be ascribed to anionic surfactants as follows:

Carboxylates: $C_nH_{2n+1}COO^-X^+$
Sulphates: $C_nH_{2n+1}OSO_3^-X^+$
Sulphonates: $C_nH_{2n+1}SO_3^-X^+$
Phosphates: $C_nH_{2n+1}OPO(OH)O^-X^+$

with n being the range 8–16 atoms and the counterion X^+ is usually Na^+.

Several other anionic surfactants are commercially available such as sulphosuccinates, isethionates (esters of isothionic acid with the general formula $RCOOCH_2-CH_2-SO_3Na$) and taurates (derivatives of methyl taurine with the general formula $RCON(R')CH_2-CH_2-SO_3Na$), sarchosinates (with the general formula $RCON(R')COONa$) and these are sometimes used for special applications.

The carboxylates are perhaps the earliest known surfactants, since they constitute the earliest soaps, e.g. sodium or potassium stearate, $C_{17}H_{35}COONa$, sodium myristate, $C_{14}H_{29}COONa$. The alkyl group may contain unsaturated portions, e.g. sodium oleate, which contains one double bond in the C_{17} alkyl chain. Most commercial soaps will be a mixture of fatty acids obtained from tallow, coconut oil, palm oil, etc. They are simply prepared by saponification of the triglycerides of oils and fats. The main attraction of these simple soaps is their low cost, their ready biodegradability and low toxicity. Their main disadvantage is their ready precipitation in water containing bivalent ions such as Ca^{2+} and Mg^{2+}. To avoid their precipitation in hard water, the carboxylates are modified by introducing some hydrophilic chains, e.g. ethoxy carboxylates with the general structure $RO(CH_2CH_2O)_nCH_2COO^-$, ester carboxylates containing hydroxyl or multi-COOH groups, sarcosinates which contain an amide group with the general structure $RCON(R')COO^-$. The addition of the ethoxylated groups results in increased water solubility and enhanced chemical stability (no hydrolysis). The modified ether carboxylates are also more compatible with electrolytes. They are also compatible with other nonionic, amphoteric and sometimes even cationic surfactants. The ester carboxylates are very soluble in water, but they suffer from the problem of hydrolysis. The sarcosinates are not very soluble in acid or neutral solutions but they are quite soluble in alkaline media. They are compatible with other anionics, nonionics, and cationics. The phosphate esters have very interesting properties being intermediate between ethoxylated nonionics and sulphated derivatives. They have good compatibility with inorganic builders and they can be good emulsifiers.

The sulphates are the largest and most important class of synthetic surfactants, which are produced by reaction of an alcohol with sulphuric acid, i.e. they are esters of sulphuric acid. In practice sulphuric acid is seldom used and chlorosulphonic or sulphur dioxide/air mixtures are the most common methods of sulphating the alcohol. The properties of sulphate surfactants depend on the nature of the alkyl chain and the sulphate group. The alkali metal salts show good solubility in water, but they tend to be affected by the presence of electrolytes. The most common sulphate surfactant is sodium dodecyl sulphate (abbreviated as SDS and sometimes referred to as sodium lauryl sulphate) which is extensively used in many cosmetic formulations. At room temperature (≈ 25 °C) this surfactant is quite soluble and 30 % aqueous solutions are fairly fluid (low viscosity). However, below 25 °C, the surfactant may separate out as a soft paste as the temperature falls below its Krafft point (the temperature above which the surfactant shows a rapid increase in solubility with a further increase of temperature). The latter depends on the distribution of chain lengths in the alkyl chain, the wider the distribution, the lower the Krafft temperature. Thus, by controlling this distribution one may achieve a Krafft temperature of ≈ 10 °C. As the surfactant concentration is increased to 30–40 % (depending on the distribution of chain lengths in the alkyl group), the viscosity of the solution increases very rapidly and may produce a gel, but then falls at about 60–70 % to give a pourable liquid, after which it increases again to a gel. The concentration at which the minimum occurs varies according to the alcohol sulphate used, and also the presence of impurities such as unsaturated alcohol. The viscosity of the aqueous solutions can be reduced by addition of short chain alcohols and glycols. The critical micelle concentration (cmc) of SDS (the concentration above which the properties of the solution show abrupt changes, see below) is 8×10^{-3} mol dm^{-3} (0.24 %). The alkyl sulphates give good foaming properties with an optimum at C_{12}–C_{14}. As with the carboxylates, the sulphate surfactants are also chemically modified to change their properties. The most common modification is to introduce some ethylene oxide units in the chain, usually referred to as alcohol ether sulphates that are commonly used in shampoos. These are made by sulphation of ethoxylated alcohols. For example, sodium dodecyl 3-mole ether sulphate, which is essentially dodecyl alcohol reacted with 3 mol EO, then sulphated and neutralized by NaOH. The presence of PEO confers improved solubility when compared with the straight alcohol sulphates. In addition, the surfactant becomes more compatible with electrolytes in aqueous solution. The ether sulphates are also more chemically stable than the alcohol sulphates. The cmc of the ether sulphates is also lower than the corresponding surfactant without the EO units. The viscosity behaviour of aqueous solutions is similar to alcohol sulphates, giving gels in the range 30–60 %. The ether sulphates show a pronounced salt effect, with significant increase in the viscosity of a dilute solution on addition of electrolytes such as NaCl. The ether sulphates are commonly used in hand dish-washing liquid and in shampoos in combination with amphoteric surfactants.

With sulphonates, the sulphur atom is directly attached to the carbon atom of the alkyl group and this gives the molecule stability against hydrolysis, when compared with the sulphates (where the sulphur atom is indirectly linked to the carbon of the hydrophobe via an oxygen atom). As with the sulphates, some chemical modification is used by introducing ethylene oxide units. These surfactants have excellent water solubility and biodegradability. They are also compatible with many aqueous ions.

Another class of sulphonates is the α-olefin sulphonates which are prepared by reacting linear α-olefin with sulphur trioxide, typically yielding a mixture of alkene sulphonates (60–70 %), 3- and 4-hydroxyalkane sulphonates ($\approx$ 30 %) and some disulphonates and other species. The two main α-olefin fractions used as starting material are C_{12}–C_{16} and C_{16}–C_{18}. Fatty acid and ester sulphonates are produced by sulphonation of unsaturated fatty acids or esters. A good example is sulphonated oleic acid,

$$CH_3(CH_2)_7\underset{\underset{SO_3H}{|}}{CH}(CH_2)_8COOH$$

A special class of sulphonates are sulphosuccinates which are esters of sulphosuccinic acid,

$$\text{HSO}_3\text{ CH COOH} \atop {\displaystyle | \atop \displaystyle \text{CH}_2\text{COOH}}$$

Both mono- and diesters are produced. A widely used diester in many formulations is sodium di(2-ethylhexyl)sulphosuccinate (that is sold commercially under the trade name Aerosol OT). The cmc of the diesters is very low, in the region of 0.06 % for C_6–C_8 sodium salts and they give a minimum in the surface tension of 26 mN m^{-1} for the C_8 diester. Thus these molecules are excellent wetting agents. The diesters are soluble both in water and in many organic solvents. They are particularly useful for preparation of water-in-oil (W/O) microemulsions.

Isethionates are esters of isethionic acid HOCH$_2$CH$_2$SO$_3$H. They are prepared by reaction of acid chloride (of the fatty acid) with sodium isethionate. The sodium salt of C_{12}–C_{14} are soluble at high temperature (70 °C) but they have very low solubility (0.01 %) at 25 °C. They are compatible with aqueous ions and hence they can reduce the formation of scum in hard water. They are stable at pH 6–8 but they undergo hydrolysis outside this range. They also have good foaming properties.

Taurates are derivatives of methyl taurine CH$_2$–NH–CH$_2$–CH$_2$–SO$_3$. The latter is made by reaction of sodium isethionate with methyl amine. The taurates are prepared by reaction of fatty acid chloride with methyl taurine. Unlike the isethionates, the taurates are not sensitive to low pH. They have good foaming properties and they are good wetting agents.

Phosphate containing anionic surfactants are also used in many cosmetic formulations. Both alkyl phosphates and alkyl ether phosphates are made by treating the fatty alcohol or alcohol ethoxylates with a phosphorylating agent, usually phosphorous pentoxide, P$_4$O$_{10}$. The reaction yields a mixture of mono- and diesters of phosphoric acid. The ratio of the two esters is determined by the ratio of the reactants and the amount of water present in the reaction mixture. The physicochemical properties of the alkyl phosphate surfactants depend on the ratio of the esters. They have properties intermediate between ethoxylated nonionics (see below) and the sulphated derivatives. They have good compatibility with inorganic builders and good emulsifying properties.

2.1.2 Cationic surfactants

The most common cationic surfactants are the quaternary ammonium compounds with the general formula R'R''R'''R''''N$^+$X$^-$, where X$^-$ is usually chloride ion and R represents alkyl groups. These quaternaries are made by reacting an appropriate tertiary amine with an organic halide or organic sulphate. A common class of cationics is the alkyl trimethyl ammonium chlorides, where R contains 8–18 C atoms, e.g. dodecyl trimethyl ammonium chloride, C$_{12}$H$_{25}$(CH$_3$)$_3$NCl. Another widely used cationic surfactant class is that containing two long chain alkyl groups, i.e. dialkyl dimethyl ammonium chloride, with the alkyl groups having a chain length of 8–18 C atoms. These dialkyl surfactants are less soluble in water than the monoalkyl quaternary compounds, but they are sometimes used as hair conditioners. A widely used cationic surfactant is alkyl dimethyl benzyl ammonium chloride (sometimes referred to as benzalkonium chloride) and widely used as bactericide), having the structure,

$$\underset{\underset{CH_2}{\diagup}\underset{CH_3}{\diagdown}}{\overset{\underset{N^+Cl^-}{\diagup\diagdown}}{\overset{C_{12}H_{25}\diagdown\diagup CH_3}{}}}$$

Imidazolines can also form quaternaries, the most common product being the ditallow derivative quaternized with dimethyl sulphate,

$$[C_{17}H_{35}\ \underset{\underset{\underset{C}{\diagdown\ \!/\!/}}{N\ CH}}{\overset{CH_3}{\overset{|}{C}}}-N-CH_2-CH_2-NH-CO-C_{17}H_{35}]^+ \qquad CH_3SO_4^-$$

Cationic surfactants can also be modified by incorporating polyethylene oxide chains, e.g. dodecyl methyl polyethylene oxide ammonium chloride having the structure,

$$\underset{\underset{CH_3}{\diagup}\underset{(CH_2CH_2O)_nH}{\diagdown}}{\overset{\underset{N^+\qquad Cl^-}{\diagup\diagdown}}{\overset{C_{12}H_{25}\ \ (CH_2CH_2O)_nH}{}}}$$

Cationic surfactants are generally water soluble when there is only one long alkyl group. When there are two or more long chain hydrophobes the product becomes dispersible in water and soluble in organic solvents. They are generally compatible with most inorganic ions and hard water, but they are incompatible with metasilicates and highly condensed phosphates. They are also incompatible with protein-like materials. Cationics are generally stable to pH changes, both acid and alkaline. They are incompatible with most anionic surfactants, but they are compatible with nonionics. These cationic surfactants are insoluble in hydrocarbon oils. In contrast, cationics with two or more long alkyl chains are soluble in hydrocarbon solvents, but they become only dispersible in water (sometimes forming bilayer vesicle type structures). They are generally chemically stable and can tolerate electrolytes. The cmc of cationic surfactants is close to that of anionics with the same alkyl chain length. For example, the cmc of benzalkonium chloride is 0.17 %. The primary use of cationic surfactants is for their tendency to adsorb at negatively charged surfaces, e.g. hair, and they can be applied as hair conditioners.

2.1.3 Amphoteric (zwitterionic) surfactants

These are surfactants containing both cationic and anionic groups. The most common amphoterics are the N-alkyl betaines which are derivatives of trimethyl glycine $(CH_3)_3NCH_2COOH$ (that was described as betaine). An example of betaine surfactant is lauryl amido propyl dimethyl betaine $C_{12}H_{25}CON(CH_3)_2CH_2COOH$. These alkyl betaines are sometimes described as alkyl dimethyl glycinates. The main characteristic of amphoteric surfactants is their dependence on the pH of the solution in which they are dissolved. In acid pH solutions, the molecule acquires a positive charge and it behaves like a cationic, whereas in alkaline pH solutions, they become negatively charged and behave like an anionic. A specific pH can be defined at which both ionic groups show equal ionization (the isoelectric point of the molecule). This can be described by the following scheme,

$$N^+ \ldots COOH \leftrightarrow N^+ \ldots COO^- \leftrightarrow NH \ldots COO^-$$
acid, pH 3   isoelectric   pH 6, alkaline

Amphoteric surfactants are sometimes referred to as zwitterionic molecules. They are soluble in water, but the solubility shows a minimum at the isoelectric point. Amphoterics show excellent compatibility with other surfactants, forming mixed micelles. They are chemically stable both in acids and alkalis. The surface activity of amphoterics varies widely and it depends on the distance between the charged groups and they show a maximum in surface activity at the isoelectric point.

Another class of amphoterics are the N-alkyl amino propionates having the structure $R-NHCH_2CH_2COOH$. The NH group is reactive and can react with another acid molecule (e.g. acrylic) to form an amino dipropoionate $R-N(CH_2CH_2COOH)_2$. Alkyl imidazoline-based product can also be produced by reacting alkyl imidazoline with a chloro acid. However, the imidazoline ring breaks down during the formation of the amphoteric.

The change in charge with pH of amphoteric surfactants affects their properties, such as wetting, foaming, etc. At the isoelectric point (IEP), the properties of amphoterics resemble those of nonionics very closely. Below and above the IEP, the properties shift towards those of cationic and anionic surfactants respectively. Zwitterionic surfactants have excellent dermatological properties. They also exhibit low eye irritation and they are frequently used in shampoos and other personal care products (cosmetics). Due to their mild characteristics, i.e. low eye and skin irritation, amphoterics are widely used in shampoos. They also provide antistatic properties to hair, good conditioning and foam booster.

2.1.4 Nonionic surfactants

The most common nonionic surfactants are those based on ethylene oxide, referred to as ethoxylated surfactants. Several classes can be distinguished: alcohol ethoxylates, fatty acid ethoxylates, monoalkaolamide ethoxylates, sorbitan ester ethoxylates, fatty amine ethoxylates and ethylene oxide-propylene oxide copolymers (sometimes referred to as polymeric surfactants). Another important class of nonionics are the multihydroxy products such as glycol esters, glycerol (and polyglycerol) esters, glucosides (and polyglucosides) and sucrose esters. Amine oxides and sulphinyl surfactants represent nonionics with a small head group.

The alcohol ethoxylates are generally produced by ethoxylation of a fatty chain alcohol such as dodecanol. Several generic names are given to this class of surfactants such as ethoxylated fatty alcohols, alkyl polyoxyethylene glycol, monoalkyl polyethylene oxide glycol ethers, etc. A typical example is dodecyl hexaoxyethylene glycol monoether with the chemical formula $C_{12}H_{25}(OCH_2CH_2O)_6OH$ (sometimes abbreviated as $C_{12}E_6$). In practice, the starting alcohol will have a distribution of alkyl chain lengths and the resulting ethoxylate will have a distribution of ethylene oxide (EO) chain length. Thus the numbers listed in the literature refer to average numbers.

The cmc of nonionic surfactants is about two orders of magnitude lower than the corresponding anionics with the same alkyl chain length. At a given alkyl chain length, the cmc decreases with a decrease in the number of EO units. The solubility of the alcohol ethoxylates depends both on the alkyl chain length and the number of ethylene oxide units in the molecule. Molecules with an average alkyl chain length of 12 C atoms and containing more than 5 EO units are usually soluble in water at room temperature. However, as the temperature of the solution is gradually raised, the solution becomes cloudy (as a result of dehydration of the PEO chain and the change in the conformation of the PEO chain) and the temperature at which this occurs is referred to as the cloud point (CP) of the surfactant. At a given alkyl chain length, the CP increases with increasing the EO chain of the molecule. The CP changes with changing concentration of the surfactant solution and the trade literature usually quotes the CP of a 1% solution. The CP is also affected by the presence of electrolyte in the aqueous solution. Most electrolytes lower the CP of a nonionic surfactant solution. Nonionics tend to have maximum surface activity near to the cloud point. The CP of most nonionics increases markedly on addition of small quantities of anionic surfactants. The surface tension of alcohol ethoxylate solutions decreases with an increase in its concentration, until it reaches its cmc, after which it remains constant with any further increase in its concentration. The minimum surface tension reached at and above the cmc decreases with decreasing the number of EO units of the chain (at a given alkyl chain). The viscosity of a nonionic surfactant solution increases gradually with an increase in its concentration, but at a critical concentration (which depends on the alkyl and EO chain length) the viscosity shows a rapid increase and ultimately a gel-like structure appears. This results from the formation of liquid crystalline structure of the hexagonal type. In many cases, the viscosity reaches a maximum after which it shows a decrease due to the formation of other structures (e.g. lamellar phases) (see below).

The fatty acid ethoxylates are produced by reaction of ethylene oxide with a fatty acid or a polyglycol and they have the general formula $RCOO-(CH_2CH_2O)_nH$. When a polyglycol is used, a mixture of mono- and diesters $(RCOO-(CH_2CH_2O)_n-OCOR)$ is produced. These surfactants are generally soluble in water provided there are enough EO units and the alkyl chain length of the acid is not too long. The monoesters are much more soluble in water than the diesters. In the latter case, a longer EO chain is required to render the molecule soluble. The surfactants are compatible with aqueous ions, provided there is not much unreacted acid. However, these surfactants undergo hydrolysis in highly alkaline solutions.

The sorbitan esters and their ethoxylated derivatives (Spans and Tweens) are perhaps one of the most commonly used nonionics. The sorbitan esters are produced by reaction of sorbitol with a fatty acid at a high temperature (> 200 °C). The sorbitol dehydrates to 1,4-sorbitan and then esterification takes place. If one mole of fatty acid is reacted with one mole of sorbitol, one obtains a monoester (some diester is also produced as a by-product). Thus, sorbitan monoester has the following general formula,

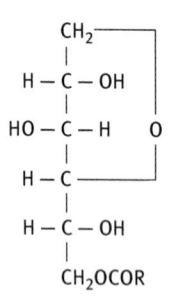

$$\begin{array}{c} CH_2- \\ | \\ H-C-OH \\ | \\ HO-C-H \quad O \\ | \\ H-C- \\ | \\ H-C-OH \\ | \\ CH_2OCOR \end{array}$$

The free OH groups in the molecule can be esterified, producing di- and tri-esters. Several products are available depending on the nature of the alkyl group of the acid and whether the product is a mono-, di- or tri-ester. Some examples are given below,

Sorbitan monolaurate: Span 20
Sorbitan monopalmitate: Span 40
Sorbitan monostearate: Span 60
Sorbitan mono-oleate: Span 80
Sorbitan tristearate: Span 65
Sorbitan trioleate: Span 85

The ethoxylated derivatives of Spans (Tweens) are produced by reaction of ethylene oxide on any hydroxyl group remaining on the sorbitan ester group. Alternatively, the sorbitol is first ethoxylated and then esterified. However, the final product has different surfactant properties to the Tweens. Some examples of Tween surfactants are given below,

Polyoxyethylene (20) sorbitan monolaurate:	Tween 20
Polyoxyethylene (20) sorbitan monopalmitate:	Tween 40
Polyoxyethylene (20) sorbitan monostearate:	Tween 60
Polyoxyethylene (20) sorbitan mono-oleate:	Tween 80
Polyoxyethylene (20) sorbitan tristearate:	Tween 65
Polyoxyethylene (20) sorbitan tri-oleate:	Tween 85

The sorbitan esters are insoluble in water, but soluble in most organic solvents (low HLB number surfactants). The ethoxylated products are generally soluble in water and they have relatively high HLB numbers. One of the main advantages of the sorbitan esters and their ethoxylated derivatives is their approval in cosmetics and some pharmaceutical preparations.

Ethoxylated fats and oils are also used in some cosmetic formulations, e.g. castor oil ethoxylates which are good solubilizers for water-insoluble ingredients.

The amine ethoxylates are prepared by addition of ethylene oxide to primary or secondary fatty amines. With primary amines both hydrogen atoms on the amine group react with ethylene oxide and therefore the resulting surfactant has the structure,

$$\text{R-N} \begin{array}{l} \diagup (CH_2CH_2O)_xH \\ \diagdown (CH_2CH_2O)_yH \end{array}$$

The above surfactants acquire a cationic character if the EO units are small in number and if the pH is low. However, at high EO levels and neutral pH they behave very similarly to nonionics. At low EO content, the surfactants are not soluble in water, but become soluble in an acid solution. At high pH, the amine ethoxylates are water soluble provided the alkyl chain length of the compound is not long (usually a C_{12} chain is adequate for reasonable solubility at sufficient EO content).

Amine oxides are prepared by oxidizing a tertiary nitrogen group with aqueous hydrogen peroxide at temperatures in the region 60–80 °C. Several examples can be quoted: N-alkyl amidopropyl-dimethyl amine oxide, N-alkyl bis(2-hydroxyethyl) amine oxide and N-alkyl dimethyl amine oxide. They have the general formula,

$$\text{Coco CONHCH}_2\text{CH}_2\text{CH}_2\text{N} \begin{array}{c} CH_3 \\ | \\ \longrightarrow O \\ | \\ CH_3 \end{array} \qquad \text{Alkyl amidopropyl-dimethyl amine oxide}$$

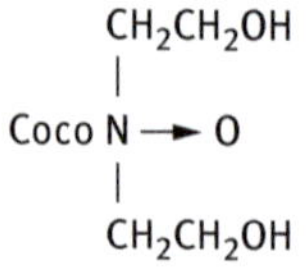

```
        CH₂CH₂OH
        |
Coco N⎯→O        Coco bis (2-hydroxyethyl) amine oxide
        |
        CH₂CH₂OH


          CH₃
          |
C₁₂H₂₅N⎯→O       Lauryl dimethyl amine oxide
          |
          CH₃
```

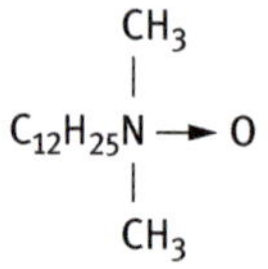

In acid solutions, the amino group is protonated and acts as a cationic surfactant. In neutral or alkaline solution the amine oxides are essentially nonionic in character. Alkyl dimethyl amine oxides are water soluble up to C_{16} alkyl chain. Above pH 9, amine oxides are compatible with most anionics. At pH 6.5 and below some anionics tend to interact and form precipitates. In combination with anionics, amine oxides can be foam boosters (e.g. in shampoos).

2.1.5 Surfactants derived from mono- and polysaccharides

Several surfactants have been synthesized starting from mono- or oligo-saccharides by reaction with the multifunctional hydroxyl groups: alkyl glucosides, alkyl polyglucosides, sugar fatty acid esters and sucrose esters, etc. The technical problem is one of joining a hydrophobic group to the multihydroxyl structure. Several surfactants have been made, e.g. esterification of sucrose with fatty acids or fatty glycerides to produce sucrose esters having the following structure,

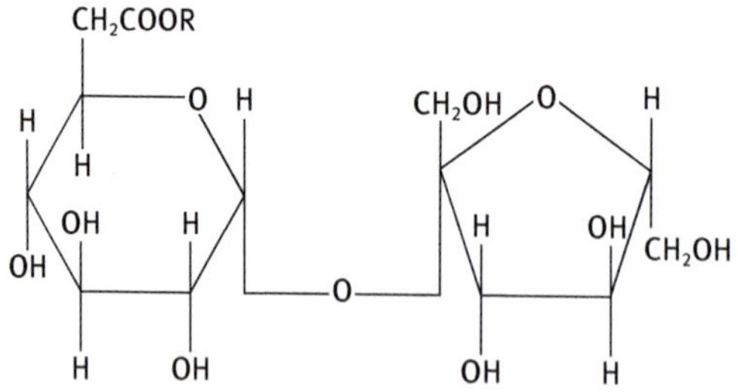

The most interesting sugar surfactants are the alkyl polyglucosides (APG) which are synthesized using a two-stage transacetalization process. In the first stage, the carbohydrate reacts with a short chain alcohol, e.g. butanol or propylene glycol. In the second stage, the short chain alkyl glucoside is transacetalized with a relatively long chain alcohol (C_{12-14}–OH) to form the required alkyl polyglucoside. This process is applied if oligo- and polyglucoses (e.g. starch, syrups with a low dextrose equivalent, DE) are used. In a simplified transacetalization process, syrups with high glucose content (DE > 96 %) or solid glucose types can react with short-chain alcohols under normal pressure. Commercial alkyl polyglucosides (APG) are complex mixtures of species varying in the degree of polymerization (DP, usually in the range 1.1–3) and in the length of the alkyl chain. When the latter is shorter than C_{14} the product is water soluble. The cmc values of APGs are comparable to nonionic surfactants and they decrease with increasing alkyl chain length.

APG surfactants have good solubility in water and they have high cloud points (> 100 °C). They are stable in neutral and alkaline solutions but are unstable in strong acid solutions. APG surfactants can tolerate high electrolyte concentrations and they are compatible with most types of surfactants. They are used in personal care products for cleansing formulations as well as for skincare and hair products.

2.1.6 Naturally occurring surfactants

The important naturally occurring class of surfactants which are widely used in cosmetic formulations are the lipids, of which phosphatidylcholine (lecithin), lysolecithin, phosphatidylethanolamine and phosphatidylinositol are the most commonly used surfactants. The structure of these lipids is given in Fig. 2.1. These lipids are used as emulsifiers as well as for production of liposomes or vesicles for skincare products. The lipids form coarse turbid dispersions of large aggregates (liposomes) which on ultrasonic irradiation form smaller units or vesicles. The liposomes are smectic mesophases of phospholipids organized into bilayers which assume a multilamellar or unilamellar structure. The multilamellar species are heterogeneous aggregates, most commonly prepared by dispersal of a thin film of phospholipid (alone or with cholesterol) into water. Sonication of the multilamellar units can produce the unilamellar liposomes, sometimes referred to as vesicles. The net charge of liposomes can be varied by incorporation of a long chain amine, such as stearyl amine (to give a positively charged vesicle) or dicetyl phosphate (giving negatively charged species). Both lipid-soluble and water-soluble actives can be entrapped in liposomes. The liposoluble actives are solubilized in the hydrocarbon interiors of the lipid bilayers, whereas the water-soluble actives are intercalated in the aqueous layers.

2.1.7 Polymeric (macromolecular) surfactants

The polymeric surfactants possess considerable advantages for use in cosmetic ingredients. They will be described in detail in Chapter 3. As will be discussed, the most commonly used materials are the ABA block copolymers, with A being poly(ethylene oxide) and B poly(propylene oxide) (Pluronics). On the whole, polymeric surfactants have much lower toxicity, sensitization and irritation potentials, provided they are not contaminated with traces of the parent monomers. As will be discussed in Chapter 3, these molecules provide greater stability and in some case they can be used to adjust the viscosity of the cosmetic formulation.

2.1.8 Silicone surfactants

In recent years, there has been a great trend towards using silicone oils for many cosmetic formulations. In particular, volatile silicone oils have found application in many cosmetic products, owing to the pleasant dry sensation they impart to the skin. These volatile silicones evaporate without unpleasant cooling effects or without leaving a residue. Due to their low surface energy, silicone help spread the various active ingredients over the surface of hair and skin.

The chemical structure of the silicone compounds used in cosmetic preparations varies according to the application. As an illustration Fig. 2.2 shows some typical structures of cyclic and linear silicones. The backbones can carry various attached "functional" groups, e.g. carboxyl, amine, sulphhydryl, etc. While most silicone oils can be emulsified using conventional hydrocarbon surfactants, there has been a trend in re-

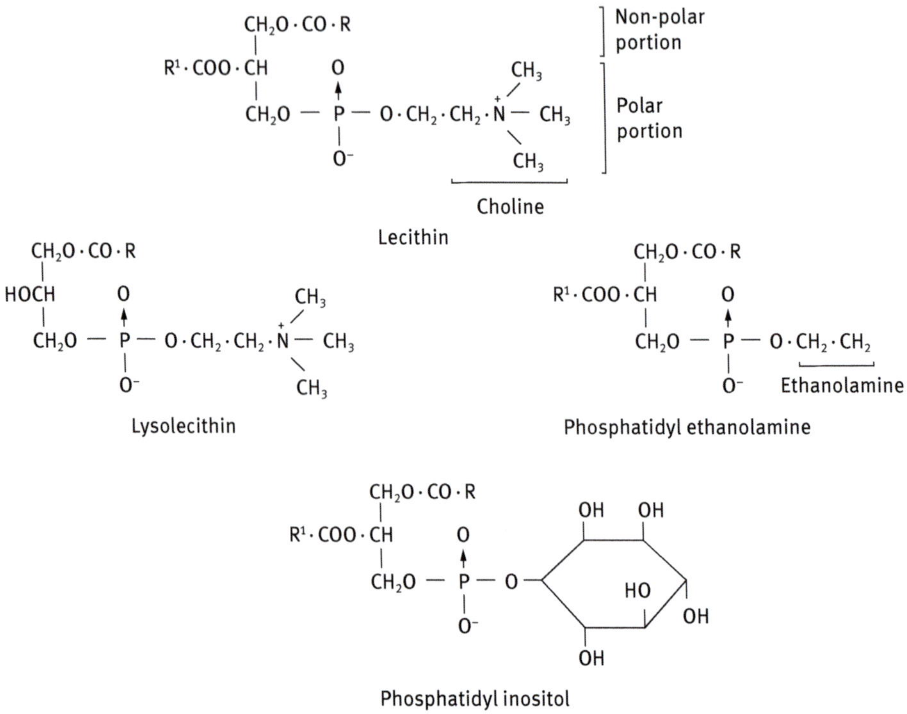

Fig. 2.1: Structure of lipids.

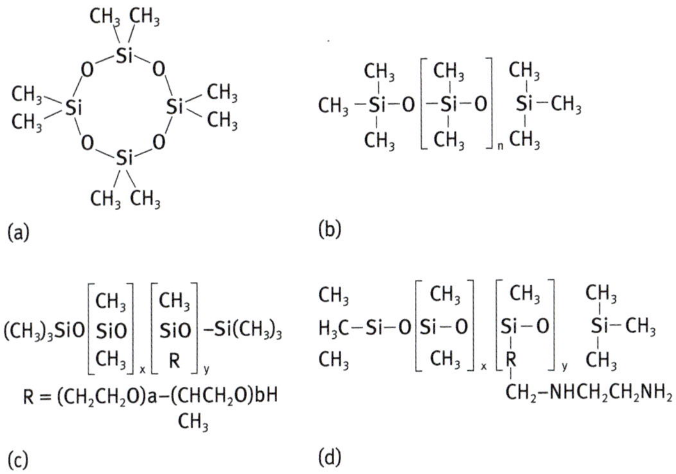

Fig. 2.2: Structural formulae of typical silicone compounds used in cosmetic formulations: (a) cyclic siloxane; (b) linear siloxane; (c) siloxane-polyethylene oxide copolymer; (d) siloxane-polyethylene amine copolymer.

cent years to use silicone surfactants for producing the emulsion. Typical structures of siloxane-polyethylene oxide and siloxane polyethylene amine copolymers are shown in Fig. 2.2. The surface activity of these block copolymers depend on the relative length of the hydrophobic silicone backbone and the hydrophilic (e.g. PEO) chains. The attraction of using silicone oils and silicone copolymers is their relatively small medical and environmental hazards, when compared to their hydrocarbon counterparts.

2.2 Physical properties of surfactant solutions and the process of micellization

The physical properties of surface active agent solutions differ from those of non-amphipathic molecule solutions (such as sugars) in one major aspect, namely the abrupt changes in their properties above a critical concentration [4]. This is illustrated in Fig. 2.3 which shows plots of several physical properties (osmotic pressure, surface tension, turbidity, solubilization, magnetic resonance, equivalent conductivity and self-diffusion) as a function of concentration for an anionic surfactant. At low concentrations, most properties are similar to those of a simple electrolyte. One notable exception is the surface tension, which decreases rapidly with increasing surfactant concentration. However, all the properties (interfacial and bulk) show an abrupt change at a particular concentration, which is consistent with the fact that at and above this concentration, surface active molecules or ions associate to form larger units. These associated units are called micelles (self-assembled structures) and the

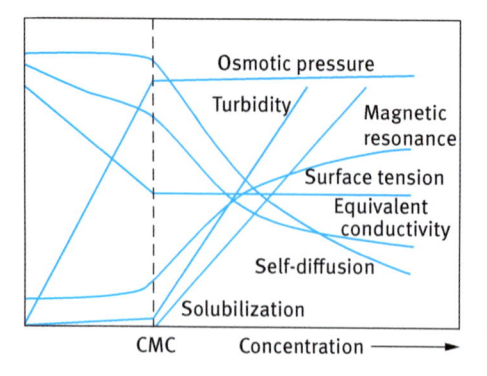

Fig. 2.3: Variation of solution properties with surfactant concentration.

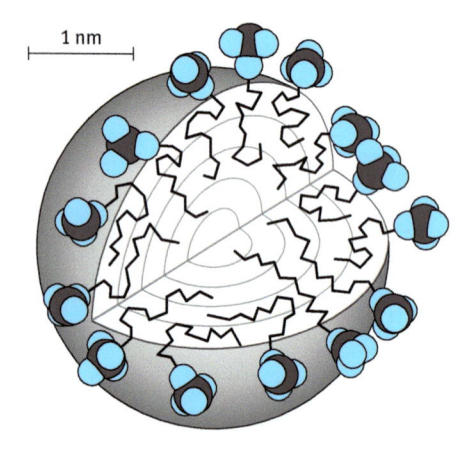

Fig. 2.4: Illustration of a spherical micelle for dodecyl sulphate [4].

first formed aggregates are generally approximately spherical in shape. A schematic representation of a spherical micelle is given in Fig. 2.4.

The concentration at which this association phenomenon occurs is known as the critical micelle concentration (cmc). Each surfactant molecules has a characteristic cmc value at a given temperature and electrolyte concentration. The most common technique for measuring the cmc is surface tension, γ, which shows break at the cmc, after which γ remains virtually constant with a further increase in concentration. However, other techniques such as self-diffusion measurements, NMR and fluorescence spectroscopy can be applied. A compilation of cmc values was given in 1971 by Mukerjee and Mysels [5], which is clearly not an up-to-date text, but is an extremely valuable reference. As an illustration, the cmc values of a number of surface active agents are given in Tab. 2.1, to show some of the general trends [1–3]. Within any class of surface active agent, the cmc decreases with increasing chain length of the hydrophobic portion (alkyl group). As a general rule, the cmc decreases by a factor of 2 for ionics (without added salt) and by a factor of 3 for nonionics on adding one methylene group to the alkyl chain. With nonionic surfactants, increasing the length of the hydrophilic group (polyethylene oxide) causes an increase in cmc.

Tab. 2.1: Critical micelle concentration of surfactant classes.

Surface active agent	cmc / mol dm^{-3}
(A) Anionic	
Sodium octyl-l-sulphate	1.30×10^{-1}
Sodium decyl-l-sulphate	3.32×10^{-2}
Sodium dodecyl-l-sulphate	8.39×10^{-3}
Sodium tetradecyl-l-sulphate	2.05×10^{-3}
(B) Cationic	
Octyl trimethyl ammonium bromide	1.30×10^{-1}
Decetryl trimethyl ammonium bromide	6.46×10^{-2}
Dodecyl trimethyl ammonium bromide	1.56×10^{-2}
Hexactecyltrimethyl ammonium bromide	9.20×10^{-4}
(C) Nonionic	
Octyl hexaoxyethylene glycol monoether C_8E_6	9.80×10^{-3}
Decyl hexaoxyethylene glycol monoether $C_{10}E_6$	9.00×10^{-4}
Decyl nonaoxyethylene glycol monoether $C_{10}E_9$	1.30×10^{-3}
Dodecyl hexaoxyethylene glycol monoether $C_{12}E_6$	8.70×10^{-5}
Octylphenyl hexaoxyethylene glycol monoether C_8E_6	2.05×10^{-4}

In general, nonionic surfactants have lower cmc values than their corresponding ionic surfactants of the same alkyl chain length. Incorporation of a phenyl group in the alkyl group increases its hydrophobicity to a much smaller extent than increasing its chain length with the same number of carbon atoms. The valency of the counterion in ionic surfactants has a significant effect on the cmc. For example, increasing the valency of the counterion from 1 to 2 causes a reduction of the cmc by roughly a factor of 4.

The cmc is, to a first approximation, independent of temperature. This is illustrated in Fig. 2.4 which shows the variation of cmc of SDS with temperature. The cmc varies in a non-monotonic way by ca. 10–20 % over a wide temperature range. The shallow minimum around 25 °C can be compared with a similar minimum in the solubility of hydrocarbon in water [1–3]. However, nonionic surfactants of the ethoxylate type show a monotonic decrease [1–3] of cmc with increasing temperature as is illustrated in Fig. 2.4 for $C_{10}E_5$. The effect of addition of cosolutes, e.g. electrolytes and non-electrolytes, on the cmc can be very striking. For example, addition of 1:1 electrolyte to a solution of anionic surfactant gives a dramatic lowering of the cmc, which may amount to an order of magnitude. The effect is moderate for short-chain surfactants, but is much larger for long-chain ones. At high electrolyte concentrations, the reduction in cmc with increasing number of carbon atoms in the alkyl chain is much stronger than without added electrolyte. This rate of decrease at high electrolyte concentrations is comparable to that of nonionics. The effect of added electrolyte also depends on the valency of the added counterions. In contrast, for nonionics, addition of electrolytes causes only small variation in the cmc.

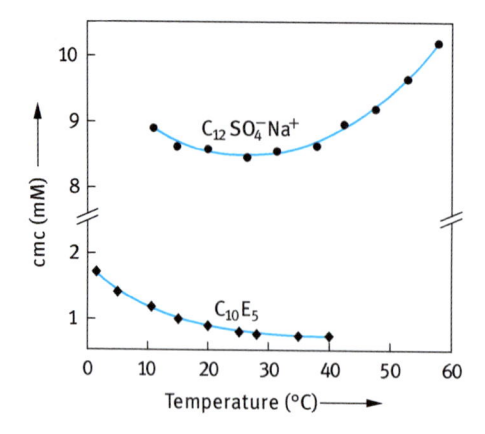

Fig. 2.5: Temperature dependence of the cmc of SDS and C$_{10}$E$_5$ [1–3].

Non-electrolytes such as alcohols can also cause a decrease in the cmc [1–3]. The alcohols are less polar than water and are distributed between the bulk solution and the micelles. The more preference they have for the micelles, the more they stabilize them. A longer alkyl chain leads to a less favourable location in water and more favourable location in the micelles

The presence of micelles can account for many of the unusual properties of solutions of surface active agents. For example, it can account for the near constant surface tension value, above the cmc (see Fig. 2.3). It also accounts for the reduction in molar conductance of the surface active agent solution above the cmc, which is consistent with the reduction in mobility of the micelles as a result of counterion association. The presence of micelles also accounts for the rapid increase in light scattering or turbidity above the cmc. The presence of micelles was originally suggested by McBain [6] who suggested that below the cmc most of the surfactant molecules are unassociated, whereas in the isotropic solutions immediately above the cmc, micelles and surfactant ions (molecules) are though to co-exist, the concentration of the latter changing very slightly as more surfactant is dissolved. However, the self-association of an amphiphile occurs in a stepwise manner with one monomer added to the aggregate at a time. For a long chain amphiphile, the association is strongly cooperative up to a certain micelle size where counteracting factors became increasingly important. Typically the micelles have a closely spherical shape in a rather wide concentration range above the cmc. Originally, it was suggested by Adam [7] and Hartley [8] that micelles are spherical in shape and have the following properties: (i) the association unit is spherical with a radius approximately equal to the length of the hydrocarbon chain; (ii) the micelle contains about 50–100 monomeric units; aggregation number generally increases with increasing alkyl chain length; (iii) with ionic surfactants, most counterions are bound to the micelle surface, thus significantly reducing the mobility from the value to be expected from a micelle with non-counterion bonding; (iv) micellization occurs over a narrow concentration range as a result of the high association number of surfactant micelles; (v) the interior of the surfactant micelle has essentially the properties of a liquid hydrocarbon. This is confirmed by the high mobility of the alkyl chains and the ability of the micelles to solubilize many water insoluble organic molecules, e.g. dyes and agrochemicals. To a first approximation, micelles can, over a wide concentration range above the cmc, be viewed as microscopic liquid hydrocarbon droplets covered with polar head groups, which interact strongly with water molecules. It appears that the radius of the micelle core constituted of the alkyl chains is close to the extended length of the alkyl chain, i.e. in the range 1.5–3.0 nm. As we will see later, the driving force for micelle formation is the elimination of the contact between the alkyl chains and water. The larger a spherical micelle, the more efficient this is, since the volume-to-area ratio increases. It should be noted that the surfactant molecules in the micelles are not all extended. Only one molecule needs to be extended to satisfy the criterion that the radius of the micelle core is close to the extended length of the alkyl chain. The majority of surfactant molecules are in a disordered state. In other words, the interior of the micelle is close to that of the corresponding alkane in a neat liquid oil. This explains the large solubilization capacity of the micelle towards a broad range of nonpolar and weakly polar substances. At the surface of the micelle, associated counterions (in the region of 50–80 % of the surfactant ions) are present. However, simple inorganic counterions are very loosely associated with the micelle. The counterions are very mobile (see below) and there is no specific complex formed with a definite counterion head group distance. In other words, the counterions are associated by long-range electrostatic interactions.

A useful concept for characterizing micelle geometry is the critical packing parameter [9], CPP. The aggregation number N is the ratio between the micellar core volume, V_{mic}, and the volume of one chain, v,

$$N = \frac{V_{mic}}{v} = \frac{(4/3)\pi R_{mic}^3}{v}, \tag{2.1}$$

where R_{mic} is the radius of the micelle.

The aggregation number, N, is also equal to the ratio of the area of a micelle, A_{mic}, to the cross-sectional area, a, of one surfactant molecule,

$$N = \frac{A_{mic}}{a} = \frac{4\pi R_{mic}^2}{a}. \tag{2.2}$$

Combining equations (2.1) and (2.2),

$$\frac{v}{R_{mic} a} = \frac{1}{3} \tag{2.3}$$

Since R_{mic} cannot exceed the extended length of a surfactant alkyl chain, l_{max},

$$l_{max} = 1.5 + 1.265 n_c. \tag{2.4}$$

This means that for a spherical micelle,

$$\frac{v}{l_{max} a} \leq \frac{1}{3}. \tag{2.5}$$

The ratio $v/(l_{max} a)$ is denoted as the critical packing parameter (CPP).

Although the spherical micelle model accounts for many of the physical properties of solutions of surfactants, a number of phenomena remain unexplained, without considering other shapes. For example, McBain [10] suggested the presence of two types of micelles, spherical and lamellar in order to account for the drop in molar conductance of surfactant solutions. The lamellar micelles are neutral and hence they account for the reduction in the conductance. Later, Harkins et al. [11] used McBain's model of lamellar micelles to interpret his X-ray results in soap solutions. Moreover, many modern techniques such as light scattering and neutron scattering indicate that in many systems the micelles are not spherical. For example, Debye and Anacker [12] proposed a cylindrical micelle to explain the light scattering results on hexadecyltrimethyl ammonium bromide in water. Evidence for disc-shaped micelles has also been obtained under certain conditions. A schematic representation of the spherical, lamellar and rod-shaped micelles, suggested by McBain, Hartley and Debye is given in Fig. 2.6. Many ionic surfactants show dramatic temperature-dependent solubility as illustrated in Fig. 2.7. The solubility first increases gradually with rising temperature, and then, above a certain temperature, there is a sudden increase in solubility with a further increase in temperature. The cmc increases gradually with increasing temperature. At a particular temperature, the solubility becomes equal to the cmc, i.e. the solubility curve intersects the cmc and this temperature is referred to as the Krafft temperature. At this temperature an equilibrium exists between solid hydrated surfactant, micelles and monomers (i.e. the Krafft temperature is a "triple point"). Surfactants with ionic head groups and long straight alkyl chains have high Krafft temperatures. The Krafft temperature increases with as the alkyl chain of the surfactant molecule increases. It can be reduced by introducing branching in the alkyl chain. The Krafft temperature is also reduced by using alkyl chains with a wide distribution of the chain lengths. Addition of electrolytes causes an increase in the Krafft temperature.

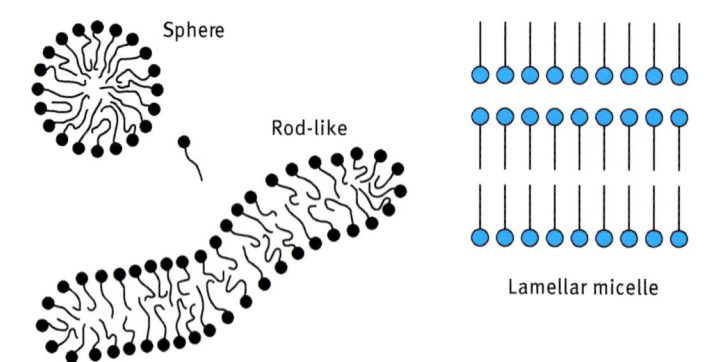

Fig. 2.6: Shape of micelles.

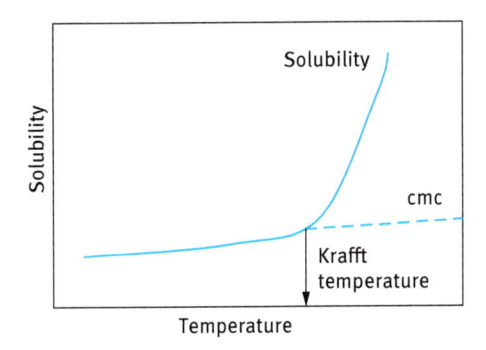

Fig. 2.7: Variation of solubility and critical micelle concentration (cmc) with temperature.

With nonionic surfactants of the ethoxylate type, an increase in temperature for a solution at a given concentration causes dehydration of the PEO chains and at a critical temperature the solution become cloudy. This is illustrated in Fig. 2.8 which shows the phase diagram of $C_{12}E_6$. Below the cloud point (CP) curve one can identify different liquid crystalline phases Hexagonal–Cubic–Lamellar which are schematically shown in Fig. 2.9.

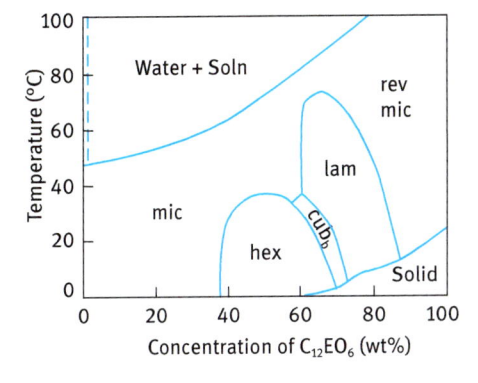

Fig. 2.8: Phase diagram of $C_{12}E_6$.

2.2.1 Thermodynamics of micellization

The process of micellization is one of the most important characteristics of surfactant solution and hence it is essential to understand its mechanism (the driving force for micelle formation). This requires analysis of the dynamics of the process (i.e. the kinetic aspects) as well as the equilibrium aspects whereby the laws of thermodynamics may be applied to obtain the free energy, enthalpy and entropy of micellization.

2.2.1.1 Kinetic aspects

Micellization is a dynamic phenomenon in which n monomeric surfactant molecules associate to form a micelle S_n, i.e.,

$$nS \Leftrightarrow S_n. \tag{2.6}$$

Hartley [8] envisaged a dynamic equilibrium whereby surface active agent molecules are constantly leaving the micelles whilst other molecules from solution enter the micelles. The same applies to the counterions with ionic surfactants, which can exchange between the micelle surface and bulk solution.

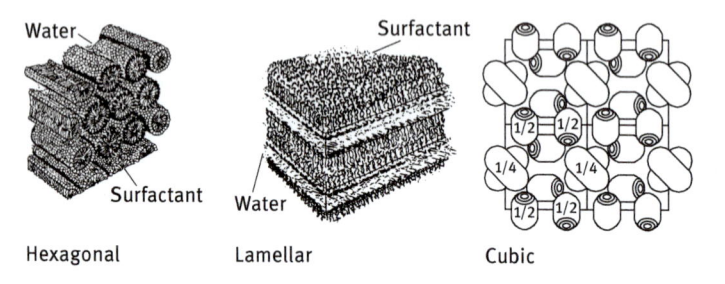

Fig. 2.9: Schematic picture of liquid crystalline phases.

Experimental investigation using fast kinetic methods such as stop flow, temperature and pressure jumps, and ultrasonic relaxation measurements have shown that there are two relaxation processes for micellar equilibrium [13–19] characterized by relaxation times τ_1 and τ_2. The first relaxation time, τ_1, is of the order of 10^{-7} s (10^{-8} to 10^{-3} s) and represents the lifetime of a surface active molecule in a micelle, i.e. it represents the association and dissociation rate for a single molecule entering and leaving the micelle, which may be represented by the equation,

$$S + S_{n-1} \underset{K^-}{\overset{K^+}{\Leftrightarrow}} S_n, \tag{2.7}$$

where K^+ and K^- represent the association and dissociation rate respectively for a single molecule entering or leaving the micelle.

The slower relaxation time τ_2 corresponds to a relatively slow process, namely the micellization-dissolution process represented by equation (2.6). The value of τ_2 is of the order of milliseconds (10^{-3}–1 s) and hence can be conveniently measured by stopped flow methods. The fast relaxation time τ_1 can be measured using various techniques depending on its range. For example, τ_1 values in the range of 10^{-8}–10^{-7} s are accessible to ultrasonic absorption methods, whereas τ_1 in the range of 10^{-5}–10^{-3} s can be measured by pressure jump methods. The value of τ_1 depends on surfactant concentration, chain length and temperature. τ_1 increases with increasing chain length of surfactants, i.e. the residence time increases with increasing chain length.

The above discussion emphasizes the dynamic nature of micelles and it is important to realize that these molecules are in continuous motion and that there is a constant interchange between micelles and solution. The dynamic nature also applies to the counterions which exchange rapidly with lifetimes in the range 10^{-9}–10^{-8} s. Furthermore, the counterions appear to be laterally mobile and not to be associated with (single) specific groups on the micelle surfaces.

2.2.1.2 Equilibrium aspects: Thermodynamics of micellization

Various approaches have been employed in tackling the problem of micelle formation. The most simple approach treats micelles as a single phase, and this is referred to as the phase separation model. In this model, micelle formation is considered as a phase separation phenomenon and the cmc is then the saturation concentration of the amphiphile in the monomeric state whereas the micelles constitute the separated pseudophase. Above the cmc, a phase equilibrium exists with a constant activity of the surfactant in the micellar phase. The Krafft point is viewed as the temperature at which solid hydrated surfactant, micelles and a solution saturated with undissociated surfactant molecules are in equilibrium at a given pressure.

Consider an anionic surfactant, in which n surfactant anions, S^-, and n counterions M^+ associate to form a micelle, i.e.,

$$nS^- + nM^+ \Leftrightarrow S_n . \tag{2.8}$$

The micelle is simply a charged aggregate of surfactant ions plus an equivalent number of counterions in the surrounding atmosphere and is treated as a separate phase. The chemical potential of the surfactant in the micellar state is assumed to be constant, at any given temperature, and this may be adopted as the standard chemical potential, μ_m^0, by analogy to a pure liquid or a pure solid. Considering the equilibrium between micelles and monomer, then,

$$\mu_m^0 = \mu_1^0 + RT \ln a_1 , \tag{2.9}$$

where μ_1 is the standard chemical potential of the surfactant monomer and a_1 is its activity which is equal to $f_1 x_1$, where f_1 is the activity coefficient and x_1 the mole fraction. Therefore, the standard free energy of micellization per mole of monomer, ΔG_m^0, is given by,

$$\Delta G_m^0 = \mu_m^0 - \mu_1^0 = RT \ln a_1 \approx RT \ln x_1 , \tag{2.10}$$

where f_1 is taken as unity (a reasonable value in very dilute solution). The cmc may be identified with x_1 so that

$$\Delta G_m^0 = RT \ln [\text{cmc}] . \tag{2.11}$$

In equation (2.10), the cmc is expressed as a mole fraction, which is equal to $C/(55.5 + C)$, where C is the concentration of surfactant in mol dm^{-3}, i.e.,

$$\Delta G_m^0 = RT \ln C - RT \ln (55.5 + C) . \tag{2.12}$$

It should be stated that ΔG^0 should be calculated using the cmc expressed as a mole fraction as indicated by equation (2.12). However, most cmc quoted in the literature are given in mol dm^{-3} and, in many cases of ΔG^0 values have been quoted when the cmc was simply expressed in mol dm^{-3}. Strictly speaking, this is incorrect, since ΔG^0 should be based on x_1 rather than on C. The value of ΔG^0 when the cmc is expressed in mol dm^{-3} is substantially different from the ΔG^0 value when the cmc is expressed in mole fraction. For example, dodecyl hexaoxyethylene glycol, the quoted cmc value is 8.7×10^{-5} mol dm^{-3} at 25 °C. Therefore,

$$\Delta G^0 = RT \ln \frac{8.7 \times 10^{-5}}{55.5 + 8.7 \times 10^{-5}} = -33.1 \text{ kJ mol}^{-1} \tag{2.13}$$

when the mole fraction scale is used. On the other hand,

$$\Delta G^0 = RT \ln 8.7 \times 10^{-5} = -23.2 \text{ kJ mol}^{-1} \tag{2.14}$$

when the molarity scale is used.

The phase separation model has been questioned for two main reasons. Firstly, according to this model a clear discontinuity in the physical property of a surfactant solution, such as surface tension, turbidity, etc. should be observed at the cmc. This is not always found experimentally and the cmc is not a sharp breakpoint. Secondly, if two phases actually exist at the cmc, then equating the chemical potential of the surfactant molecule in the two phases would imply that the activity of the surfactant in the aqueous phase would be constant above the cmc. If this was the case, the surface tension of a surfactant solution should remain constant above the cmc. However, careful measurements have shown that the surface tension of a surfactant solution decreases slowly above the cmc, particularly when using purified surfactants.

A convenient solution for relating ΔG_m^0 to [cmc] was given by Phillips [18] for ionic surfactants; he arrived at the following expression,

$$\Delta G_m^0 = \{2 - (p/n)\} RT \ln [\text{cmc}], \tag{2.15}$$

where p is the number of free (unassociated) surfactant ions and n is the total number of surfactant molecules in the micelle. For many ionic surfactants, the degree of dissociation (p/n) is ≈ 0.2 so that,

$$\Delta G_m^0 = 1.8 RT \ln [\text{cmc}]. \tag{2.16}$$

Comparison with equation (2.11) clearly shows that for similar ΔG_m, the [cmc] is about two orders of magnitude higher for ionic surfactants when compared with nonionic surfactant of the same alkyl chain length (see Tab. 2.1).

In the presence of excess added electrolyte, with mole fraction x, the free energy of micellization is given by the expression,

$$\Delta G_m^0 = RT \ln [\text{cmc}] + \{1 - (p/n)\} \ln x. \tag{2.17}$$

Equation (2.17) shows that as x increases, the [cmc] decreases.

It is clear from equation (2.15) that as p → 0, i.e. when most charges are associated with counterions,

$$\Delta G_m^0 = 2RT \ln [cmc], \quad (2.18)$$

whereas when p ≈ n, i.e. the counterions are bound to micelles,

$$\Delta G_m^0 = RT \ln [cmc], \quad (2.19)$$

which is the same equation for nonionic surfactants.

2.2.1.3 Enthalpy and entropy of micellization

The enthalpy of micellization can be calculated from the variation of cmc with temperature. This follows from,

$$-\Delta H^0 = RT^2 \frac{d\ln [cmc]}{dT}. \quad (2.20)$$

The entropy of micellization can then be calculated from the relationship between ΔG^0 and ΔH^0, i.e.,

$$\Delta G^0 = \Delta H^0 - T\Delta S^0. \quad (2.21)$$

Therefore, ΔH^0 may be calculated from the surface tension – log C plots at various temperatures. Unfortunately, the errors in locating the cmc (which in many cases is not a sharp point) leads to a large error in the value of ΔH^0. A more accurate and direct method of obtaining ΔH^0 is microcalorimetry. As an illustration, the thermodynamic parameters, ΔG^0, ΔH^0, and $T\Delta S^0$ for octylhexaoxyethylene glycol monoether (C_8E_6) are given in Tab. 2.2.

Tab. 2.2: Thermodynamic quantities for micellization of octylhexaoxyethylene glycol monoether.

Temp / °C	ΔG^0 / kJ mol^{-1}	ΔH^0 / kJ mol^{-1} (from cmc)	ΔH^0 / kJ mol^{-1} (from calorimetry)	$T\Delta S^0$ / kJ mol^{-1} K^{-1}
25	−21.3 ± 2.1	8.0 ± 4.2	20.1 ± 0.8	41.8 ± 1.0
40	−23.4 ± 2.1		−14.6 ± 0.8	38.0 ± 1.0

It can be seen from Tab. 2.2 that ΔG^0 is large and negative. However, ΔH^0 is positive, indicating that the process is endothermic. In addition, $T\Delta S^0$ is large and positive which implies that in the micellization process there is a net increase in entropy. This positive enthalpy and entropy points to a different driving force for micellization from that encountered in many aggregation processes.

The influence of alkyl chain length of the surfactant on the free energy, enthalpy and entropy of micellization was demonstrated by Rosen [20] who listed these parameters as a function of alkyl chain length for sulphoxide surfactants. The results are given in Tab. 2.3 it can be seen that the standard free energy of micellization becomes increasingly negative as the chain length increases. This is to be expected since the cmc decreases with increasing alkyl chain length. However, ΔH^0 becomes less positive and $T\Delta S^0$ becomes more positive with an increase in chain length of the surfactant. Thus, the large negative free energy of micellization is made up of a small positive enthalpy (which decreases slightly with increasing chain length of the surfactant) and a large positive entropy term $T\Delta S^0$, which becomes more positive as the chain is lengthened. As we will see in the next section, these results can be accounted for in terms of the hydrophobic effect which will be described in some detail.

Tab. 2.3: Change in thermodynamic parameters of micellization of alkyl sulphoxide with increasing chain length of the alkyl group.

Surfactant	ΔG / kJ mol^{-1}	ΔH^0 / kJ mol^{-1}	$T\Delta S^0$ / kJ mol^{-1} K^{-1}
$C_6H_{13}S(CH_3)O$	−12.0	10.6	22.6
$C_7H_{15}S(CH_3)O$	−15.9	9.2	25.1
$C_8H_{17}S(CH_3)O$	−18.8	7.8	26.4
$C_9H_{19}S(CH_3)O$	−22.0	7.1	29.1
$C_{10}H_{21}S(CH_3)O$	−25.5	5.4	30.9
$C_{11}H_{23}S(CH_3)O$	−28.7	3.0	31.7

2.2.1.4 Driving force for micelle formation

Until recently, the formation of micelles was regarded primarily as an interfacial energy process, analogous to the process of coalescence of oil droplets in an aqueous medium. If this were the case, micelle formation would be a highly exothermic process, as the interfacial free energy has a large enthalpy component. As mentioned above, experimental results have clearly shown that micelle formation involves only a small enthalpy change and is often endothermic. The negative free energy of micellization is the result of a large positive entropy. This led to the conclusion that micelle formation must be a predominantly entropy driven process.

Two main sources of entropy have been suggested. The first is related to the so-called "hydrophobic effect". This effect was first established from a consideration of the free energy enthalpy and entropy of transfer of hydrocarbon from water to a liquid hydrocarbon. Some results are listed in Tab. 2.4. This table also lists the heat capacity change ΔC_p on transfer from water to a hydrocarbon, as well as C_p^0, gas, i.e. the heat capacity in the gas phase. It can be seen from Tab. 2.4 that the principal contribution to the value of ΔG^0 is the large positive value of ΔS^0, which increases with increasing hydrocarbon chain length, whereas ΔH^0 is positive, or small and negative.

To account for this large positive entropy of transfer several authors [21–23] suggest that the water molecules around a hydrocarbon chain are ordered, forming "clusters" or "icebergs". On transfer of an alkane from water to a liquid hydrocarbon, these clusters are broken, thus releasing water molecules which now have a higher entropy. This accounts for the large entropy of transfer of an alkane from water to a hydrocarbon medium. This effect is also reflected in the much higher heat capacity change on transfer, ΔC_p^0, when compared with the heat capacity in the gas phase, C_p^0. This effect is also operative on transfer of surfactant monomer to a micelle, during the micellization process. The surfactant monomers will also contain "structured" water around their hydrocarbon chain. On transfer of such monomers to a micelle, these water molecules are released and they have a higher entropy.

Tab. 2.4: Thermodynamic parameters for transfer of hydrocarbons from water to liquid hydrocarbon at 25 °C.

Hydrocarbon	ΔG^0 kJ mol^{-1}	ΔH^0 kJ mol^{-1}	ΔS^0 kJ mol^{-1} K^{-1}	ΔC_p^0 kJ mol^{-1} K^{-1}	$C_p^{0,gas}$ kJ mol^{-1} K^{-1}
C_2H_6	−16.4	10.5	88.2	—	—
C_3H_8	−20.4	7.1	92.4	—	—
C_4H_{10}	−24.8	3.4	96.6	−273	−143
C_5H_{12}	−28.8	2.1	105.0	−403	−172
C_6H_{14}	−32.5	0	109.2	−441	−197
C_6H_6	−19.3	−2.1	58.8	−227	−134
$C_6H_5CH_3$	−22.7	−1.7	71.4	−265	−155
$C_6H_5C_2H_5$	−26.0	−2.0	79.8	−319	−185
$C_6H_5C_3H_8$	−29.0	−2.3	88.2	−395	—

The second source of entropy increase on micellization may arise from the increase in flexibility of the hydrocarbon chains on their transfer from an aqueous to a hydrocarbon medium [21]. The orientations and bendings of an organic chain are likely to be more restricted in an aqueous phase compared to an organic phase. It should be mentioned that with ionic and zwitterionic surfactants, an additional entropy contribution associated with the ionic head groups, must be considered. Upon partial neutralization of the ionic charge by the counterions when aggregation occurs, water molecules are released. This will be associated with an entropy increase which should be added to the entropy increase resulting from the hydrophobic effect mentioned above. However, the relative contribution of the two effects is difficult to assess in a quantitative manner.

2.3 Micellization in surfactant mixtures (mixed micelles)

In most cosmetic and personal care applications, more than one surfactant molecule is used in the formulation. It is, therefore, necessary to predict the type of possible interactions and whether these lead to some synergistic effects. Two general cases may be considered: surfactant molecules with no net interaction (with similar head groups) and systems with net interaction [1–3]. The first case is that when mixing two surfactants with the same head group but with different chain lengths. In analogy with the hydrophilic-lipophilic balance (HLB) for surfactant mixtures, one can also assume the cmc of a surfactant mixture (with no net interaction) to be an average of the two cmcs of the single components [1–3],

$$cmc = x_1 cmc_1 + x_2 cmc_2, \qquad (2.22)$$

where x_1 and x_2 are the mole fractions of the respective surfactants in the system. However, the mole fractions should not be those in the whole system, but those inside the micelle. This means that equation (2.22) should be modified,

$$cmc = x_1^m cmc_1 + x_2^m cmc_2. \qquad (2.23)$$

The superscript m indicates that the values are inside the micelle. If x_1 and x_2 are the solution composition, then,

$$\frac{1}{cmc} = \frac{x_1}{cmc_1} + \frac{x_2}{cmc_2}. \qquad (2.24)$$

The molar composition of the mixed micelle is given by,

$$x_1^m = \frac{x_1 cmc_2}{x_1 cmc_2 + x_2 cmc_1}. \qquad (2.25)$$

Fig. 2.10 shows the calculated cmc and the micelle composition as a function of solution composition using equations (2.24) and (2.25) for three cases where $cmc_2/cmc_1 = 1, 0.1,$ and 0.01. As can be seen, the cmc and micellar composition change dramatically with solution composition when the cmcs of the two surfactants vary considerably, i.e. when the ratio of cmcs is far from 1. This fact is used when preparing microemulsions where the addition of medium chain alcohol (like pentanol or hexanol) changes the properties considerably. If component 2 is much more surface active, i.e. $cmc_2/cmc_1 \ll 1$, and it is present in low concentrations (x_2 is of the order of 0.01), then from equation (2.25) $x_1^m \approx x_2^m \approx 0.5$, i.e. at the cmc of the systems the micelles are up to 50 % composed of component 2. This illustrates the role of contaminants in surface activity, e.g. dodecyl alcohol in sodium dodecyl sulphate (SDS).

Fig. 2.11 shows the cmc as a function of molar composition of the solution and in the micelles for a mixture of SDS and nonylphenol with 10 mol ethylene oxide (NP–E_{10}). If the molar composition of the micelles is used as the x-axis, the cmc is more or less the arithmetic mean of the cmcs of the two surfactants. If, on the other hand, the molar composition in the solution is used as the x-axis (which at the cmc is equal to the total molar concentration), then the cmc of the mixture shows a dramatic decrease at low fractions of NP–E_{10}. This decrease is due to the preferential absorption of NP–E_{10} in the micelle. This higher absorption is due to the higher hydrophobicity of the NP–E_{10} surfactant when compared with SDS.

2.3 Micellization in surfactant mixtures (mixed micelles)

With many cosmetic and personal care formulations, surfactants of different kinds are mixed together, for example anionics and nonionics. The nonionic surfactant molecules shield the repulsion between the negative head groups in the micelle and hence there will be a net interaction between the two types of molecules. Another example is the case when anionic and cationic surfactants are mixed, whereby very strong interaction will take place between the oppositely charged surfactant molecules. To account for this interaction, equation (2.25) has to be modified by introducing activity coefficients of the surfactants, f_1^m and f_2^m in the micelle,

$$\text{cmc} = x_1^m f_1^m \text{cmc}_1 + x_2^m f_2^m \text{cmc}_2 . \tag{2.26}$$

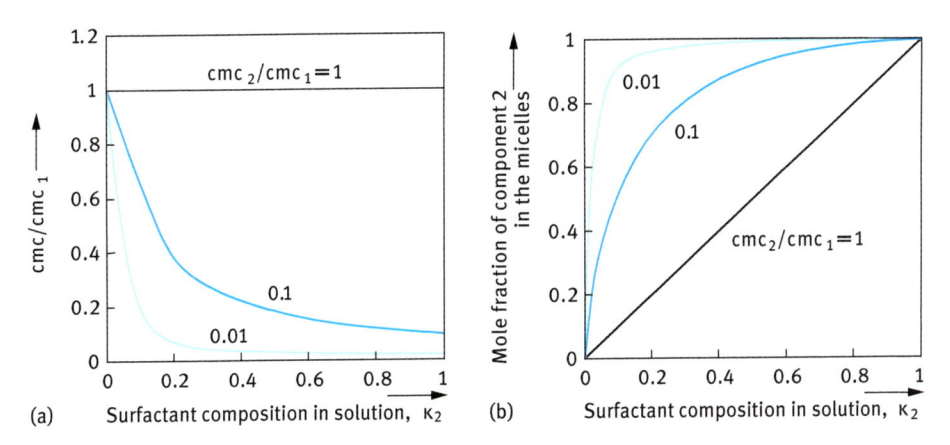

Fig. 2.10: Calculated cmc (a) and micellar composition (b) as a function of solution composition for three ratios of cmcs.

An expression for the activity coefficients can be obtained using the regular solutions theory [1–3],

$$\ln f_1^m = (x_2^m)^2 \beta , \tag{2.27}$$

$$\ln f_2^m = (x_2^m)^2 \beta , \tag{2.28}$$

where β is an interaction parameter between the surfactant molecules in the micelle. A positive β value means that there is a net repulsion between the surfactant molecules in the micelle, whereas a negative β value means a net attraction.

The cmc of the surfactant mixture and the composition x_1 are given by the following equations,

$$\frac{1}{\text{cmc}} = \frac{x_1}{f_1^m \text{cmc}_1} + \frac{x_2}{f_2^m \text{cmc}_2} \tag{2.29}$$

$$x_1^m = \frac{x_1 f_2^m \text{cmc}_2}{x_1 f_2^m \text{cmc}_2 + x_2 f_2^m \text{cmc}_1} . \tag{2.30}$$

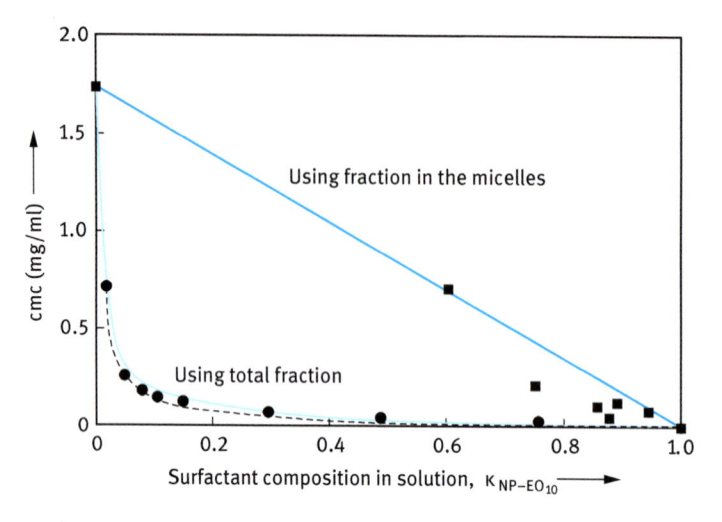

Fig. 2.11: cmc as a function of surfactant composition, x_1, or micellar surfactant composition, x_1^m for the system SDS + NP–E$_{10}$.

Fig. 2.12 shows the effect of increasing the β parameter on the cmc and micellar composition for two surfactant with a cmc ratio of 0.1. This figure shows that as β becomes more negative, the cmc of the mixture decreases. β values in the region of –2 are typical for anionic/nonionic mixtures, whereas values in the region of –10 to –20 are typical of anionic/cationic mixtures. With increasing the negative value of β, the mixed micelles tend towards a mixing ratio of 50 : 50, which reflects the mutual electrostatic attraction between the surfactant molecules. The predicted cmc and micellar composition depends both on the ratio of the cmcs as well as the value of β.

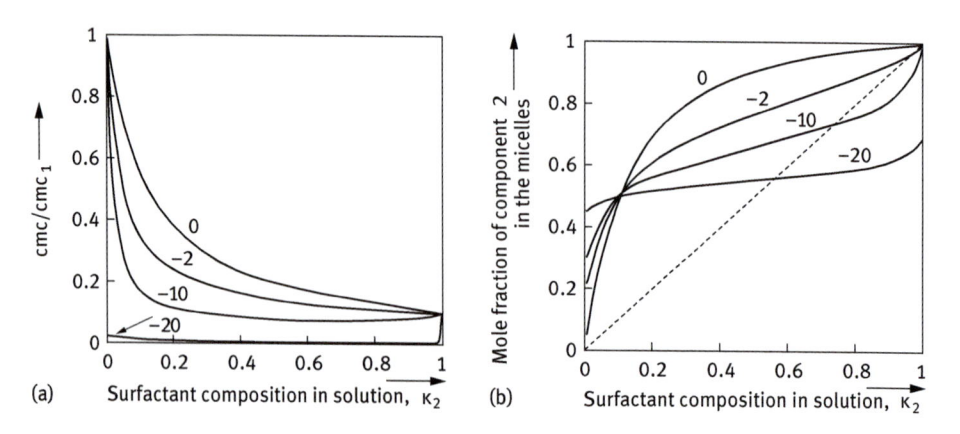

Fig. 2.12: cmc (a) and micellar composition (b) for various values of β for a system with a cmc ratio cmc$_2$/cmc$_1$ of 0.1.

When the cmcs of the single surfactants are similar, the predicted value of the cmc is very sensitive to small variation in β. On the other hand, when the ratio of the cmcs is large, the predicted value of the mixed cmc and the micellar composition are insensitive to variation of the β parameter. For mixtures of nonionic and ionic surfactants, the β decreases with increasing electrolyte concentration. This is due to the screening of the electrostatic repulsion on the addition of electrolyte. With some surfactant mixtures, the β decreases with increasing temperature, i.e. the net attraction decreases with increasing temperature.

2.4 Surfactant–polymer interaction

Mixtures of surfactants and polymers are very common in many cosmetic and personal care formulations. With many suspension and emulsion systems stabilized with surfactants, polymers are added for a number of reasons. For example, polymers are added as suspending agents ("thickeners") to prevent sedimentation or creaming of these systems. Water soluble polymers are added for enhancement of the function of the system, e.g. in shampoos, hair sprays, lotions, and creams. The interaction between surfactants and water soluble polymers results in some synergistic effects, e.g. enhancing the surface activity, stabilizing foams and emulsions, etc. It is therefore important to study the interaction between surfactants and water soluble polymers in a systematic way.

One of the earliest studies of surfactant/polymer interaction came from surface tension measurements. Fig. 2.13 shows some typical results for the effect of addition of polyvinylpyrrolidone (PVP) on the γ–log C curves of SDS [24].

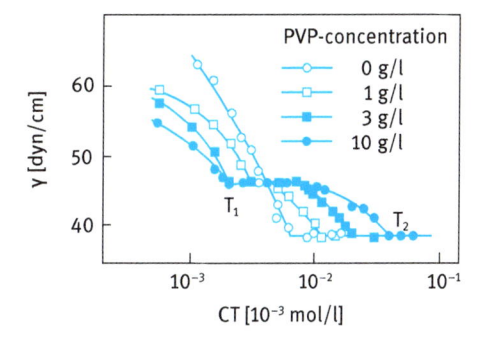

Fig. 2.13: γ–log C curves for SDS solutions in the presence of different concentrations of PVP.

In a system of fixed polymer concentration and varying surfactant concentrations, two critical concentrations appear, denoted by T_1 and T_2. T_1 represents the concentration at which interaction between the surfactant and polymer first occurs. This is sometimes termed the critical aggregation concentration (CAC), i.e. the onset of association of surfactant to the polymer. Because of this there is no further increase in surface activity and thus no lowering of surface tension. T_2 represents the concentration at which the polymer becomes saturated with surfactant. Since T_1 is generally lower than the cmc of the surfactant in the absence of polymer, then "adsorption" or "aggregation" of SDS on or with the polymer is more favourable than normal micellization. As the polymer is saturated with surfactant (i.e. beyond T_2) the surfactant monomer concentration and the activity start to increase again and there is lowering of γ until the monomer concentration reaches the cmc, after which γ remains virtually constant and normal surfactant micelles begin to form.

The above picture is confirmed if the association of surfactant is directly monitored (e.g. by using surfactant selective electrodes, by equilibrium dialysis or by some spectroscopic technique). The binding isotherms are illustrated in Fig. 2.14.

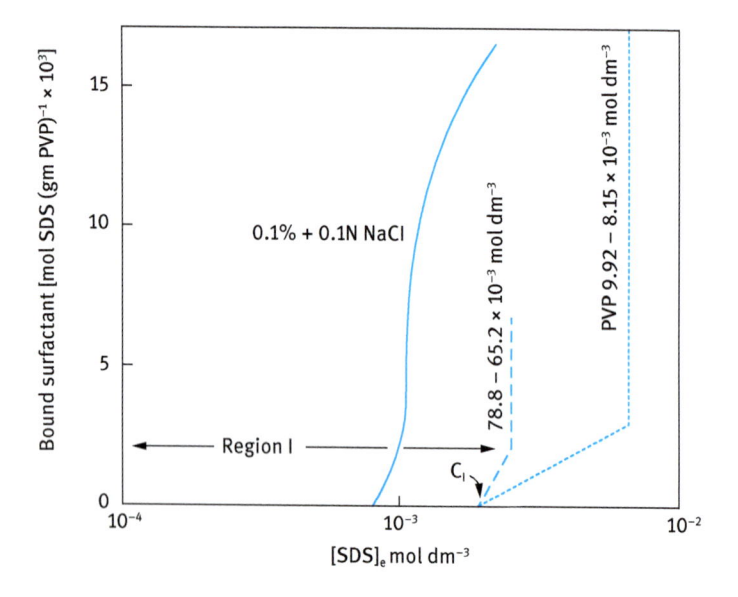

Fig. 2.14: Binding isotherms of a surfactant to a water soluble polymer.

At low surfactant concentration, there is no significant interaction (binding). At the CAC, a strongly co-operative binding is indicated and at higher concentrations a plateau is reached. A further increase in surfactant concentration produces "free" surfactant molecules until the surfactant activity or concentration joins the curve obtained in the absence of polymer. The binding isotherms of Fig. 2.14 show the strong analogy with micelle formation and the interpretation of these isotherms in terms of a depression of the cmc.

Several conclusions could be drawn from the experimental binding isotherms of mixed surfactant/polymer solutions: (i) the CAC/cmc is only weakly dependent on polymer concentration over wide ranges; (ii) CAC/cmc is, to a good approximation, independent of polymer molecular weight down to low values. For very low molecular weight the interaction is weakened; (iii) the plateau binding increases linearly with polymer concentration; (iv) anionic surfactants show a marked interaction with most homopolymers (e.g. PEO and PVP) while cationic surfactants show a weaker but still significant interaction. Nonionic and zwitterionic surfactants only rarely show a distinct interaction with homopolymers.

A schematic representation of the association between surfactants and polymers for a wide range of concentration of both components [25] is shown in Fig. 2.15. It can be seen that at low surfactant concentration (region I) there is no significant association at any polymer concentration. Above the CAC (region II) association increases up to a surfactant concentration which increases linearly with increasing polymer concentration. In region III, association is saturated and the surfactant monomer concentration increases until region IV is reached where there is co-existence of surfactant aggregates at the polymer chains and free micelles.

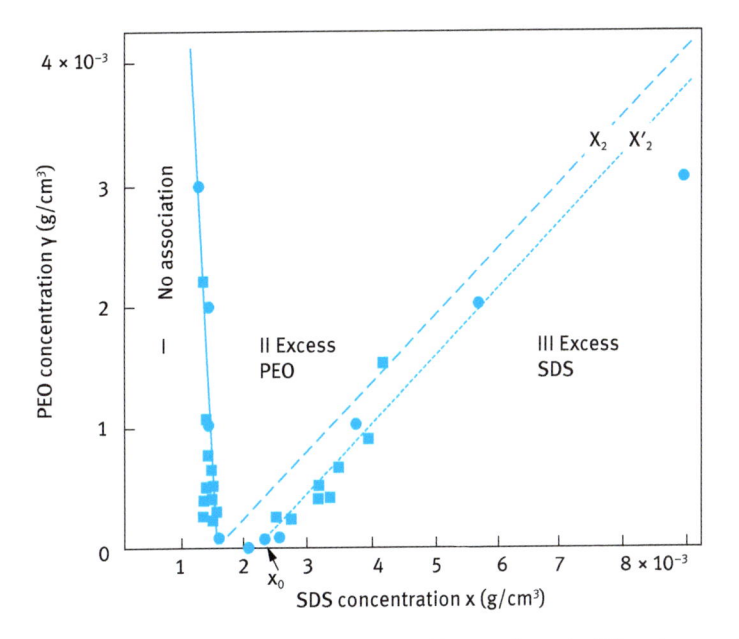

Fig. 2.15: Association between surfactant and homopolymer in different concentration domains [25].

2.4.1 Factors influencing the association between surfactant and polymer

Several factors influence the interaction between surfactant and polymer and these are summarized as follows. (i) Temperature; increasing temperature generally increases the CAC, i.e. the interaction becomes less favourable. (ii) Addition of electrolyte; this generally decreases the CAC, i.e. it increases the binding. (iii) Surfactant chain length; an increase in the alkyl chain length decreases the CAC, i.e. it increases association. A plot of log CAC versus the number of carbon atoms, n, is linear (similar to the log cmc–n relationship obtained for surfactants alone). (iv) Surfactant structure; alkyl benzene sulphonates are similar to SDS, but introduction of EO groups in the chain weakens the interaction. (v) Surfactant classes; weaker interaction is generally observed with cationics when compared to anionics. However, the interaction can be promoted by using a strongly interacting counterion for the cationic (e.g. CNS^-). Interaction between ethoxylated surfactants and nonionic polymers is weak. The interaction is stronger with alkyl phenol ethoxylates. (vi) Polymer molecular weight; a minimum molecular weight of ≈ 4000 for PEO and PVP is required for "complete" interaction. (vii) Amount of polymer; the CAC seems to be insensitive to (or slightly lower) with increasing polymer concentration. T_2 increases linearly with increasing polymer concentration. (viii) Polymer structure and hydrophobicity; several uncharged polymers such as PEO, PVP and polyvinyl alcohol (PVOH) interact with charged surfactants. Many other uncharged polymers interact weakly with charged surfactants, e.g. hydroxyethyl cellulose (HEC), dextran and polyacrylamide (PAAm). For anionic surfactants, the following order of increased interaction has been listed: PVOH < PEO < MEC (methyl cellulose) < PVAc (partially hydrolysed polyvinyl acetate) < PPO ≈ PVP. For cationic surfactants, the following order was listed: PVP < PEO < PVOH < MEC < PVAc < PPO. The position of PVP can be explained by the slight positive charge on the chain which causes repulsion with cations and attraction with anionics.

2.4.2 Interaction models

NMR data showed that every "bound" surfactant molecule experienced the same environment, i.e. the surfactant molecules might be bound in micelle-like clusters, but with smaller size. Assuming that each polymer molecule consists of a number of "effective segments" of mass M_s (minimum molecular weight for interaction to occur), then each segment will bind a cluster of n surfactant anions, D^-, and the binding equilibrium may be represented by,

$$P + nD^- \leftrightarrow PD_n^{n-} \tag{2.31}$$

and the equilibrium constant is given by,

$$K = \frac{[PD_n^{n-}]}{[P][D^-]^n}, \tag{2.32}$$

K is obtained from the half saturation condition,

$$K = [D^-]_{1/2}^n. \tag{2.33}$$

By varying n and using the experimental binding isotherms, one obtains the following values: $M_s = 1830$ and $n = 15$. The free energy of binding is given by the expression,

$$\Delta G^0 = -RT \ln K^{1/n} . \tag{2.34}$$

ΔG^0 was found to be -5.07 kcal mol^{-1} which is close to that for surfactants.

Najaragan [26] introduced a comprehensive thermodynamic treatment of surfactant/polymer interaction. The aqueous solution of surfactant and polymer was assumed to contain both free micelles and "micelles" bound to the polymer molecule. The total surfactant concentration, X_t, is partitioned into single dispersed surfactant, X_1, surfactant in free micelles, X_f, and surfactant bound as aggregates, X_b,

$$X_t = X_1 + g_f(K_f X_1)^{g_f} + g_b n X_p \left[\frac{(K_b X_1)^{g_b}}{1 + (K_b X_1)^{g_b}} \right] . \tag{2.35}$$

g_f is the average aggregation number of free micelles, K_f is the intrinsic equilibrium constant for formation of free micelles, n is the number of binding sites for surfactant aggregates of average size g_b, K_b is the intrinsic equilibrium constant for binding surfactant on the polymer and X_p is the total polymer concentration (mass concentration is nX_p).

The polymer-micelle complexation may affect the conformation of the polymer, but this was assumed not to affect K_b and g_b. The relative magnitudes of K_b, K_f and g_b determine whether complexation with the polymer occurs as well as the critical surfactant concentration exhibited by the system. If $K_f > K_b$ and $g_b = g_f$ then free micelles occur in preference to complexation. If $K_f < K_b$ and $g_b = g_f$, then micelles bound to polymer occur first. If $K_f < K_b$, but g_b is much smaller than g_f, then the free micelles can occur prior to saturation of the polymer. A first critical surfactant concentration (CAC) occurs close to $X_1 = K_b^{-1}$. A second critical concentration occurs near $X_1 = K_f^{-1}$. Depending on the magnitude of nX_p, one may observe only one critical concentration over a finite range of surfactant concentrations.

Fig. 2.16 shows the relationship between X_1 and X_f for different polymer concentrations (SDS/PEO system) using the following values for K_b, K_f, g_f and g_b: $K_b = 319$, $K_f = 120$, $g_b = 51$ and $g_f = 54$.

In the region from 0 to A, the surfactant molecules remain singly dispersed. In the region from A to B, polymer-bound micelles occur; X_1 increases very little in this region (large size of polymer-bound micelles). If g_b is small (say 10), then X_1 should increase more significantly in this region. If nX_p is small, the region AB is confined to a narrow surfactant concentration range. If nX_p is very large, the saturation point B may not be reached. At B the polymer is saturated with surfactant. In the region AC, an increase in X_t is accompanied by an increase in X_1. At C the formation of free micelles becomes possible; CD denotes the surfactant concentration range over which any further addition of surfactant results in formation of free micelles. The point C depends on the polymer mass concentration (nX_p).

The above theoretical predictions were verified by the results of Guilyani and Wolfram [27] using specific ion electrodes. This is illustrated in Fig. 2.17.

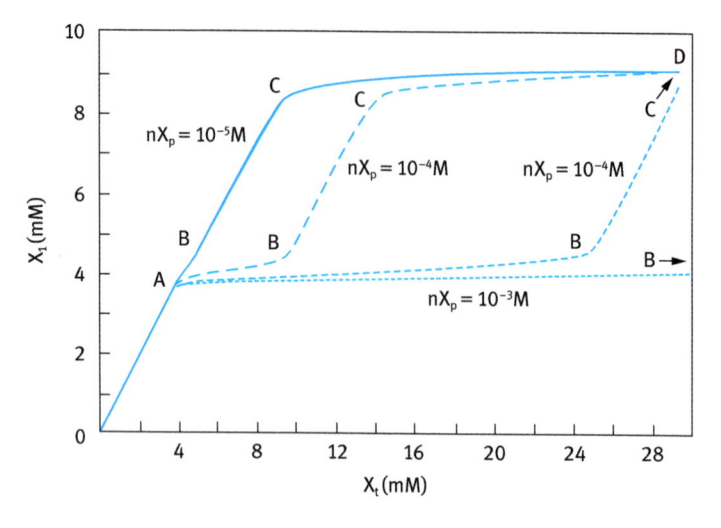

Fig. 2.16: Variation of X_1 with X_t for the SDS/PEO system.

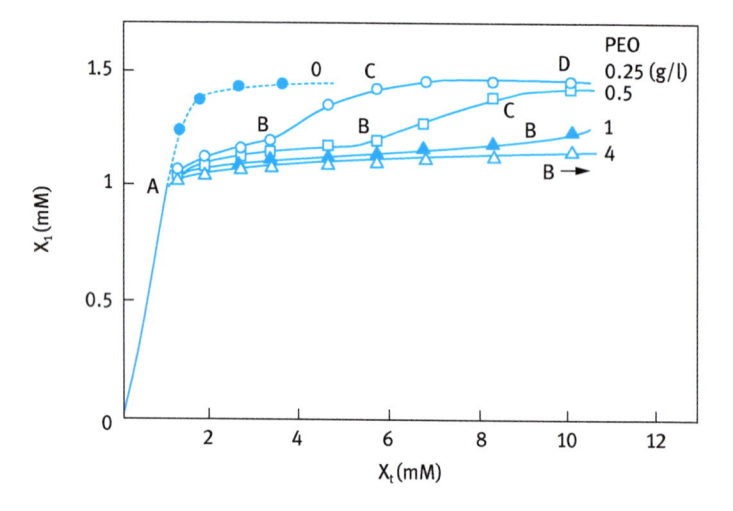

Fig. 2.17: Experimentally measured values of X_1 versus X_t for SDS/PEO system.

2.4.3 Driving force for surfactant/polymer interaction

The driving force for polymer/surfactant interaction is the same as that for the process of micellization (see above). As with micelles the main driving force is the reduction of hydrocarbon/water contact area of the alkyl chain of the dissolved surfactant. A delicate balance between several forces is responsible for the surfactant/polymer association. For example, aggregation is resisted by the crowding of the ionic head groups at the surface of the micelle. Packing constraints also resist association. Molecules that screen the repulsion between the head groups, e.g. electrolytes and alcohol, promote association. A polymer molecule with hydrophobic and hydrophilic segments (which is also flexible) can enhance association by ion-dipole association between the dipole of the hydrophilic groups and the ionic head groups of the surfactant. In addition, contact between the hydrophobic segments of the polymer and the exposed hydrocarbon areas of the micelles can enhance association. With SDS/PEO and SDS/PVP, the association complexes are approximately three monomer units per molecule of aggregated surfactant.

2.4.4 Structure of surfactant/polymer complexes

Generally speaking there are two alternative pictures of mixed surfactant/polymer solutions, one describing the interaction in terms of a strongly co-operative association or binding of the surfactant to the polymer chain and one in terms of a micellization of surfactant on or in the vicinity of the polymer chain. For polymers with hydrophobic groups the binding approach is preferred, whereas for hydrophilic homopolymers the micelle formation picture is more likely. The latter picture has been suggested by Cabane [28] who proposed a structure in which the aggregated SDS is surrounded by macromolecules in a loopy configuration. A schematic picture of this structure, that is sometimes referred to as "pearl-necklace model" is given in Fig. 2.18.

The consequences of the above model are: (i) More favourable free energy of association (CAC < cmc) and increased ionic dissociation of the aggregates. (ii) An altered environment of the CH_2 groups of the surfactant near the head group. The micelle sizes are similar with polymer present and without, and the aggregation numbers are typically similar or slightly lower than those of the micelles forming in the absence of a polymer. In the presence of a polymer, the surfactant chemical potential is lowered with respect to the situation without polymer [29].

Fig. 2.18: Schematic representation of the topology of surfactant/polymer complexes according to Cabane [28].

2.4.5 Surfactant/hydrophobically modified polymer interaction

Water soluble polymers are modified by grafting a low amount of hydrophobic groups (of the order of 1 % of the monomers reacted in a typical molecule) resulting in the formation of "associative structures". These molecules are referred to as associative thickeners and are used as rheology modifiers in many cosmetic and personal care products. An added surfactant will interact strongly with the hydrophobic groups of the polymer, leading to a strengthened association between the surfactant molecules and the polymer chain. A schematic picture for the interaction between SDS and hydrophobically modified hydroxyethyl cellulose (HM-HEC) is shown in Fig. 2.19 which shows the interaction at various surfactant concentrations [1–3].

Initially the surfactant monomers interact with the hydrophobic groups of the HM polymer and at some surfactant concentration (CAC), the micelles can crosslink the polymer chains. At higher surfactant concentrations, the micelles which are now abundant will no longer be shared between the polymer chains, i.e. the crosslinks are broken. These effects are reflected in the variation of viscosity with surfactant concentration for HM polymer as is illustrated in Fig. 2.19. The viscosity of the polymer increases with increasing surfactant concentration, reaching a maximum at an optimum concentration (maximum crosslinks) and then decreases with a further increase in surfactant concentration (Fig. 2.20). For the unmodified polymer, the changes in viscosity are relatively small.

2.4.6 Interaction between surfactants and polymers with opposite charge (surfactant/polyelectrolyte interaction)

The case of surfactant polymer pairs in which the polymer is a polyion and the surfactant is also ionic, but of opposite charge is of special importance in many cosmetic formulations, e.g. hair conditioners. As an illustration, the interaction between SDS and cationically modified cellulosic polymer (Polymer JR, Union Carbide), that is used as a hair conditioner, is shown in Fig. 2.21 using surface tension γ measurements [30]. The γ–log C curves for SDS in the presence and absence of the polyelectrolyte are shown in Fig. 2.22 which also shows the appearance of the solutions. At low surfactant concentration, there is a synergistic lowering of surface tension, i.e. the surfactant/polyelectrolyte complex is more surface active. The low surface tension is also present in the precipitation zone. At high surfactant concentrations, γ approaches that of the polymer-free surfactant in the micellar region. These trends are schematically illustrated in Fig. 2.23

2.4 Surfactant–polymer interaction — 47

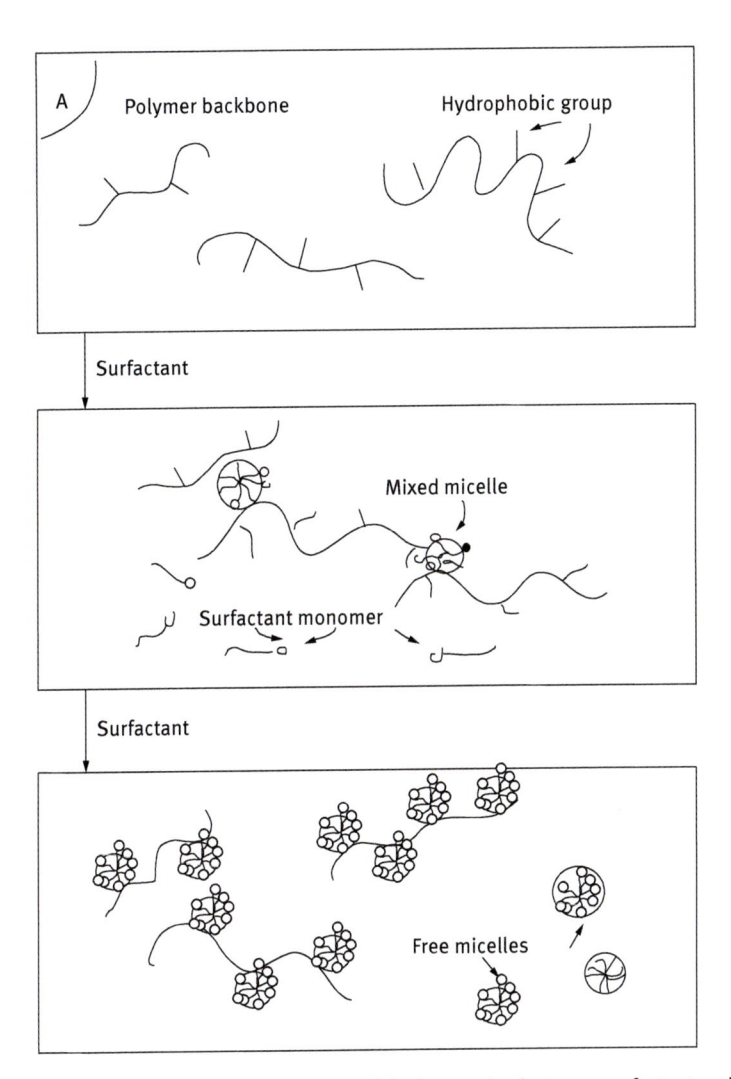

Fig. 2.19: Schematic representation of the interaction between surfactant and HM polymer.

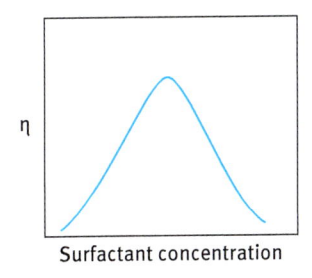

Fig. 2.20: Viscosity-surfactant concentration relationship for HM modified and unmodified polymer solutions.

48 — 2 Surfactants in cosmetics

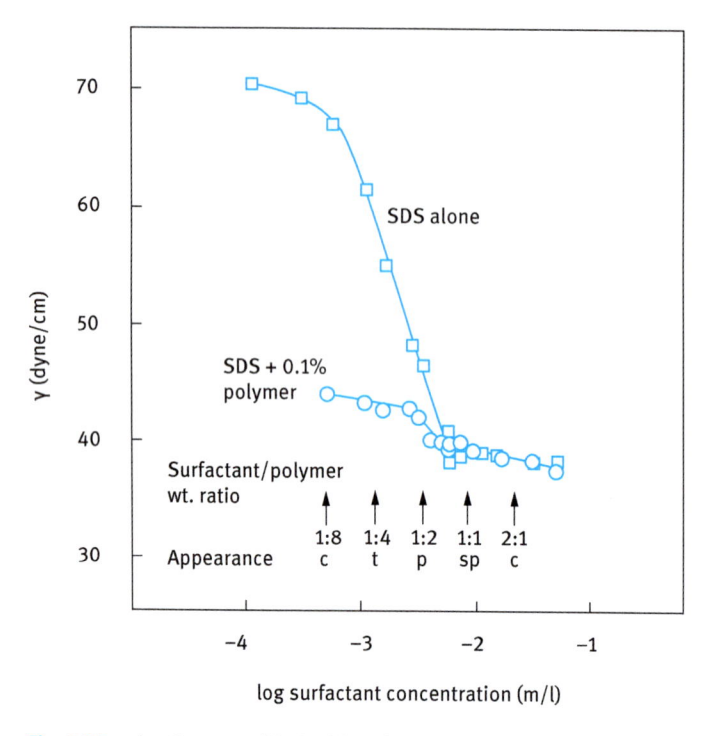

Fig. 2.21: γ–log C curves of SDS with and without addition of polymer (0.1 % JR 400) – c (clear), t (turbid), p (precipitate), sp (slight precipitate).

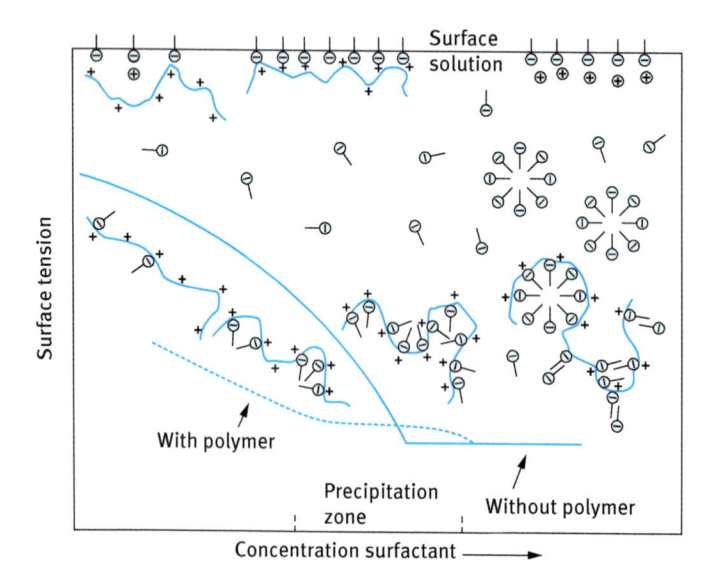

Fig. 2.22: Schematic representation of surfactant/polyelectrolyte interaction.

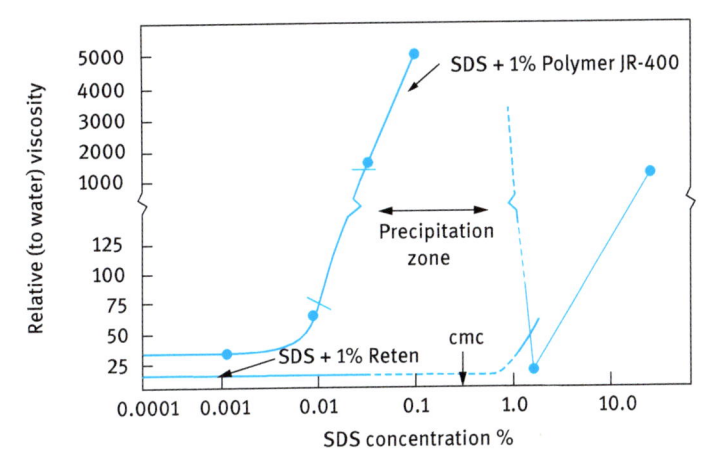

Fig. 2.23: Relative viscosity of 1% JR and 1% Reten as a function of SDS concentration.

The surfactant/polyelectrolyte interaction has several consequences on application in hair conditioners. The most important effect is the high foaming power of the complex. Maximum foaming occurs in the region of highest precipitation, i.e. maximum hydrophobization of the polymer. It is likely that the precipitate stabilize the foam. Direct determination of the amount of surfactant bound to the polyelectrolyte chains revealed a number of interesting features. The binding occurs at very low surfactant concentration (1/20th of the cmc). The degree of binding β reached a value of 0.5 ($\beta = 1$ corresponds to a bound DS^- ion for each ammonium group). β versus SDS concentration curves were identical for polymeric homologues with a degree of cationic substitution (CS) > 0.23. Precipitation occurred when $\beta = 1$.

The binding of cationic surfactants to anionic polyelectrolytes also showed a number of interesting features. The binding affinity depends on the nature of the polyanion. Addition of electrolytes increases the steepness of binding, but the binding occurs at higher surfactant concentration as the electrolyte concentration is increased. Increasing the alkyl chain length of the surfactant increases binding, a process that is similar to micellization.

Viscometric measurements showed a rapid increase in the relative viscosity at a critical surfactant concentration. However, the behaviour depends on the type of polyelectrolyte used. As an illustration, Fig. 2.20 shows the viscosity-SDS concentration curves for two types of cationic polyelectrolyte: JR 400 (cationically modified cellulosic) and Reteen (an acrylamide/β-methylacryloxytrimethyl ammonium chloride copolymer, ex Hercules).

The difference between the two polyelectrolytes is striking and it suggests little change in the conformation of Reten on addition of SDS, but strong intermolecular association between polymer JR 400 and SDS.

References

[1] Tadros, Th. F., "Applied Surfactants", Wiley-VCH, Germany (2005).
[2] Tadros, Th. F., "Cosmetics", in "Encyclopedia of Colloid and Interface Science", Th. F. Tadros (ed.), Springer, Germany (2013).
[3] Tadros, Th. F., "Introduction to Surfactants", De Gruyter, Germany (2014).
[4] Holmberg, K., Jonsson, B., Kronberg, B. and Lindman, B., "Surfactants and Polymers in Aqueous Solution", 2nd Edition, John Wiley & Sons, USA (2003).
[5] Mukerjee, P. and Mysels, K. J., "Critical Micelle Concentrations of Aqueous Surfactant Systems", National Bureau of Standards Publication, Washington DC (1971).
[6] McBain, J. W., Trans. Faraday Soc., **9**, 99 (1913).
[7] Adam, N. K., J. Phys. Chem., **29**, 87 (1925).
[8] Hartley, G. S., "Aqueous Solutions of Paraffin Chain Salts", Hermann and Cie, Paris (1936).
[9] Israelachvili, J. N., "Intermolecular and Surface Forces, with Special applications to Colloidal and Biological Systems", Academic Press, London (1985), p. 251.
[10] McBain, J. W., "Colloid Science", Heath, Boston (1950).
[11] Harkins, W. D., Mattoon, W. D. and Corrin, M. L., J. Amer. Chem. Soc., **68**, 220 (1946); J. Colloid Sci., **1**, 105 (1946).
[12] Debye, P. and Anaker, E. W., J. Phs. and Colloid Chem., **55**, 644 (1951).
[13] Anainsson, E. A. G. and Wall, S. N., J. Phys. Chem., **78**, 1024 (1974); **79**, 857 (1975).
[14] Aniansson, E. A. G., Wall, S. N., Almagren, M., Hoffmann, H., Ulbricht, W., Zana, R., Lang, J. and Tondre, C., J. Phys. Chem., **80**, 905 (1976).
[15] Rassing, J., Sams, P.J. and Wyn-Jones, E., J. Chem. Soc., Faraday II, **70**, 1247 (1974).
[16] Jaycock, M. J. and Ottewill, R. H., Fourth Int. Congress Surface Activity, **2**, 545 (1964).
[17] Okub, T., Kitano, H., Ishiwatari, T. and Isem, N., Proc. Royal Soc., **A36**, 81 (1979).
[18] Phillips, J. N., Trans. Faraday Soc., **51**, 561 (1955).
[19] Kahlweit, M. and Teubner, M., Adv. Colloid Interface Sci., **13**, 1 (1980).
[20] Rosen, M. L., "Surfactants and Interfacial Phenomena", Wiley-Interscience, NY (1978).
[21] Tanford, "The Hydrophobic Effect", 2nd Edition, Wiley, NY (1980).
[22] Stainsby, G. and Alexander, A. E., Trans. Faraday Soc., **46**, 587 (1950).
[23] Arnow, R. H. and Witten, L., J. Phys. Chem., **64**, 1643 (1960).
[24] Robb, I. D., Chemistry and Industry, 530535 (1972).
[25] Cabane, B. and Duplessix, R., J. Phys. (Paris), **43**, 1529 (1982).
[26] Nagaarajan, R., Colloids and Surfaces, **13**, 1 (1885).
[27] Gilyani, T. and Wolfram, E., Colloids and Surfaces, **3**, 181 (1981).
[28] Cabane, B., J. Phys. Chem., **81**, 1639 (1977).
[29] Evans, D. F. and Winnerstrom, H., "The Colloidal Domain. Where Physics, Chemistry, Biology and Technology Meet", John Wiley and Sons INC. VCH, NY (1994). p. 312.
[30] Goddard, E. D., Colloids and Surfaces, **19**, 301 (1986).

3 Polymeric surfactants in cosmetic formulations

3.1 Introduction

As mentioned in Chapter 2, polymeric (macromolecular) surfactants possess considerable advantages for use in cosmetic ingredients. The most commonly used materials are the ABA block copolymers, with A being poly(ethylene oxide) and B poly(propylene oxide) (Pluronics). On the whole, polymeric surfactants have much lower toxicity, sensitization and irritation potentials, provided they are not contaminated with traces of the parent monomers. These molecules provide greater stability and in some case they can be used to adjust the viscosity of the cosmetic formulation [1–3].

In this chapter, I will give a section on the general classification of polymeric surfactants. This is followed by a section on the adsorption and conformation of polymeric surfactants at interfaces, with particular reference to the solid/liquid interface. The last section gives two examples of the high stability obtained using model oil-in-water (O/W) and water-in-oil (W/O) emulsions.

3.2 General classification of polymeric surfactants

Perhaps the simplest type of polymeric surfactant is a homopolymer [4] that is formed from the same repeating units, such as poly(ethylene oxide) or poly(vinylpyrrolidone). These homopolymers have little surface activity at the O/W interface, since the homopolymer segments (ethylene oxide or vinylpyrrolidone) are highly water soluble and have little affinity to the interface. However, such homopolymers may adsorb significantly at the S/L interface. Even if the adsorption energy per monomer segment to the surface is small (fraction of kT, where k is the Boltzmann constant and T is the absolute temperature), the total adsorption energy per molecule may be sufficient to overcome the unfavourable entropy loss of the molecule at the S/L interface.

Clearly, homopolymers are not the most suitable emulsifiers or dispersants. A small variant is to use polymers that contain specific groups that have high affinity to the surface. This is exemplified by partially hydrolysed poly(vinyl acetate) (PVAc), technically referred to as poly(vinyl alcohol) (PVA). The polymer is prepared by partial hydrolysis of PVAc, leaving some residual vinyl acetate groups. Most commercially available PVA molecules contain 4–12 % acetate groups. These acetate groups, which are hydrophobic, give the molecule its amphipathic character. On a hydrophobic surface such as polystyrene, the polymer adsorbs with preferential attachment of the acetate groups on the surface, leaving the more hydrophilic vinyl alcohol segments dangling in the aqueous medium. These partially hydrolysed PVA molecules also exhibit surface activity at the O/W interface [1–3].

The most convenient polymeric surfactants are those of the block and graft copolymer type [4]. A block copolymer is a linear arrangement of blocks of variable monomer composition. The nomenclature for a diblock is poly-A-block–poly-B and for a triblock is poly-A-block–poly-B–poly-A. One of the most widely used triblock polymeric surfactants are the "Pluronics" (BASF, Germany), which consist of two poly-A blocks of poly(ethylene oxide) (PEO) and one block of poly(propylene oxide) (PPO). Several chain lengths of PEO and PPO are available. More recently triblocks of PPO–PEO–PPO (inverse Pluronics) became available for some specific applications. These polymeric triblocks can be applied as emulsifiers or dispersants, whereby the assumption is made that the hydrophobic PPO chain resides at the hydrophobic surface, leaving the two PEO chains dangling in aqueous solution and hence providing steric repulsion. The reason for the surface activity of the PEO–PPO–PEO triblock copolymers at the O/W interface may stem from a process of "rejection" anchoring of the PPO chain since it is not soluble in both oil and water. Several other di- and triblock copolymers have been synthesized, although these are of limited commercial availability. Typical examples are diblocks of polystyrene-block–polyvinyl alcohol, triblocks of poly(methyl methacrylate)-block poly(ethylene oxide)-block poly(methyl methacrylate), diblocks of polystyrene block–polyethylene oxide and triblocks of polyethylene oxide-block polystyrene-polyethylene oxide [4]. An alternative (and perhaps more efficient) polymeric surfactant is the amphipathic graft copolymer consisting of a polymeric backbone B (polystyrene or polymethyl methacrylate) and several A chains ("teeth") such as polyethylene oxide [1–3] This graft copolymer is sometimes referred to as a "comb" stabilizer. This copolymer is usually prepared by grafting a macromonomer such as methoxy polyethylene oxide methacrylate with polymethyl methacrylate. The "grafting onto" technique has also been used to synthesize polystyrene-polyethylene oxide graft copolymers.

Recently, graft copolymers based on polysaccharides [5–7] have been developed for stabilization of disperse systems [8]. One of the most useful graft copolymers are those based on inulin obtained from chicory roots. It is a linear polyfructose chain with a glucose end. When extracted from chicory roots, inulin has a wide range of chain lengths ranging from 2–65 fructose units. It is fractionated to obtain a molecule with narrow molecular weight distribution with a degree of polymerization > 23 and this is commercially available as INUTEC®N25. The latter molecule is used to prepare a series of graft copolymers by random grafting of alkyl chains (using alky isocyanate) on the inulin backbone. The first molecule of this series is INUTEC®SP1 (Beneo-Remy, Belgium) that is obtained by random grafting of C_{12} alkyl chains. It has an average molecular weight of ≈ 5000 Daltons and its structure is given in Fig. 3.1. The molecule is schematically illustrated in Fig. 3.2 which shows the hydrophilic polyfructose chain (backbone) and the randomly attached alkyl chains. The main advantages of INUTEC®SP1 as stabilizer for disperse systems are: (i) Strong adsorption to the particle or droplet by multipoint attachment with several alkyl chains. This ensures lack of desorption and displacement of the molecule from the interface. (ii) Strong

Fig. 3.1: Structure of INUTEC® SP1.

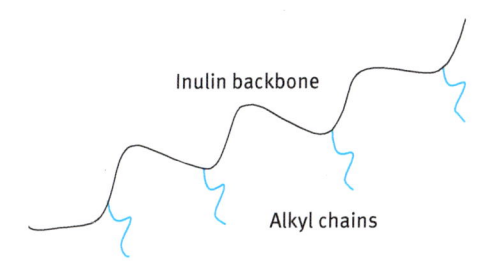

Fig. 3.2: Schematic representation of INUTEC® SP1 polymeric surfactant.

hydration of the linear polyfructose chains both in water and in the presence of high electrolyte concentrations and high temperatures. This ensures effective steric stabilization (see Chapter 5).

3.3 Polymeric surfactant adsorption and conformation

Understanding the adsorption and conformation of polymeric surfactants at interfaces is key to knowing how these molecules act as stabilizers. Most basic ideas on adsorption and conformation of polymers have been developed for the solid/liquid interface [9]. The adsorption and conformation of polymers at the solid/liquid interface requires understanding of the various interactions involved (polymer/surface, chain/solvent and surface/solvent). The most important problem to be solved is the conformation of the polymer molecule at the solid surface. This was recognized by

Jenkel and Rumbach in 1951 [10] who found that the amount of polymer adsorbed per unit area of the surface would correspond to a layer more than 10 molecules thick if all the segments of the chain are attached. They suggested a model in which each polymer molecule is attached in sequences separated by bridges which extend into solution. In other words, not all the segments of a macromolecule are in contact with the surface. The segments which are in direct contact with the surface are termed "trains"; those in between and extended into solution are termed "loops"; the free ends of the macromolecule also extending in solution are termed "tails". This is illustrated in Fig. 3.1 (a) for a homopolymer. Examples of homopolymers that are formed from the same repeating units are poly(ethylene oxide) or poly(vinylpyrrolidone). These homopolymers have little surface activity at the O/W interface, since the homopolymer segments (ethylene oxide or vinylpyrrolidone) are highly water soluble and have little affinity to the interface. However, such homopolymers may adsorb significantly at the S/L interface. Even if the adsorption energy per monomer segment to the surface is small (fraction of kT, where k is the Boltzmann constant and T is the absolute temperature), the total adsorption energy per molecule may be sufficient to overcome the unfavourable entropy loss of the molecule at the S/L interface.

Clearly, homopolymers are not the most suitable emulsifiers or dispersants. A small variant is to use polymers that contain specific groups that have high affinity to the surface. This is exemplified by partially hydrolysed poly(vinyl acetate) (PVAc),

(a) Homopolymer sequence of loop stails and trains

(b) Chains with "blocks" that have higher affinity to the surface

(c) Chain lying flat on the surface

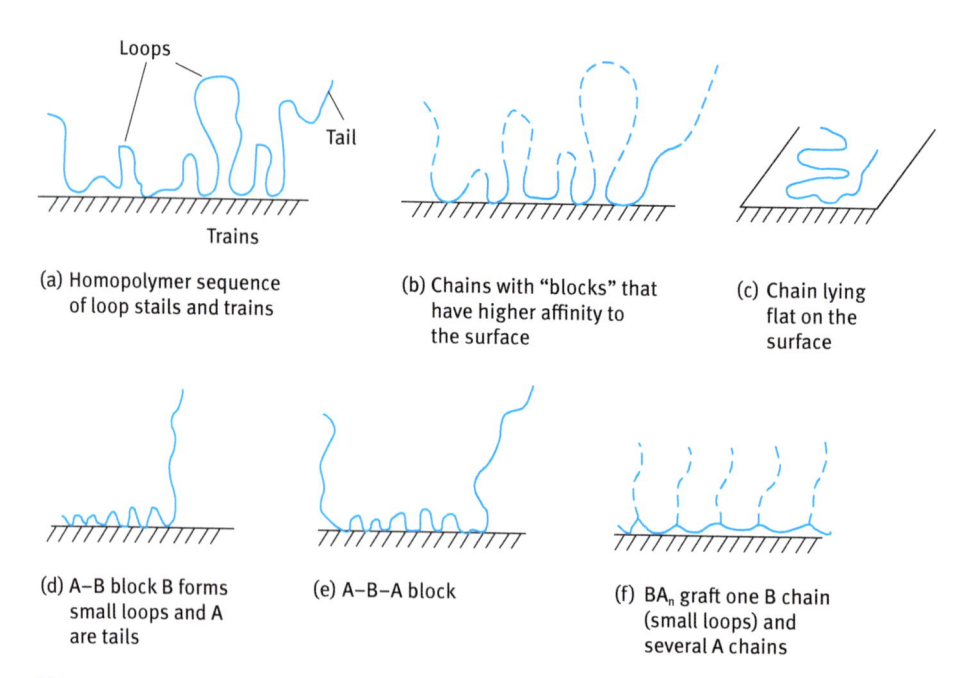

(d) A–B block B forms small loops and A are tails

(e) A–B–A block

(f) BA$_n$ graft one B chain (small loops) and several A chains

Fig. 3.3: Various conformations of macromolecules on a plane surface.

technically referred to as poly(vinyl alcohol) (PVA). The polymer is prepared by partial hydrolysis of PVAc, leaving some residual vinyl acetate groups. Most commercially available PVA molecules contain 4–12 % acetate groups. These acetate groups, which are hydrophobic, give the molecule its amphipathic character. On a hydrophobic surface such as polystyrene, the polymer adsorbs with preferential attachment of the acetate groups on the surface, leaving the more hydrophilic vinyl alcohol segments dangling in the aqueous medium. The configuration of such "blocky" copolymers is illustrated in Fig. 3.1 (b). Clearly, if the molecule is made fully from hydrophobic segments, the chain will adopt a flat configuration as is illustrated in Fig. 3.1 (c). The most convenient polymeric surfactants are those of the block and graft copolymer type. A block copolymer is a linear arrangement of blocks of variable monomer composition. The nomenclature for a diblock is poly-A-block–poly-B and for a triblock is poly-A-block–poly-B–poly-A. An example of an A–B diblock is polystyrene block–polyethylene oxide and its conformation is represented in Fig. 3.1 (d).

One of the most widely used triblock polymeric surfactants are the "Pluronics" (BASF, Germany) which consists of two poly-A blocks of poly(ethylene oxide) (PEO) and one block of poly(propylene oxide) (PPO). Several chain lengths of PEO and PPO are available. More recently triblocks of PPO–PEO–PPO (inverse Pluronics) became available for some specific applications. These polymeric triblocks can be applied as emulsifiers or dispersants, whereby the assumption is made that the hydrophobic PPO chain resides at the hydrophobic surface, leaving the two PEO chains dangling in aqueous solution and hence providing steric repulsion. Several other triblock copolymers have been synthesized, although these are of limited commercial availability. Typical examples are triblocks of poly(methyl methacrylate)-block poly(ethylene oxide)-block poly(methyl methacrylate). The conformation of these triblock copolymers is illustrated in Fig. 3.1 (e). An alternative (and perhaps more efficient) polymeric surfactant is the amphipathic graft copolymer consisting of a polymeric backbone B (polystyrene or polymethyl methacrylate) and several A chains ("teeth") such as polyethylene oxide. This graft copolymer is sometimes referred to as a "comb" stabilizer. Its configuration is illustrated in Fig. 3.1 (f).

The adsorption and conformation of the polymer at the surface can be predicted using theoretical treatments. The most successful theories are based on statistical thermodynamics [11–21]. More recently a step-weighted random walk [22–27] was developed to predict the adsorption isotherm, the fraction of segments in direct contact with the surface (trains) and the layer extension (loops and tails). For full characterization of polymer adsorption one needs to measure the amount of polymer adsorbed per unit area Γ, the fraction of segments in direct contact with the surface p and the adsorbed layer thickness δ. The various methods that can be applied for such measurements are described. Examples of adsorption of homopolymers and block copolymers are given.

The polymer/surface interaction is described in terms of adsorption energy per segment χ^s. The polymer/solvent interaction is described in terms of the Flory–

Huggins interaction parameter χ. For adsorption to occur, a minimum energy of adsorption per segment χ^s is required. When a polymer molecule adsorbs on a surface, it loses configurational entropy and this must be compensated by an adsorption energy χ^s per segment. This is schematically shown in Fig. 3.4, where the adsorbed amount Γ is plotted versus χ^s. The minimum value of χ^s can be very small (< 0.1 kT) since a large number of segments per molecule are adsorbed. For a polymer with say 100 segments and 10 % of these are in trains, the adsorption energy per molecule now reaches 1 kT (with χ^s = 0.1 kT). For 1000 segments, the adsorption energy per molecule is now 10 kT.

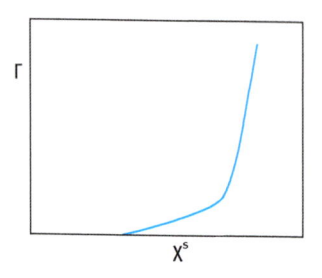

Fig. 3.4: Variation of adsorption amount Γ with adsorption energy per segment χ^s.

As mentioned above, homopolymers are not the most suitable for stabilization of dispersions. For strong adsorption, one needs the molecule to be "insoluble" in the medium and to have a strong affinity ("anchoring") to the surface. For stabilization, one needs the molecule to be highly soluble in the medium and strongly solvated by its molecules; this requires a Flory–Huggins interaction parameter less than 0.5. The above opposing effects can be resolved by introducing "short" blocks in the molecule which are insoluble in the medium and have a strong affinity to the surface, for example partially hydrolysed polyvinyl acetate (88 % hydrolysed, i.e. with 12% acetate groups), usually referred to as polyvinyl alcohol (PVA),

$$-(CH_2-CH)_x-(CH_2-CH^-)_y-(CH_2-CH)_x-$$
$$\;\;\;\;\;\;\;\;\;|\;\;\;\;\;\;\;\;\;\;\;\;\;\;\;\;|\;\;\;\;\;\;\;\;\;\;\;\;\;\;\;\;\;|$$
$$\;\;\;\;\;\;\;\;OH\;\;\;\;\;\;\;\;OCOCH_3\;\;\;\;OH$$

As mentioned above, these requirements are better satisfied using A–B, A–B–A and BA$_n$ graft copolymers. B is chosen to be highly insoluble in the medium and it should have high affinity to the surface. This is essential to ensure strong "anchoring" to the surface (irreversible adsorption). A is chosen to be highly soluble in the medium and strongly solvated by its molecules. The Flory–Huggins parameter χ can be applied in this case. For a polymer in a good solvent χ has to be lower than 0.5; the smaller the χ value the better the solvent for the polymer chains. Examples of B for a hydrophobic particle in aqueous media are polystyrene, polymethylmethacrylate. Examples of A in aqueous media are polyethylene oxide, polyacrylic acid, polyvinylpyrrolidone and polysaccharides. For nonaqueous media such as hydrocarbons, the A chain(s) could be poly(12-hydroxystearic acid).

For a full description of polymer adsorption one needs to obtain information on the following: (i) The amount of polymer adsorbed Γ (in mg or mol) per unit area of the particles. It is essential to know the surface area of the particles in the suspension. Nitrogen adsorption on the powder surface may give such information (by application of the BET equation) provided there will be no change in area on dispersing the particles in the medium. For many practical systems, a change in surface area may occur on dispersing the powder, in which case one has to use dye adsorption to measure the surface area (some assumptions have to be made in this case). (ii) The fraction of segments in direct contact with the surface, i.e. the fraction of segments in trains p (p = (Number of segments in direct contact with the surface)/(Total Number)). (iii) The distribution of segments in loops and tails, $\rho(z)$, which extend in several layers from the surface. $\rho(z)$ is usually difficult to obtain experimentally although recently application of small angle neutron scattering could obtain such information. An alternative and useful parameter for assessing "steric stabilization" is the hydrodynamic thickness, δ_h (thickness of the adsorbed or grafted polymer layer plus any contribution from the hydration layer). Several methods can be applied to measure δ_h as will be discussed below.

3.3.1 Measurement of the adsorption isotherm

This is by far the easiest to obtain [28]. One measures the polymeric surfactant concentration before ($C_{initial}$, C_1) and after ($C_{equilibrium}$, C_2)

$$\Gamma = \frac{(C_1 - C_2)V}{A}, \qquad (3.1)$$

where V is the total volume of the solution and A is the specific surface area (m² g⁻¹). It is necessary in this case to separate the particles from the polymer solution after adsorption. This could be carried out by centrifugation and/or filtration. One should make sure that all particles are removed. To obtain this isotherm, one must develop a sensitive analytical technique for determination of the polymeric surfactant concentration in the ppm range. It is essential to follow the adsorption as a function of time to determine the time required to reach equilibrium. For some polymer molecules such as polyvinyl alcohol, PVA, and polyethylene oxide, PEO, (or blocks containing PEO), analytical methods based on complexation with iodine/potassium iodide or iodine/boric acid potassium iodide have been established. For some polymers with specific functional groups spectroscopic methods may be applied, e.g. UV, IR or fluorescence spectroscopy. A possible method is to measure the change in refractive index of the polymer solution before and after adsorption. This requires very sensitive refractometers. High resolution NMR has been recently applied since the polymer molecules in the adsorbed state are in different environment than those in the bulk. The chemical shifts of functional groups within the chain are different in these two environments. This has the attraction of measuring the amount of adsorption without separating the particles.

3.3.2 Measurement of the fraction of segments p

The fraction of segments in direct contact with the surface can be directly measured using spectroscopic techniques: (i) IR if there is specific interaction between the segments in trains and the surface, e.g. polyethylene oxide on silica from nonaqueous solutions [26]. (ii) Electron spin resonance (ESR); this requires labelling of the molecule. (iii) NMR, pulse gradient or spin ECO NMR. This method is based on the fact that the segments in trains are "immobilized" and hence they have lower mobility than those in loops and tails [26].

An indirect method of determination of p is to measure the heat of adsorption ΔH using microcalorimetry [26]. One should then determine the heat of adsorption of a monomer H_m (or molecule representing the monomer, e.g. ethylene glycol for PEO); p is then given by the equation,

$$p = \frac{\Delta H}{H_m n}, \qquad (3.2)$$

where n is the total number of segments in the molecule.

The above indirect method is not very accurate and can only be used in a qualitative sense. It also requires very sensitive enthalpy measurements (e.g. using an LKB microcalorimeter).

3.3.3 Determination of the segment density distribution $\rho(z)$ and adsorbed layer thickness δ_h

The segment density distribution $\rho(z)$ is given by the number of segments parallel to the surface in the z-direction. Three direct methods can be applied for determination of adsorbed layer thickness: ellipsometry, attenuated total reflection (ATR) and neutron scattering. Both ellipsometry and ATR [26] depend on the difference between refractive indices between the substrate, the adsorbed layer and bulk solution and they require flat reflecting surface. Ellipsometry [26] is based on the principle that light undergoes a change in polarizability when it is reflected at a flat surface (whether covered or uncovered with a polymer layer).

The above limitations when using ellipsometry or ATR are overcome by the application technique of neutron scattering, which can be applied to both flat surfaces as well as particulate dispersions. The basic principle of neutron scattering is to measure the scattering due to the adsorbed layer when the scattering length density of the particle is matched to that of the medium (the so-called "contrast matching" method). Contrast matching of particles and medium can be achieved by changing the isotopic composition of the system (using deuterated particles and mixture of D_2O and H_2O). Apart from obtaining δ, one can also determine the segment density distribution $\rho(z)$.

The above technique of neutron scattering gives a clear quantitative picture of the adsorbed polymer layer. However, its application in practice is limited since one needs to prepare deuterated particles or polymers for the contrast matching procedure. The practical methods for determination of the adsorbed layer thickness are mostly based on hydrodynamic methods. Several methods may be applied to determine the hydrodynamic thickness of adsorbed polymer layers of which viscosity, sedimentation coefficient (using an ultracentrifuge) and dynamic light scattering measurements are the most convenient. The most commonly used technique is based on dynamic light scattering that is referred to a Photon Correlation Spectroscopy (PCS) which allows one to obtain the diffusion coefficient of the particles with and without the adsorbed layer (D_δ and D respectively). This is obtained from measurement of the intensity fluctuation of scattered light as the particles undergo Brownian diffusion [29, 30]. When a light beam (e.g. monochromatic laser beam) passes through a dispersion, an oscillating dipole is induced in the particles, thus re-radiating the light. Due to the random arrangement of the particles (which are separated by a distance comparable to the wavelength of the light beam, i.e. the light is coherent with the interparticle distance), the intensity of the scattered light will, at any instant, appear as random diffraction or a "speckle" pattern. As the particles undergo Brownian motion, the random configuration of the speckle pattern changes. The intensity at any one point in the pattern will, therefore, fluctuate such that the time taken for an intensity maximum to become a minimum (i.e. the coherence time) corresponds approximately to the time required for a particle to move one wavelength. Using a photomultiplier of active area about the size of a diffraction maximum, i.e. approximately one coherence area, this intensity fluctuation can be measured. A digital correlator is used to measure the photocount or intensity correlation function of the scattered light. The photocount correlation function can be used to obtain the diffusion coefficient D of the particles. For monodisperse non-interacting particles (i.e. at sufficient dilution), the normalized correlation function $[g^{(1)}(\tau)]$ of the scattered electric field is given by the equation,

$$[g^{(1)}(\tau)] = \exp-(\Gamma\tau), \tag{3.3}$$

where τ is the correlation delay time and Γ is the decay rate or inverse coherence time. Γ is related to D by the equation,

$$\Gamma = DK^2, \tag{3.4}$$

where K is the magnitude of the scattering vector that is given by,

$$K = \left(\frac{4n}{\lambda_0}\right)\sin\left(\frac{\theta}{2}\right), \tag{3.5}$$

where n is the refractive index of the solution, λ is the wavelength of light in vacuum and θ is the scattering angle.

From D, the particle radius R is calculated using the Stokes–Einstein equation,

$$D = \frac{kT}{6\pi \rho \eta R}, \qquad (3.6)$$

where k is the Boltzmann constant and T is the absolute temperature. For a polymer coated particle R is denoted R_δ which is equal to $R + \delta_h$. Thus, by measuring D_δ and D, one can obtain δ_h. It should be mentioned that the accuracy of the PCS method depends on the ratio of δ/R, since δ_h is determined by difference. Since the accuracy of the measurement is +1 %, δ_h should be at least 10 % of the particle radius. This method can only be used with small particles and reasonably thick adsorbed layers.

3.4 Examples of the adsorption results of nonionic polymeric surfactant

3.4.1 Adsorption isotherms

Fig. 3.5 shows the adsorption isotherms for PEO with different molecular weights on PS (at room temperature) [31]. It can be seen that the amount adsorbed in mg m^{-2} increases with increasing polymer molecular weight. Fig. 3.6 shows the variation of the hydrodynamic thickness δ_h with molecular weight M. δ_h shows a linear increase with log M. δ_h increases with n, the number of segments in the chain according to,

$$\delta_h \approx n^{0.8}. \qquad (3.7)$$

Fig. 3.7 shows the adsorption isotherms of PVA with various molecular weights on PS latex (at 25 °C) [28]. The polymers were obtained by fractionation of a commercial sample of PVA with an average molecular weight of 45 000. The polymer also contained

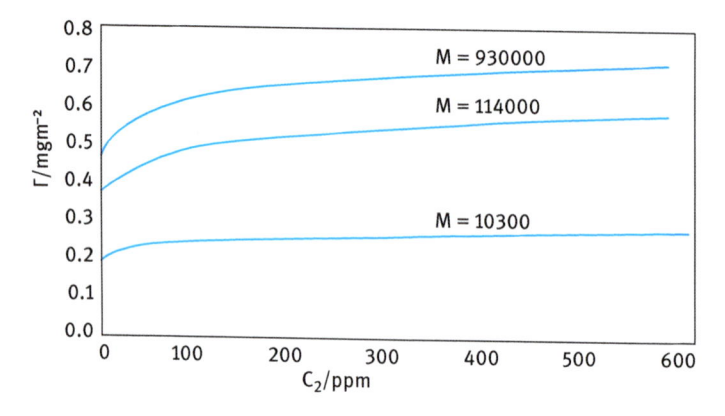

Fig. 3.5: Adsorption isotherms for PEO on PS.

3.4 Examples of the adsorption results of nonionic polymeric surfactant — 61

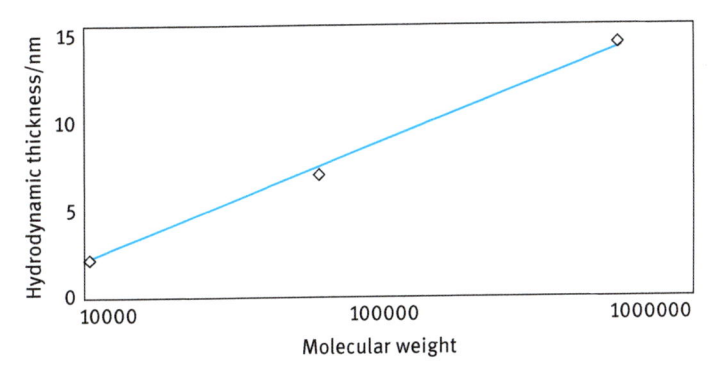

Fig. 3.6: Hydrodynamic thickness of PEO on PS as a function of the molecular weight.

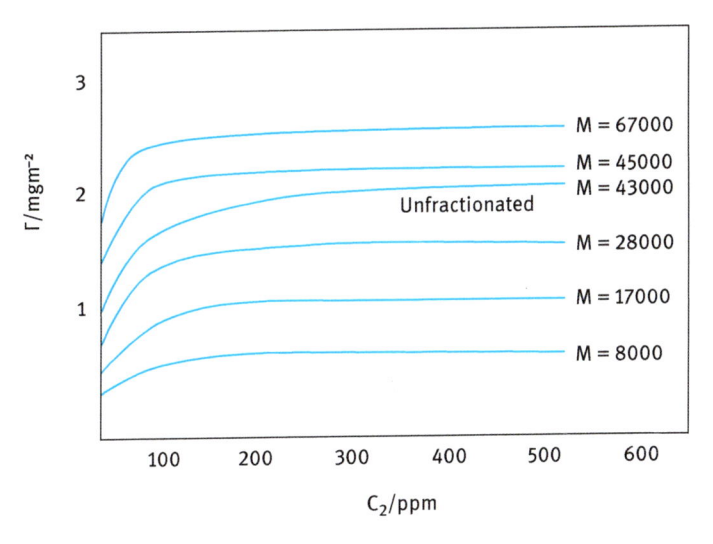

Fig. 3.7: Adsorption isotherms of PVA with different molecular weights on polystyrene latex at 25 °C.

12 % vinyl acetate groups. As with PEO, the amount of adsorption increases with increasing M. The isotherms are also of the high affinity type. Γ at the plateau increases linearly with $M^{1/2}$.

The hydrodynamic thickness was determined using PCS and the results are given below.

M	67 000	43 000	28 000	17 000	8 000
δ_h / nm	25.5	19.7	14.0	9.8	3.3

δ_h seems to increase linearly with increasing molecular weight.

The effect of solvency on adsorption was investigated by increasing the temperature (the PVA molecules are less soluble at higher temperature) or addition of electrolyte (KCl) [32]. The results are shown in Figs. 3.8 and 3.9 for M = 65 100. As can be seen from

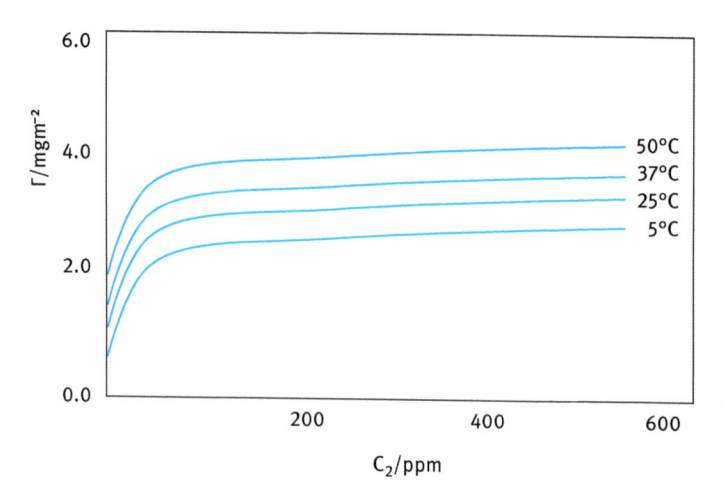

Fig. 3.8: Influence of temperature on adsorption.

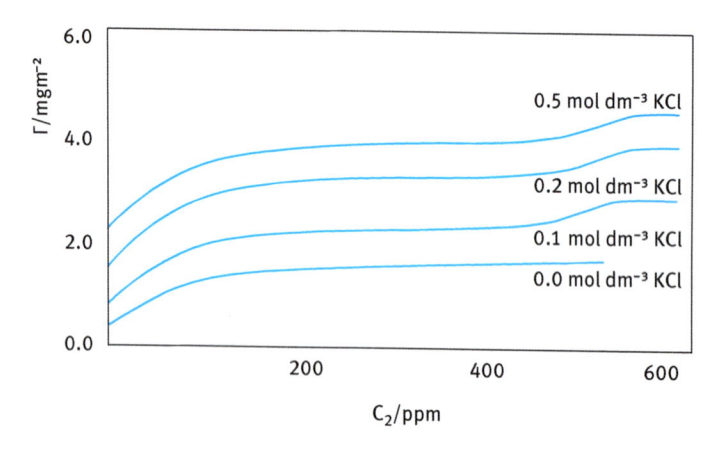

Fig. 3.9: Influence of addition of KCl on adsorption.

Fig. 3.8, an increase in temperature results in a reduction of solvency of the medium for the chain (due to breakdown of hydrogen bonds) and this results in an increase in the amount adsorbed. Addition of KCl (which reduces the solvency of the medium for the chain) results in an increase in adsorption (as predicted by theory).

The adsorption of block and graft copolymers is more complex since the intimate structure of the chain determines the extent of adsorption. Random copolymers adsorb in an intermediate way to that of the corresponding homopolymers. Block copolymers retain the adsorption preference of the individual blocks. The hydrophilic block (e.g. PEO), the buoy, extends away from the particle surface into the bulk solution, whereas the hydrophobic anchor block (e.g. PS or PPO) provides firm attachment to the surface. Fig. 3.10 shows the theoretical prediction of diblock copolymer adsorp-

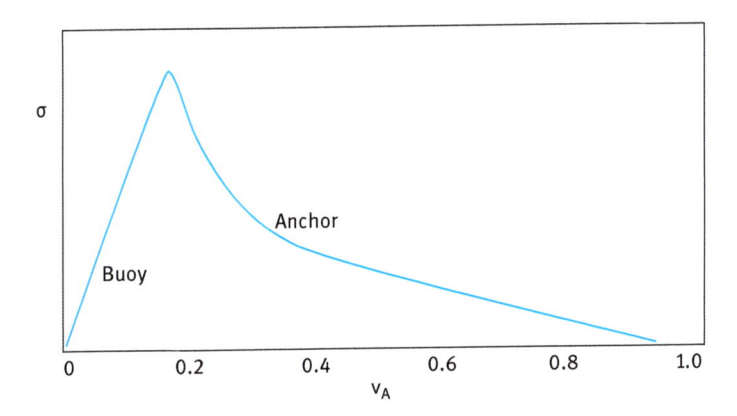

Fig. 3.10: Prediction of adsorption of diblock copolymer.

tion according to the Scheutjens and Fleer (SF) theory [23–25]. The surface density σ is plotted versus the fraction of anchor segments ν_A. The adsorption depends on the anchor/buoy composition.

The amount of adsorption is higher than for homopolymers and the adsorbed layer thickness is more extended and dense. For a triblock copolymer A–B–A, with two buoy chains and one anchor chain, the behaviour is similar to that of diblock copolymers. This is shown in Fig. 3.11 for PEO–PPO–PEO block (Pluronic).

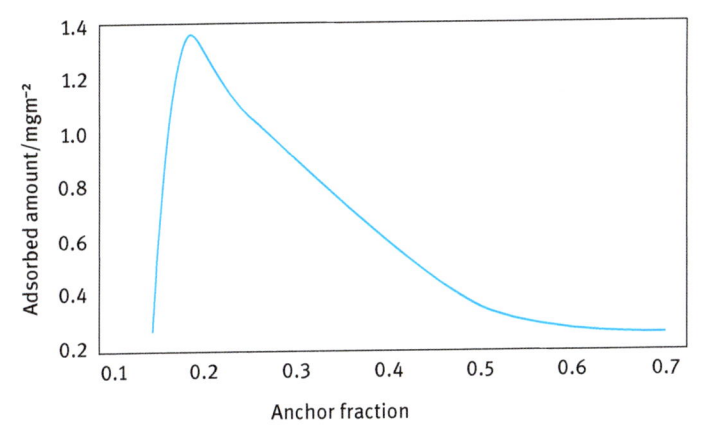

Fig. 3.11: Adsorbed amount (mg m-2) versus fraction of anchor segment for an A–B–A triblock copolymer (PEO–PPO–PEO).

3.4.2 Adsorbed layer thickness results

Fig. 3.12 shows a plot of ρ(z) against z for PVA (M = 37000) adsorbed on deuterated PS latex in D_2O/H_2O [33].

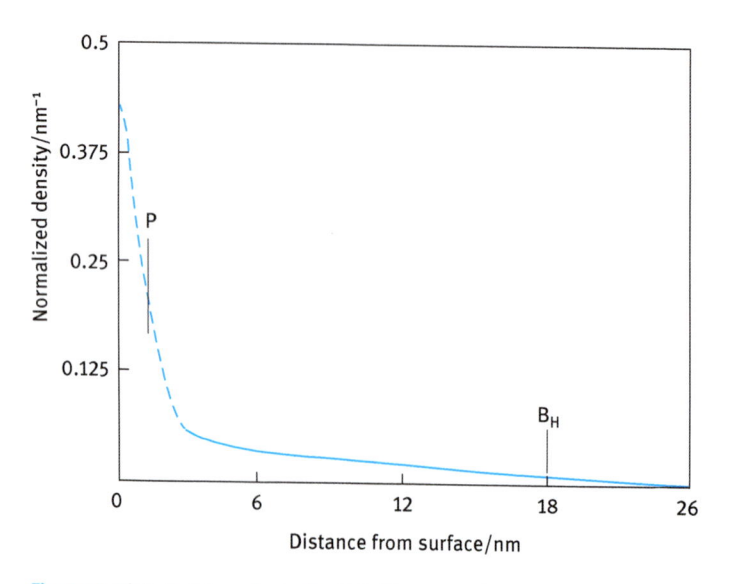

Fig. 3.12: Plot of ρ(z) against z for PVA (M = 37000) adsorbed on deuterated PS latex in D_2O/H_2O.

The results shows a monotonic decay of ρ(z) with distance z from the surface and several regions may be distinguished. Close to the surface (0 < z < 3 nm), the decay in ρ(z) is rapid and assuming a thickness of 1.3 nm for the bound layer, p was calculated to be 0.1, which is in close agreement with the results obtained using NMR measurements. In the middle region, ρ(z) shows a shallow maximum followed by a slow decay which extends to 18 nm, i.e. close to the hydrodynamic layer thickness δ_h of the polymer chain (see below). δ_h is determined by the longest tails and is about 2.5 times the radius of gyration in bulk solution (≈ 7.2 nm). This slow decay of ρ(z) with z at long distances is in qualitative agreement with Scheutjens and Fleer's theory [23] which predicts the presence of long tails. The shallow maximum at intermediate distances suggests that the observed segment density distribution is a summation of a fast monotonic decay due to loops and trains together with the segment density for tails which have a maximum density away from the surface. The latter maximum was clearly observed for a sample which had PEO grafted to a deuterated polystyrene latex [34] (where the configuration is represented by tails only).

The hydrodynamic thickness of block copolymers shows different behaviour from that of homopolymers (or random copolymers). Fig. 3.13 shows the theoretical prediction of the adsorbed layer thickness δ that is plotted as a function of v.

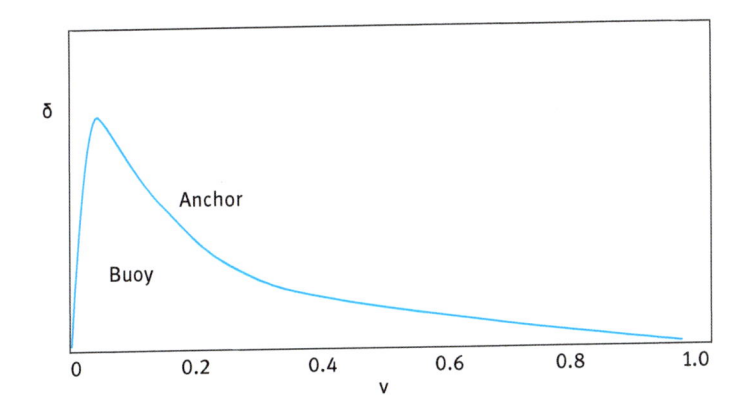

3.4 Examples of the adsorption results of nonionic polymeric surfactant — 65

Fig. 3.13: Theoretical predictions of the adsorbed layer thickness for a diblock copolymer.

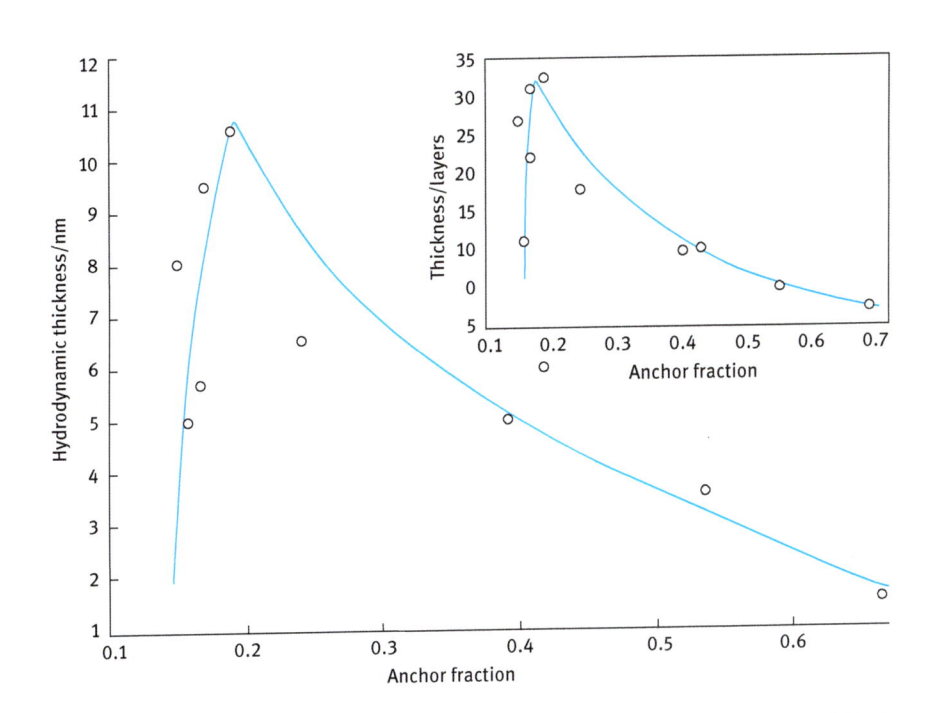

Fig. 3.14: Hydrodynamic thickness versus fraction of anchor segment v_A for PEO–PPO–PEO block copolymer onto polystyrene latex. Insert shows the mean field calculation of thickness versus anchor fraction using the SF theory.

Fig. 3.14 shows the hydrodynamic thickness versus fraction of anchor segment for an A–B–A block copolymer of (polyethylene oxide)–poly(propylene oxide)–poly(ethylene oxide) (PEO–PPO–PEO) [31] versus fraction of anchor segment. The theoretical (Scheutjens and Fleer) prediction of adsorbed amount and layer thickness versus fraction of anchor segment are shown in the inserts of Fig. 3.14. When there are two buoy blocks and a central anchor block, as in the above example, the A–B–A block shows similar behaviour to that of an A–B block. However, if there are two anchor blocks and a central buoy block, surface precipitation of the polymer molecule at the particle surface is observed and this is reflected in a continuous increase of adsorption with increasing polymer concentration as has been show for an A–B–A block of PPO–PEO–PPO [31].

3.5 Kinetics of polymer adsorption

The kinetics of polymer adsorption is a highly complex process. Several distinct processes can be distinguished, each with a characteristic time scale [31]. These processes may occur simultaneously and hence it is difficult to separate them. The first process is the mass transfer of the polymer to the surface, which may be either diffusion or convection. Having reached the surface, the polymer must then attach itself to a surface site, which depends on any local activation energy barrier. Finally, the polymer will undergo large-scale rearrangements as it changes from its solution conformation to a "tail-train-loop" conformation. Once the polymer has reached the surface, the amount of adsorption increases with time. The increase is rapid at the beginning but subsequently slows as the surface becomes saturated. The initial rate of adsorption is sensitive to the bulk polymer solution concentration and molecular weight as well as the solution viscosity. Nevertheless, all the polymer molecules arriving at the surface tend to adsorb immediately. The concentration of unadsorbed polymer around the periphery of the forming layer (the surface polymer solution) is zero and, therefore the concentration of polymer in the interfacial region is significantly greater than the bulk polymer concentration. Mass transport is found to dominate the kinetics of adsorption until 75 % of full surface coverage. At higher surface coverage, the rate of adsorption decreases since the polymer molecules arriving at the surface cannot immediately adsorb. Over time, equilibrium is set up between this interfacial concentration of polymer and the concentration of polymer in the bulk. Given that the adsorption isotherm is of the high affinity type, no significant change in adsorbed amount is expected, even over decades of polymer concentration. If the surface polymer concentration increases toward that of the bulk solution, the rate of adsorption decreases because the driving force for adsorption (the difference in concentration between the surface and bulk solutions) decreases. Adsorption processes tend to be very rapid and an equilibrated polymer layer can form within several 1000 s. However, desorption is a much slower process and this can take several years!

3.6 Emulsions stabilized by polymeric surfactants

Two examples are given to illustrate the effective stabilization when using polymeric surfactants. The first example is an O/W emulsion based on a graft copolymer AB$_n$ with A being polyfructose and B are several alkyl groups grafted on the polyfructose chain, i.e. INUTEC®SP1 described above [8]. This polymeric surfactant produces enhanced steric stabilization both in water and high electrolyte concentrations as will be discussed later. The second example is a W/O emulsion stabilized using an A–B–A block copolymer of poly(12-hydroxystearic acid) (PHS) (the A chains) and poly(ethylene oxide) (PEO) (the B chain): PHS–PEO–PHS. The PEO chain (that is soluble in the water droplets) forms the anchor chain, whereas the PHS chains form the stabilizing chains. PHS is highly soluble in most hydrocarbon solvents and is strongly solvated by its molecules. The structure of the PHS–PEO–PHS block copolymer is schematically shown in Fig. 3.15.

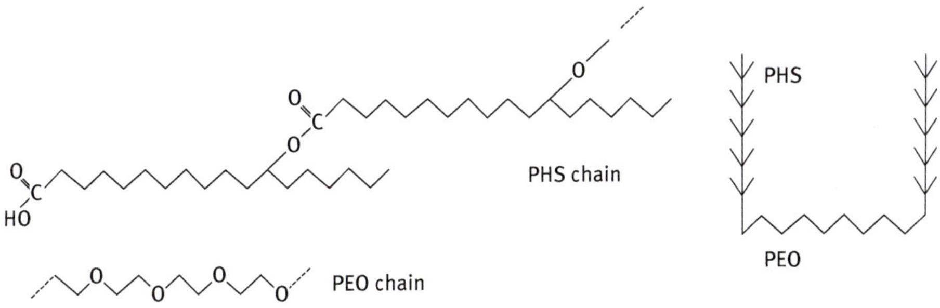

Fig. 3.15: Schematic representation of the structure of PHS-PEO-PHS block copolymer.

The conformation of the polymeric surfactant at the W/O interface is schematically shown in Fig. 3.16.

Emulsions of Isopar M/water and cyclomethicone/water were prepared using INUTEC®SP1. 50/50 (v/v) O/W emulsions were prepared and the emulsifier concentration was varied from 0.25 to 2 (w/v) % based on the oil phase. 0.5 (w/v) % emulsifier was sufficient for stabilization of these 50/50 (v/v) emulsions [8]. The emulsions were stored at room temperature and 50 °C and optical micrographs were taken at intervals of time (for a year) in order to check the stability. Emulsions prepared in water were very stable showing no change in droplet size distribution over more than a one year period and this indicated absence of coalescence. Any weak flocculation that occurred was reversible and the emulsion could be redispersed by gentle shaking.

Fig. 3.17 shows an optical micrograph for a dilute 50/50 (v/v) emulsion that was stored for 1.5 and 14 weeks at 50 °C. No change in droplet size was observed after storage for more than one year at 50 °C, indicating absence of coalescence. The same result was obtained when using different oils. Emulsions were also stable against coales-

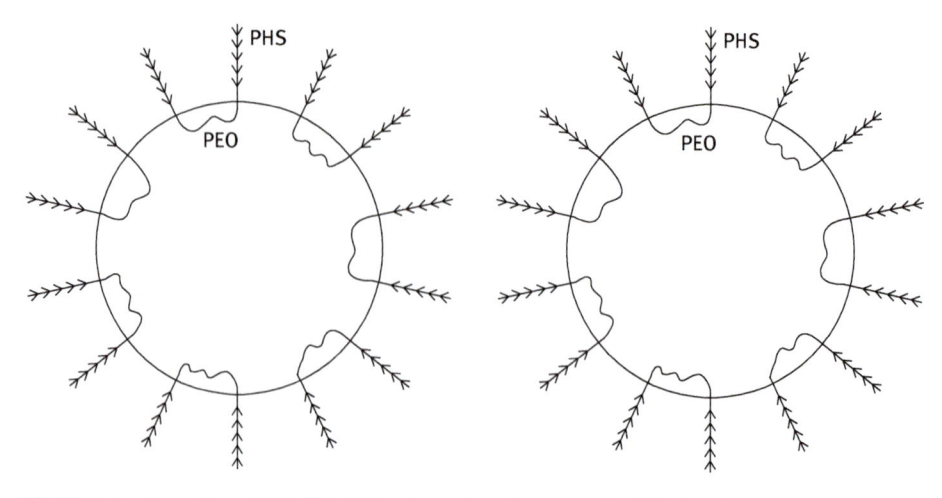

Fig. 3.16: Conformation of PHS-PEO-PHS polymeric surfactant at the W/O interface.

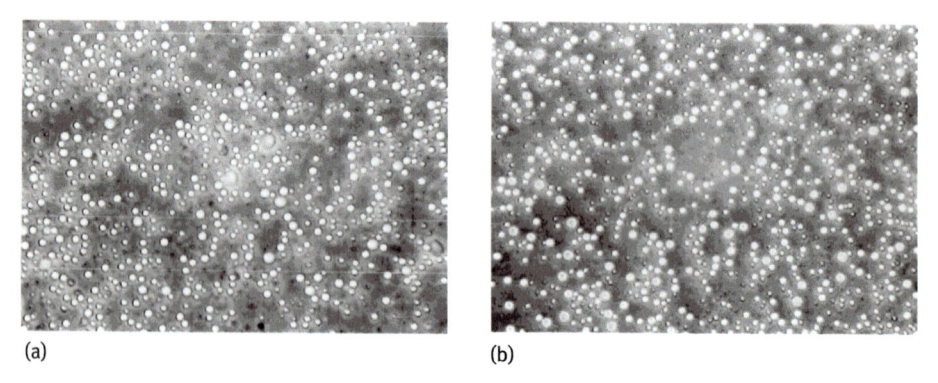

Fig. 3.17: Optical micrographs of O/W emulsions stabilized with INUTEC®SP1 stored at 50 °C for 1.5 weeks (a) and 14 weeks (b).

cence in the presence of high electrolyte concentrations (up to 4 mol dm^{-3} or ≈ 25 % NaCl). This stability in high electrolyte concentrations is not observed with polymeric surfactants based on polyethylene oxide.

The high stability observed using INUTEC®SP1 is related to its strong hydration both in water and in electrolyte solutions. The hydration of inulin (the backbone of HMI) could be assessed using cloud point measurements. A comparison was also made with PEO with two molecular weights, namely 4000 and 20 000. Solutions of PEO 4000 and 20 000 showed a systematic decrease in the cloud point with an increase in NaCl or MgSO$_4$ concentration. In contrast, inulin showed no cloud point up to 4 mol dm^{-3} NaCl and up to 1 mol dm^{-3} MgSO$_4$. These results explain the difference between PEO and inulin. With PEO, the chains show dehydration when the NaCl concentration is increased above 2 mol dm^{-3} or 0.5 mol dm^{-3} MgSO$_4$. The inulin

3.6 Emulsions stabilized by polymeric surfactants

chains remain hydrated at much higher electrolyte concentrations. It seems that the linear polyfructose chains remain strongly hydrated at high temperature and high electrolyte concentrations.

The high emulsion stability obtained when using INUTEC®SP1 can be accounted for by the following factors: (i) The multipoint attachment of the polymer by several alkyl chains that are grafted on the backbone. (ii) The strong hydration of the polyfructose "loops" both in water and high electrolyte concentrations (χ remains below 0.5 under these conditions). (iii) The high volume fraction (concentration) of the loops at the interface. (iv) Enhanced steric stabilization; this is the case with multipoint attachment which produces strong elastic interaction.

Evidence for the high stability of the liquid film between emulsion droplets when using INUTEC®SP1 was obtained by Exerowa et al. [31] using disjoining pressure measurements. This is illustrated in Fig. 3.18 which shows a plot of disjoining pressure versus separation distance between two emulsion droplets at various electrolyte concentrations. The results show that by increasing the capillary pressure a stable Newton Black Film (NBF) is obtained at a film thickness of ≈ 7 nm. The lack of rupture of the film at the highest pressure applied of 4.5×10^4 Pa indicates the high stability of the film in water and in high electrolyte concentrations (up to $2.0 \, \text{mol dm}^{-3}$ NaCl). This result is consistent with the high emulsion stability obtained at high electrolyte concentrations and high temperature. Emulsions of Isopar M-in-water are very stable

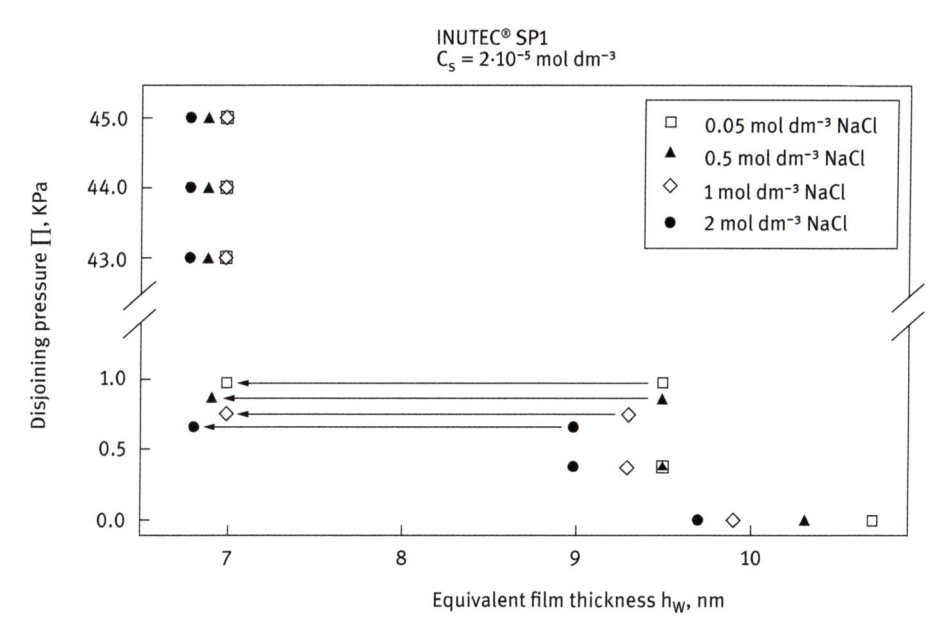

Fig. 3.18: Variation of disjoining pressure with equivalent film thickness at various NaCl concentrations.

under such conditions and this could be accounted for by the high stability of the NBF. The droplet size of 50 : 50 O/W emulsions prepared using 2 % INUTEC®SP1 is in the region of 1–10 μm. This corresponds to a capillary pressure of $\approx 3 \times 10^4$ Pa for the 1 μm drops and $\approx 3 \times 10^3$ Pa for the 10 μm drops. These capillary pressures are lower than those to which the NBF have been subjected to and this clearly indicates the high stability obtained against coalescence in these emulsions.

W/O emulsions (the oil being Isopar M) were prepared using PHS-PEO-PHS block copolymer at high water volume fractions (> 0.7). The emulsions have a narrow droplet size distribution with a z-average radius of 183 nm. They also remained fluid up to high water volume fractions (> 0.6). This could be illustrated from viscosity–volume fraction curves as is shown in Fig. 3.19. The effective volume fraction ϕ_{eff} of the emulsions (the core droplets plus the adsorbed layer) could be calculated from the relative viscosity and using the Dougherty–Krieger equation [36],

$$\eta_r = \left[1 - \frac{\phi_{\text{eff}}}{\phi_p}\right]^{-[\eta]\phi_p}, \tag{3.8}$$

where η_r is the relative viscosity, ϕ_p is the maximum packing fraction (≈ 0.7) and $[\eta]$ is the intrinsic viscosity that is equal to 2.5 for hard spheres.

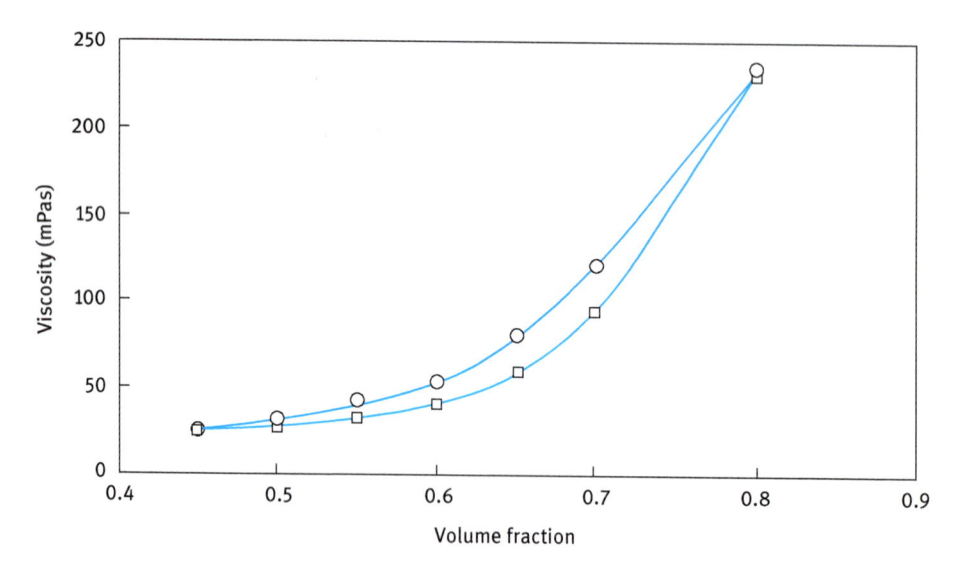

Fig. 3.19: Viscosity–volume fraction for W/O emulsion stabilized with PHS–PEO–PHS block copolymer.

The calculations based on equation (3.8) are shown in Fig. 3.17 (square symbols). From the effective volume fraction ϕ_{eff} and the core volume fraction ϕ, the adsorbed layer thickness could be calculated. This was found to be in the region of 10 nm at $\phi = 0.4$ and it decreased with increasing ϕ.

The W/O emulsions prepared using the PHS–PEO–PHS block copolymer remained stable both at room temperature and 50 °C. This is consistent with the structure of the block copolymer: the B chain (PEO) is soluble in water and it forms a very strong anchor at the W/O interface. The PHS chains (the A chains) provide effective steric stabilization since the chains are highly soluble in Isopar M and are strongly solvated by its molecules.

References

[1] Tadros, Th. F., "Polymeric Surfactants", in "Encyclopedia of Colloid and Interface Science", Th. F. Tadros (ed.), Springer, Germany (2013).
[2] Tadros, Th. F., in "Principles of Polymer Science and Technology in Cosmetics and Personal Care", E. D. Goddard and J. V. Gruber (eds.), Marcel Dekker, NY (1999).
[3] Tadros, Th. F., in "Novel Surfactants", K. Holmberg (ed.), Marcel Dekker, NY (2003).
[4] Piirma, I., "Polymeric Surfactants", Surfactant Science Series, No. 42, NY, Marcel Dekker, (1992).
[5] Stevens, C. V., Meriggi, A., Peristerpoulou, M., Christov, P. P., Booten, K., Levecke, B., Vandamme, A., Pittevils, N. and Tadros, Th. F., Biomacromolecules, **2**, 1256 (2001).
[6] Hirst, E. L., McGilvary, D. I., and Percival, E. G., J. Chem. Soc., 1297 (1950).
[7] Suzuki, M., in "Science and Technology of Fructans", M. Suzuki and N. J. Chatterton (eds.), CRC Press, Boca Raton, FL. (1993), p. 21.
[8] Tadros, Th. F., Vandamme, A., Levecke, B., Booten, K. and Stevens, C. V., Advances Colloid Interface Sci., **108-109**, 207 (2004).
[9] Tadros, Th. F., in "Polymer Colloids", R. Buscall, T. Corner and J. F. Stageman (eds.), Applied Sciences, Elsevier, London (1985) p. 105.
[10] Jenkel E., and Rumbach, R., Z. Elekrochem., **55**, 612 (1951).
[11] Simha, R., Frisch, L. and Eirich, F. R., J. Phys. Chem., **57**, 584 (1953).
[12] Silberberg, A., J. Phys. Chem., **66**, 1872 (1962).
[13] Di Marzio, E. A., J. Chem. Phys., **42**, 2101 (1965).
[14] Di Marzio, E. A. and McCrakin, F. L., J. Chem. Phys., **43**, 539 (1965).
[15] Hoeve, C. A., Di Marzio, E. A. and Peyser, P., J. Chem. Phys., **42**, 2558 (1965).
[16] Silberberg, A., J. Chem. Phys., **48**, 2835 (1968).
[17] Hoeve, C. A. J., J. Chem. Phys., **44**, 1505 (1965); J. Chem. Phys., **47**, 3007 (1966).
[18] Hoeve, C. A., J. Polym. Sci., **30**, 361 (1970); **34**, 1 (1971).
[19] Silberberg, A., J. Colloid Interface Sci., **38**, 217 (1972)
[20] Hoeve, C. A. J., J. Chem. Phys., **44**, 1505 (1965).
[21] Hoeve, C. A. J., J. Chem. Phys., **47**, 3007 (1966).
[22] Roe, R. J., J. Chem. Phys., **60**, 4192 (1974).
[23] Scheutjens, J. M. H. M. and Fleer, G. J., J. Phys. Chem., **83**, 1919 (1979).
[24] Scheutjens, J. M. H. M. and Fleer, G. J., J. Phys. Chem., **84**, 178 (1980).
[25] 25 Scheutjens, J. M. H. M. and Fleer, G. J., Adv. Colloid Interface Sci., **16**, 341 (1982).
[26] Fleer, G. J., Cohen-Stuart, M. A., Scheutjens, J. M. H. M. Cosgrove, T. and Vincent, B., "Polymers at Interfaces", Chapman and Hall, London (1993).
[27] Cohen-Stuart, M. A., Scheutjens, J. M. H. M. and Fleer, G. J., J. Polym. Sci., Polym. Phys. Ed., **18**, 559 (1980).
[28] Garvey, M. J., Tadros, Th. F. and Vincent, B., J. Colloid Interface Sci., **55**, 440 (1976).

[29] Pusey, P. N., in "Industrial Polymers: Characterisation by Molecular Weights", J. H. S. Green and R. Dietz (eds.), London, Transcripta Books (1973).
[30] Garvey, M. J., Tadros, Th. F. and Vincent, B., J. Colloid Interface Sci., **49**, 57 (1974).
[31] Obey, T. M. and Griffiths, P. C., in "Principles of Polymer Science and Technology in Cosmetics and Personal Care", E. D. Goddard and J. V. Gruber (eds.), Marcel Dekker, NY (1999), Chapter 2.
[32] van den Boomgaard, Th., King, T. A., Tadros, Th. F., Tang, H. and Vincent, B., J. Colloid Interface Sci., **61**, 68 (1978).
[33] Barnett, K. G., Cosgrove, T., Vincent, B., Burgess, A., Crowley, T. L., Kims, J., Turner J. D. and Tadros, Th. F., Disc. Faraday Soc. **22**, 283 (1981).
[34] Cosgrove, T., Crowley T. L. and Ryan, T., Macromolecules, **20**, 2879 (1987).
[35] Exerowa, D., Gotchev, G., Kolarev, T., Khristov, Khr., Levecke, B. and Tadros, Th. F., Langmuir, **23**, 1711 (2007).
[36] Krieger, I. M., Advances Colloid and Interface Sci., **3**, 111 (1972).

4 Self-assembly structures in cosmetic formulations

4.1 Introduction

With most surfactants, the spherical micelles described in Chapter 2 occur over a limited range of concentration and temperature [1]. With some specific surfactant structures these spherical micelles may not form at all [1] and other self-assembly structures such as rod-shaped or lamellar structures are produced. In general one can distinguish between three types of behaviour of a surfactant as the concentration is increased: (i) Surfactants with high water solubility and their physicochemical properties such as viscosity and light scattering vary in a smooth way from the critical micelle concentration (cmc) region up to saturation. This suggests no major changes in the micelle structure, whereby the micelles remain small and spherical. (ii) Surfactants with high water solubility but as the concentration is increased there are quite dramatic changes in certain properties that are consistent with marked changes in the self-assembly structure. (iii) Surfactants with low water solubility that show phase separation at low concentration.

4.2 Self-assembly structures

Surfactant micelles and bilayers are the building blocks of most self-assembly structures. One can divide the phase structures into two main groups [1]: (i) those that are built of limited or discrete self-assemblies, which may be characterized roughly as spherical, prolate or cylindrical; (ii) infinite or unlimited self-assemblies whereby the aggregates are connected over macroscopic distances in one, two or three dimensions. The hexagonal phase (see below) is an example of one-dimensional continuity, the lamellar phase of two-dimensional continuity, whereas the bicontinuous cubic phase and the sponge phase (see later) are examples of three-dimensional continuity. These two types are schematically illustrated in Fig. 4.1.

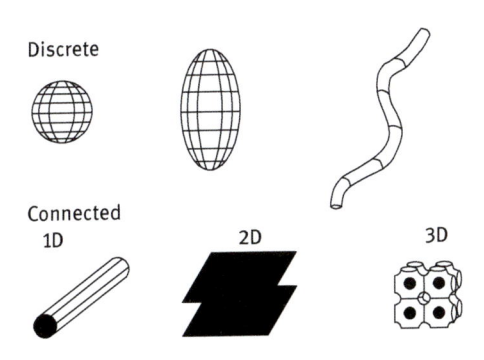

Fig. 4.1: Schematic representation of self-assembly structures.

4.3 Structure of liquid crystalline phases

The above mentioned unlimited self-assembly structures in 1D, 2D or 3D are referred to as liquid crystalline structures. These behave as fluids and are usually highly viscous. At the same time, X-ray studies of these phases yield a small number of relatively sharp lines which resemble those produced by crystals [2–6]. Since they are fluids then they are less ordered than crystals, but because of the X-ray lines and their high viscosity it is also apparent that they are more ordered than ordinary liquids. Thus, the term liquid crystalline phase is very appropriate for describing these self-assembled structures. A brief description of the various liquid crystalline structures that can be produced with surfactants is given below and Tab. 4.1 shows the most commonly used notation to describe these systems.

Tab. 4.1: Notation of the most common liquid crystalline structures.

Phase Structure	Abbreviation	Notation
Micellar	mic	L_1, S
Reversed micellar	rev mic	L_2, S
Hexagonal	hex	H_1, E, M_1, middle
Reversed hexagonal	rev hex	H_2, F, M_2
Cubic (normal micellar)	cubm	I_1, S_{1c}
Cubic (reversed micelle)	cubm	I_2
Cubic (normal bicontinuous)	cubb	I_1, V_1
Cubic (reversed bicontinuous)	cubb	I_2, V_2
Lamellar	lam	L_α, D, G, neat
Gel	gel	L_β
Sponge phase (reversed)	spo	L_3 (normal), L_4

4.3.1 Hexagonal phase

This phase is built up of (infinitely) long cylindrical micelles arranged in a hexagonal pattern, with each micelle being surrounded by six other micelles, as schematically shown in Fig. 4.2. The radius of the circular cross section (which may be somewhat deformed) is again close to the surfactant molecule length [2].

4.3.2 Micellar cubic phase

This phase is built up of regular packing of small micelles, which have similar properties to small micelles in the solution phase. However, the micelles are short prolates (axial ratio 1–2) rather than spheres since this allows a better packing. The micellar cubic phase is highly viscous. A schematic representation of the micellar cubic phase [2] is shown in Fig. 4.3.

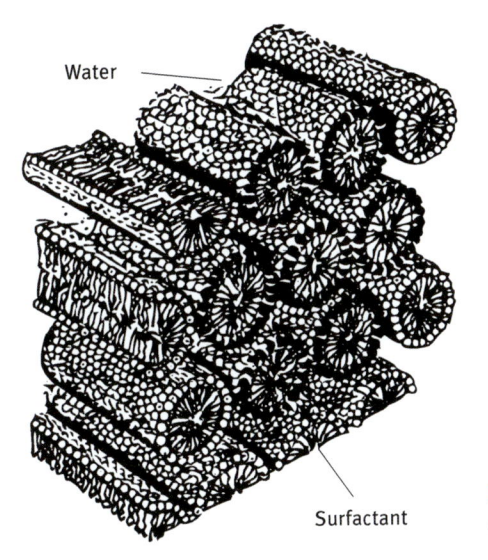

Fig. 4.2: Schematic representation of the hexagonal phase.

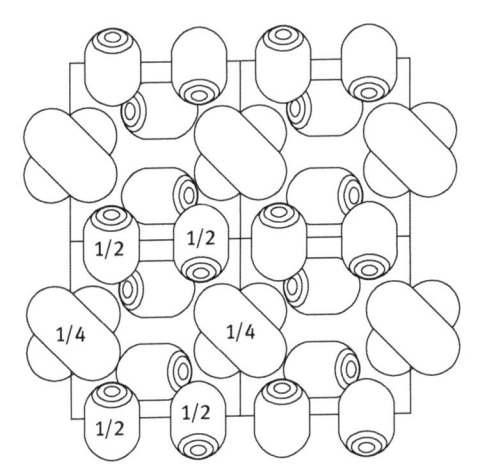

Fig. 4.3: Representation of the micellar cubic phase.

4.3.3 Lamellar phase

This phase is built of bilayers of surfactant molecules alternating with water layers. The thickness of the bilayers is somewhat lower than twice the surfactant molecule length. The thickness of the water layer can vary over wide ranges, depending on the nature of the surfactant. The surfactant bilayer can range from being stiff and planar to being very flexible and undulating. A schematic representation of the lamellar phase [2] is shown in Fig. 4.4.

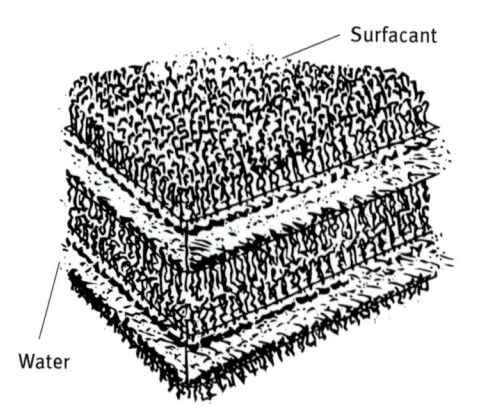

Fig. 4.4: Schematic representation of the lamellar phase [2].

4.3.4 Discontinuous cubic phases

These phases can be a number of different structures, where the surfactant molecules form aggregates that penetrate space, forming a porous connected structure in three dimensions. They can be considered as structures formed by connecting rod-like micelles (branched micelles) or bilayer structures [2].

4.3.5 Reversed structures

Except for the lamellar phase, which is symmetrical around the middle of the bilayer, the different structures have a reversed counterpart in which the polar and nonpolar parts have changed roles. For example, a hexagonal phase is built up of hexagonally packed water cylinders surrounded by the polar head groups of the surfactant molecules and a continuum of the hydrophobic parts. Similarly, reversed (micellar-type) cubic phases and reversed micelles consist of globular water cores surrounded by surfactant molecules. The radii of the water cores are typically in the range 2–10 nm.

4.4 Driving force for liquid crystalline phase formation

One of the simplest methods for predicting the shape of an aggregated structure is based on the critical packing parameter P [7]. This is a dimensionless number that relates the cross-sectional area of the hydrophobic part of the surfactant molecule a_0 (given by the ratio of the volume of the hydrophobic part of the molecule to its extended length l_c) and the head group area a,

$$P = \frac{v}{l_c a}. \tag{4.1}$$

A schematic representation of the critical packing parameter is shown in Fig. 4.5.

4.4 Driving force for liquid crystalline phase formation

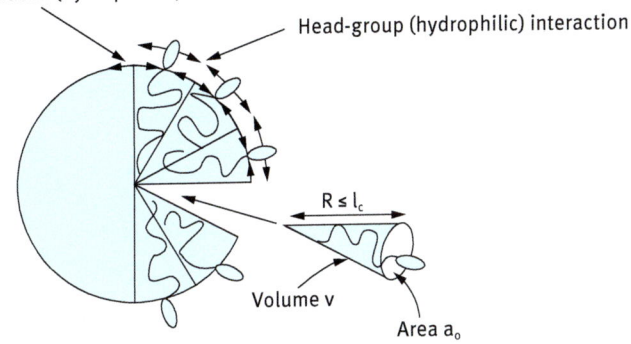

Fig. 4.5: Schematic representation of the critical packing parameter concept.

For a spherical micelle with radius r and containing n molecules each with volume v and head group cross-sectional area a,

$$n = \frac{4\pi r^3}{3v} = \frac{4\pi r^2}{a} \tag{4.2}$$

$$a = \frac{3v}{r}. \tag{4.3}$$

The cross-sectional area of the hydrocarbon tail a is given by

$$a_0 = \frac{v}{l_c}, \tag{4.4}$$

where l_c is the extended length of the hydrocarbon tail

$$P = \frac{v}{l_c a} = \frac{a_0}{a} = \frac{1}{3}\frac{r}{l_c}. \tag{4.5}$$

Since $r < l_c$, the $P \leq (1/3)$.

For a cylindrical micelle with radius r and length d,

$$n = \frac{\pi r^2 d}{v} = \frac{2\pi r d}{a} \tag{4.6}$$

$$a = \frac{2v}{r} \tag{4.7}$$

$$P = \frac{a_0}{a} = \frac{v}{l_c a} = \frac{1}{2}\frac{r}{l_c}. \tag{4.8}$$

Since $r < l_c$, $(1/3) \leq P \leq (1/2)$.
- For liposomes and vesicles $1 \geq P \geq (2/3)$.
- For lamellar micelles $P \approx 1$.
- For Reverse micelles $P > 1$.

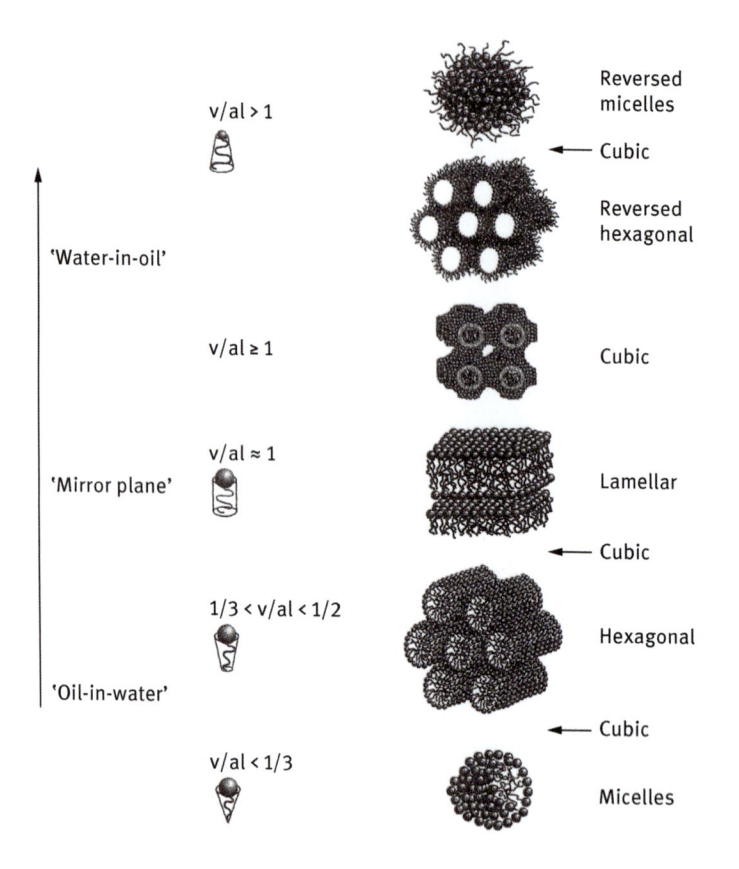

Fig. 4.6: Critical packing parameter values of surfactant molecules and preferred aggregate structure formed.

The packing parameter can be controlled by using mixtures of surfactants to arrive at the desirable structure. Fig. 4.6 shows a schematic representation of the critical packing parameter.

It should be mentioned that the simple geometrical model above does not take into account the interaction between the head groups. A strongly repulsive interaction between the head groups will drive the aggregates to the left in the model shown in Fig. 4.5, while the opposite applies for the attractive interaction. This problem can be circumvented by estimating an "effective" head group area. For example for ionic surfactants, the head group interactions will be strongly affected by the electrolyte concentration so that a will decrease on addition of electrolytes. For nonionic surfactants, temperature is very important for the interaction between the head groups.

4.5 Identification of the liquid crystalline phases and investigation of their structure

The most common procedure to establish the various liquid crystalline structures in a surfactant solution is to establish the phase diagram. The determination of a complete phase diagram involves considerable work and skill and strongly increases in difficulty as the number of components increases. As an illustration, Fig. 2.6 in Chapter 2 shows the phase diagram for a nonionic surfactant consisting of C_{12} alkyl chain and 6 mol ethylene oxide, i.e. $C_{12}E_6$. This phase diagram shows the presence of hexagonal, cubic and lamellar phases as the surfactant concentration is increased.

The distinction between solution and liquid crystalline phase is best made from studies of diffraction properties, either light, X-ray or neutrons. Liquid crystalline phases have a repetitive arrangement of aggregates and observation of a diffraction pattern can firstly give evidence for a long-range order and, secondly, distinguish between attractive structures.

The scattering of normal and polarized light is very useful for identification of different structures. Isotropic phases, i.e. solutions and cubic liquid crystals, are clear and transparent, while the anisotropic phases (hexagonal and lamellar) scatter light and appear more or less cloudy. Using polarized light and viewing the samples through crossed polarizers give a black picture for isotropic phase, while anisotropic ones can give bright images. The patterns in a polarization microscope are different for different anisotropic phase. This is illustrated in Fig. 4.7 for the hexagonal and lamellar phases.

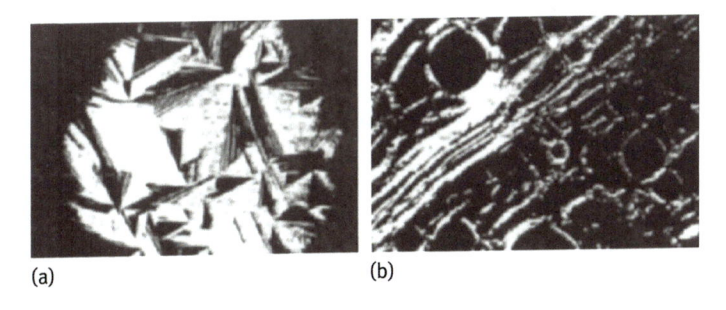

Fig. 4.7: Texture of hexagonal (a) and lamellar (b) phases.

Another very useful technique for identification of the various liquid crystalline phases is NMR spectroscopy by observing the quadrupole splittings in deuterium NMR. Different patterns are observed for different phases as illustrated in Fig. 4.8.

For an isotropic phase (micellar solution, sponge phase, cubic phase) (Fig. 4.8 (a)) a narrow singlet is observed. For a single anisotropic phase (Fig. 4.8 (b)) a doublet is observed. The magnitude of the splitting depends on the degree of anisotropy, which

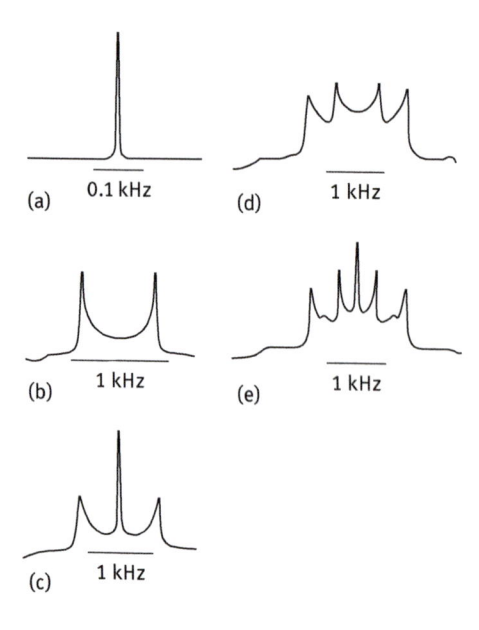

Fig. 4.8: ^{2}H NMR spectra of heavy water for different phases: (a) isotropic phases, (b) single anisotropic phase, (c) one isotropic and one anisotropic phase, (d) two anisotropic (hexagonal and lamellar) phases, (e) three-phase region with two anisotropic and one isotropic phase coexisting.

is larger for a lamellar phase when compared with a hexagonal phase. In a two-phase region, with coexistence of two phases, one observes the spectra of the two phases superimposed. For one isotropic and one anisotropic phase (Fig. 4.8 (c)) one observes one singlet and one doublet. For two anisotropic phases (hexagonal and lamellar) (Fig. 4.8 (d)) one observes two doublets. In a three-phase region with two anisotropic and one isotropic phase (Fig. 4.8 (e)) coexisting, one observes two doublets and one singlet.

4.6 Formulation of liquid crystalline phases

The formulation of liquid crystalline phases is based on the application of the above concepts. However, one must take into account the penetration of the oil between the hydrocarbon tails (which affects the volume and hence a of the chain) as well as hydration of the head group that affects a_0.

The most useful liquid crystalline phases are those of lamellar structure which can bend around the droplets producing an energy barrier against coalescence and Ostwald ripening. These lamellar liquid crystals can also extend in the bulk phase forming a "gel" network that prevents creaming or sedimentation. These liquid crystalline phases also provide the optimum consistency for sensorial application. Due to the high water content of the liquid crystalline structure (water incorporated between several bilayers) it can also provide increased skin hydration. They can influence the delivery of active ingredients both of the lipophilic and hydrophilic types. Since they mimic the skin structure (in particular the stratum corneum) they can offer prolonged hydration potential.

The key to produce lamellar liquid crystals is to use mixtures of surfactants with different P values (different HLB numbers) whose composition can be adjusted to produce the right units.

Using the above concepts, two different types of liquid crystals in oil-in-water (O/W) emulsions can be developed, namely oleosomes and hydrosomes. These structures are obtained by using several surfactant mixtures whose concentration ratio and total concentration are carefully adjusted to produce the desirable effect. These systems are described below.

4.6.1 Oleosomes

These are multilayers of lamellar liquid crystals surrounding the oil droplets that become randomly distributed as they progress into the continuous phase. The rest of the liquid crystals produce the "gel" phase that is viscoelastic. The oleosomes are produced using a mixture of Brij 72 (Steareth-2), Brij 721 (Steareth-21), a fatty alcohol and a minimum of a specific emollient such as isohexadecane or PPG-15 stearyl ether. The nature of the emollient is very crucial; it should be a medium to polar oil such as Arlamol E (PPG-15 stearyl ether), or Estol 3609 (triethylhexanoin). Very polar oils such as Prisorine 2034 (propylene glycol isostearate) or Prisonine 2040 (glyceryl isostearate) disturb the oleosome structure. Nonpolar oils such as paraffinic oils inhibit the formation of oleosomes.

The oleosomes are anisotropic and they can be identified using polarizing microscopy. Fig. 4.9 shows a schematic picture of the oleosomes.

4.6.2 Hydrosomes

In this case a "gel" network is produced in the aqueous phase by the lamellar liquid crystals. The surfactant mixture (sorbitan stearate and sucrose cocoate or sorbityl laurate) is dispersed in water at high temperature (80 °C) and this creates the lamellar phase which becomes swollen with water in between the bilayers. The oil is then emulsified and the droplets become entrapped in the "holes" of the "gel" network. The viscoelastic nature of the "gel" prevents close approach of the oil droplets. The hydrosomes can be obtained using Arlatone 2121 (sorbitan stearate and sucrose cocoate) or Arlatone LC (sorbitan stearate and sorbityl laurate). A schematic representation of hydrosomes is shown in Fig. 4.9.

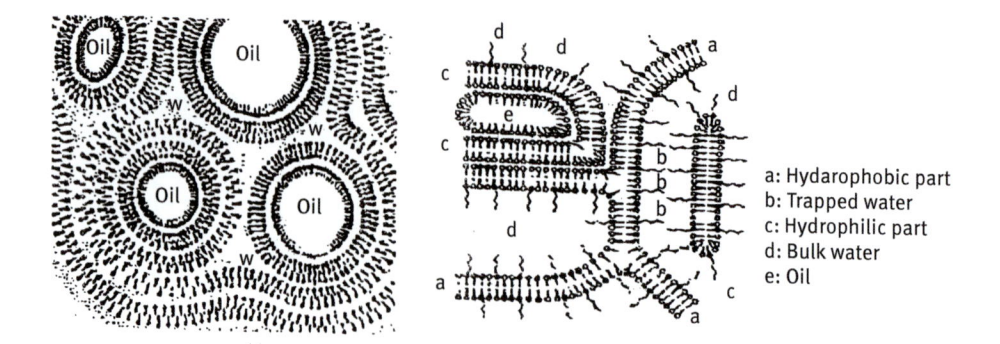

Fig. 4.9: Schematic representation of oleosomes (left) and hydrosomes (right).

a: Hydarophobic part
b: Trapped water
c: Hydrophilic part
d: Bulk water
e: Oil

References

[1] Holmberg, K., Jonsson, B., Kronberg, B. and Lindman, B., "Surfactants and Polymers in Aqueous Solution", John Wiley & Sons, USA. (2003).
[2] Laughlin, R. G., "The Aqueous Phase Behaviour of Surfactants", Academic Press, London (1994).
[3] Fontell, K., Mol. Cryst. Liquid Cryst., **63**, 59 (1981).
[4] Fontell, K., Fox, C. and Hanson, E., Mol. Cryst. Liquid Cryst., **1**, 9 (1985).
[5] Evans, D. F. and Wennerstrom, H., "The Colloid Domain, where Physics, Chemistry and Biology Meet", John Wiley & Sons, VCH, New York (1994).
[6] Tadros, Th. F., Leonard, S., Verboom, C., Wortel, V., Taelman, M. C. and Roschzttardtz, F., "Cosmetic Emulsions: Based on Surfactant Liquid Crystalline Phases: Structure, Rheology and Sensory Evaluation", in: Th. F. Tadros (ed.), "Colloids and Personal Care", Wiley-VCH, Germany (2008), Chapter 6.
[7] Israelachvili, J., "Intermolecular and Surface Forces with Special applications to Colloidal and Biological Systems", Academic Press, London (1985), p. 247.
[8] Khan, A., Fontell, K., Lindblom, G. and Lindman, B., J. Phys. Chem. **86**, 4266 (1982).

5 Interaction forces between particles or droplets in cosmetic formulations and their combination

Three main interaction forces can be distinguished: (i) van der Waals attraction; (ii) double Layer repulsion; (iii) steric interaction. These interaction forces and their combination are summarized below.

5.1 Van der Waals attraction

As is well known, atoms or molecules always attract each other at short distances of separation. The attractive forces are of three different types: dipole–dipole interaction (Keesom), dipole-induced dipole interaction (Debye) and London dispersion force. The London dispersion force is the most important, since it occurs for polar and nonpolar molecules. It arises from fluctuations in the electron density distribution.

At small distances of separation r in vacuum, the attractive energy between two atoms or molecules is given by,

$$G_{aa} = -\frac{\beta_{11}}{r^6}. \tag{5.1}$$

β_{11} is the London dispersion constant.

For particles or emulsion droplets which are made of atom or molecular assemblies, the attractive energies have to be compounded. In this process, only the London interactions have to be considered, since large assemblies have neither a net dipole moment nor a net polarization. The result relies on the assumption that the interaction energies between all molecules in one particle with all others are simply additive [1]. For the interaction between two identical spheres in vacuum the result is,

$$G_A = -\frac{A_{11}}{6}\left(\frac{2}{s^2 - 4} + \frac{2}{s^2} + \ln\frac{s^2 - 4}{s^2}\right). \tag{5.2}$$

A_{11} is known as the Hamaker constant and is defined by [1],

$$A_{11} = \pi^2 q_{11}^2 \beta_{ii}. \tag{5.3}$$

q_{11} is the number of atoms or molecules of type 1 per unit volume, and $s = (2R + h)/R$. Equation (5.2) shows that A_{11} has the dimension of energy.

For very short distances ($h \ll R$), equation (5.2) may be approximated by,

$$G_A = -\frac{A_{11}R}{12h}. \tag{5.4}$$

When the droplets are dispersed in a liquid medium, the van der Waals attraction has to be modified to take into account the medium effect. When two droplets are brought from infinite distance to h in a medium, an equivalent amount of medium has to be transported the other way round. Hamaker forces in a medium are excess forces.

Consider two identical spheres 1 at a large distance apart in a medium 2 as is illustrated in Fig. 5.1 (a). In this case the attractive energy is zero. Fig. 5.1 (b) gives the same situation with arrows indicating the exchange of 1 against 2. Fig. 5.1 (c) shows the complete exchange which now shows the attraction between the two droplets 1 and 1 and equivalent volumes of the medium 2 and 2.

The effective Hamaker constant for two identical droplets 1 and 1 in a medium 2 is given by,

$$A_{11(2)} = A_{11} + A_{22} - 2A_{12} = (A_{11}^{1/2} - A_{22}^{1/2})^2 . \tag{5.5}$$

Equation (5.5) shows that two particles of the same material attract each other unless their Hamaker constants exactly match each other. Equation (5.4) now becomes,

$$G_A = -\frac{A_{11(2)}R}{12h}, \tag{5.6}$$

where $A_{11(2)}$ is the effective Hamaker constant of two identical droplets with Hamaker constant A_{11} in a medium with Hamaker constant A_{22}.

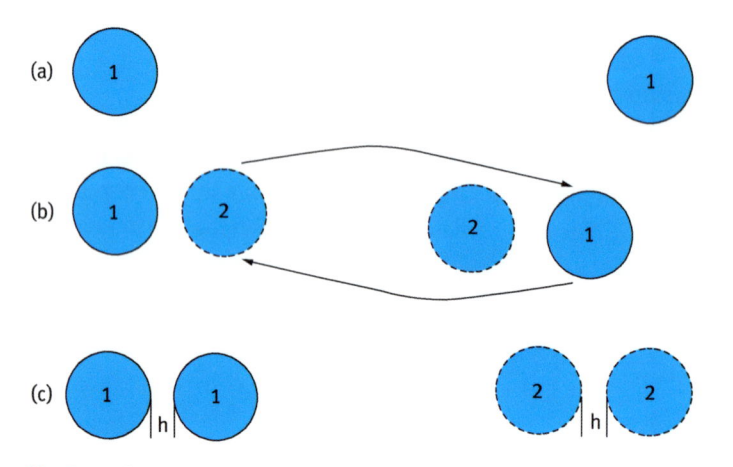

Fig. 5.1: Schematic representation of interaction of two droplets in a medium.

In most cases the Hamaker constant of the particles or droplets is higher than that of the medium. Examples of Hamaker constants for some liquids are given in Tab. 5.1. Generally speaking the effect of the liquid medium is to reduce the Hamaker constant of the droplets below its value in vacuum (air).

G_A decreases with increasing h as schematically shown in Fig. 5.2. This shows the rapid increase of attractive energy with decreasing h reaching a deep minimum at short h values. At extremely short h, Born repulsion operates due to the overlap of the electronic clouds at such very small distance (few Ångströms). Thus in the absence of any repulsive mechanism, the emulsion droplets become strongly aggregated due to the very strong attraction at short distances of separation.

To counteract the van der Waals attraction, it is necessary to create a repulsive force. Two main types of repulsion can be distinguished depending on the nature of the emulsifier used: electrostatic (due to the creation of double layers) and steric (due to the presence of adsorbed surfactant or polymer layers).

Tab. 5.1: Hamaker constants of some liquids.

Liquid	$A_{22} \times 10^{20}$ J
Water	3.7
Ethanol	4.2
Decane	4.8
Hexadecane	5.2
Cyclohexane	5.2

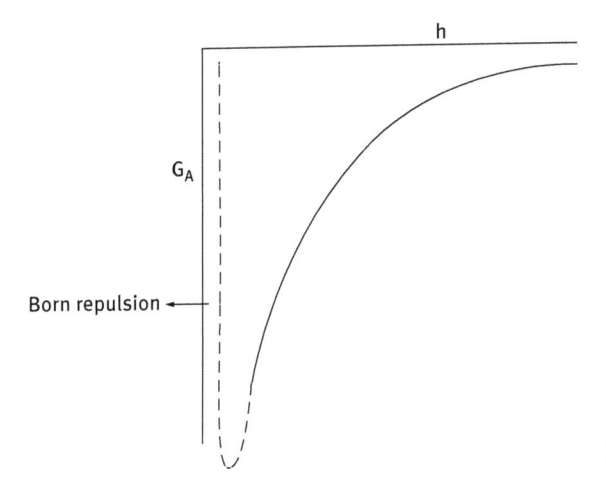

Fig. 5.2: Variation of G_A with h.

5.2 Electrostatic repulsion

This can be produced by adsorption of an ionic surfactant producing an electrical double layer whose structure was described by Gouy–Chapman, Stern and Grahame [2–5]. A schematic representation of the diffuse double layer according to Gouy and Chapman [2] is shown in Fig. 5.3.

The surface charge σ_0 is compensated by unequal distribution of counterions (opposite in charge to the surface) and co-ions (same sign as the surface) which extend to some distance from the surface [2, 3]. The potential decays exponentially with dis-

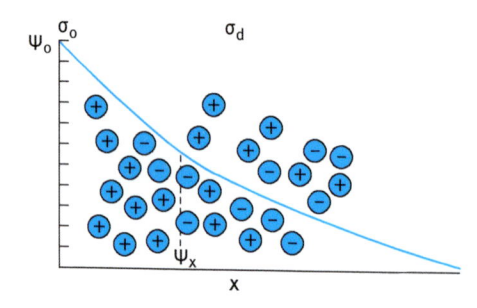

Fig. 5.3: Schematic representation of the diffuse double layer according to Gouy and Chapman.

tance x. At low potentials,

$$\psi = \psi_0 \exp{-(\kappa x)}. \tag{5.7}$$

Note that when $x = 1/\kappa$, $\psi x = \psi_0/e - 1/\kappa$ is referred to as the "thickness of the double layer".

The double layer extension depends on electrolyte concentration and valency of the counterions,

$$\left(\frac{1}{\kappa}\right) = \left(\frac{\varepsilon_r \varepsilon_0 kT}{2n_0 Z_i^2 e^2}\right)^{1/2}. \tag{5.8}$$

ε_r is the permittivity (dielectric constant); 78.6 for water at 25 °C, ε_0 is the permittivity of free space, k is the Boltzmann constant and T is the absolute temperature, n_0 is the number of ions per unit volume of each type present in bulk solution, Z_i is the valency of the ions and e is the electronic charge.

Values of $(1/\kappa)$ at various 1 : 1 electrolyte concentrations are given below.

C / mol dm^{-3}	10^{-5}	10^{-4}	10^{-3}	10^{-2}	10^{-1}
$(1/\kappa)$ / nm	100	33	10	3.3	1

The double layer extension increases with decreasing electrolyte concentration.

Stern [4] introduced the concept of the non-diffuse part of the double layer for specifically adsorbed ions, the rest being diffuse in nature. This is schematically illustrated in Fig. 5.4

The potential drops linearly in the Stern region and then exponentially. Grahame [5] distinguished two types of ions in the Stern plane, physically adsorbed counterions (outer Helmholtz plane) and chemically adsorbed ions (that lose part of their hydration shell) (inner Helmholtz plane).

When charged particles or droplets in a cosmetic formulation approach each other such that the double layers begin to overlap (droplet separation becomes less than twice the double layer extension), repulsion occurs. The individual double layers can no longer develop unrestrictedly, since the limited space does not allow complete potential decay [6]. This is illustrated in Fig. 5.5 for two flat plates.

The potential $\psi_{H/2}$ half way between the plates is no longer zero (as would be the case for isolated particles at $x \to \infty$). The potential distribution at an interdroplet

$\sigma_0 = \sigma_s + \sigma_d$

σ_s = Charge due to specifically adsorbed counter ions

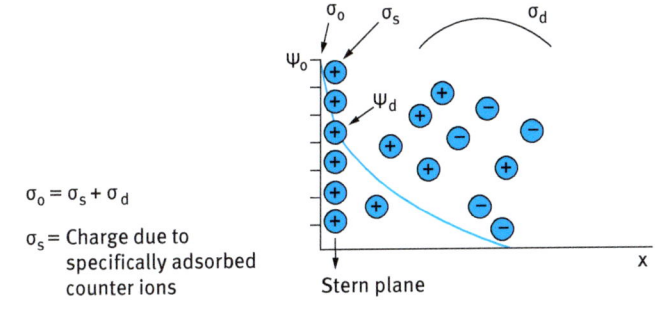

Fig. 5.4: Schematic representation of the double layer according to Stern and Grahame.

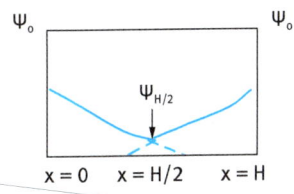

Fig. 5.5: Schematic representation of double layer interaction for two flat plates.

distance H is schematically depicted by the full line in Fig. 5.5. The Stern potential ψ_d is considered to be independent of the interdroplet distance. The dashed curves show the potential as a function of distance x to the Helmholtz plane, had the particles been at infinite distance.

For two spherical particles or droplets of radius R and surface potential ψ_0 and condition $\kappa R < 3$, the expression for the electrical double layer repulsive interaction is given by [6],

$$G_{elec} = \frac{4\pi\varepsilon_r\varepsilon_0 R^2 \psi_0^2 \exp-(\kappa h)}{2R + h}, \qquad (5.9)$$

where h is the closest distance of separation between the surfaces.

Equation (5.9) shows the exponential decay of G_{elec} with h. The higher the value of κ (i.e., the higher the electrolyte concentration), the steeper the decay, as schematically shown in Fig. 5.6. This means that at any given distance h, the double layer repulsion decreases with increasing electrolyte concentration.

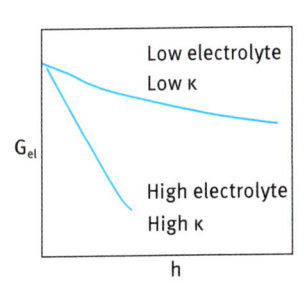

Fig. 5.6: Variation of G_{elec} with h at different electrolyte concentrations.

An important aspect of double layer repulsion is the situation during particle or droplet approach. If at any stage the assumption is made that the double layers adjust to new conditions, so that equilibrium is always maintained, then the interaction takes place at constant potential. This would be the case if the relaxation time of the surface charge is much shorter than the time the particles are in each other's interaction sphere as a result of Brownian motion. However, if the relaxation time of the surface charge is appreciably longer than the time particles are in each other's interaction sphere, the charge rather than the potential will be the constant parameter. The constant charge leads to larger repulsion than the constant potential case [7, 8].

Combination of G_{elec} and G_A results in the well-known theory of stability of colloids due to Deryaguin–Landau–Verwey–Overbeek (DLVO Theory) [9, 10],

$$G_T = G_{elec} + G_A . \qquad (5.10)$$

A plot of G_T versus h is shown in Fig. 5.7, which represents the case at low electrolyte concentrations, i.e. strong electrostatic repulsion between the particles. G_{elec} decays exponentially with h, i.e. $G_{elec} \to 0$ as h becomes large. G_A is $\propto 1/h$, i.e. G_A does not decay to 0 at large h. At long distances of separation, $G_A > G_{elec}$, resulting in a shallow minimum (secondary minimum). At very short distances, $G_A \gg G_{elec}$, resulting in a deep primary minimum. At intermediate distances, $G_{elec} > G_A$, resulting in energy maximum, G_{max}, whose height depends on ψ_0 (or ψ_d) and the electrolyte concentration and valency.

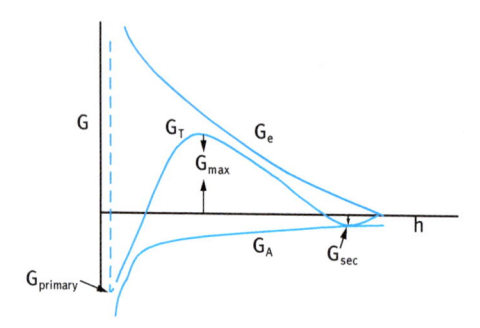

Fig. 5.7: Schematic representation of the variation of G_T with h according to the DLVO theory.

At low electrolyte concentrations ($< 10^{-2}$ mol dm^{-3} for a 1:1 electrolyte), G_{max} is high ($> 25\,kT$) and this prevents particle aggregation into the primary minimum. The higher the electrolyte concentration (and the higher the valency of the ions), the lower the energy maximum. Under some conditions (depending on electrolyte concentration and particle size), flocculation into the secondary minimum may occur. This flocculation is weak and reversible. By increasing the electrolyte concentration, G_{max} decreases until at a given concentration it vanishes and particle or droplet coagulation occurs. This is illustrated in Fig. 5.8 which shows the variation of G_T with h at various electrolyte concentrations.

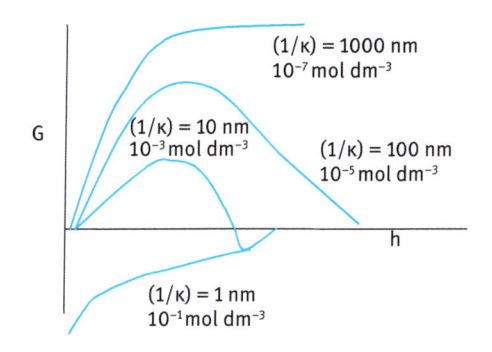

Fig. 5.8: Variation of G with h at various electrolyte concentrations.

Since approximate formulae are available for G_{elec} and G_A, quantitative expressions for $G_T(h)$ can also be formulated. These can be used to derive expressions for the coagulation concentration, which is that concentration that causes every encounter between two emulsion droplets to lead to destabilization. Verwey and Overbeek [10] introduced the following criteria for transition between stability and instability,

$$G_T (= G_{elec} + G_A) = 0 \tag{5.11}$$

$$\frac{dG_T}{dh} = 0 \tag{5.12}$$

$$\frac{dG_{elec}}{dh} = -\frac{dG_A}{dh}. \tag{5.13}$$

Using the equations for G_{elec} and G_A, the critical coagulation concentration (ccc) could be calculated. The theory predicts that the ccc is directly proportional to the surface potential ψ_0 and inversely proportional to the Hamaker constant A and the electrolyte valency Z. As will be shown below, the ccc is inversely proportional to Z^6 at high surface potential and inversely proportional to Z^2 at low surface potential.

5.3 Flocculation of electrostatically stabilized dispersions

As discussed above, the condition for kinetic stability is $G_{max} > 25$ kT. When $G_{max} <$ 5 kT, flocculation occurs. Two types of flocculation kinetics may be distinguished: fast flocculation with no energy barrier and slow flocculation when an energy barrier exists.

Fast flocculation kinetics was treated by Smoluchowki [11], who considered the process to be represented by second order kinetics and the process is simply diffusion controlled. The number of particles n at any time t may be related to the initial number (at t = 0) n_0 by the following expression,

$$n = \frac{n_0}{1 + k_0 n_0 t}, \tag{5.14}$$

where k_0 is the rate constant for fast flocculation that is related to the diffusion coefficient of the particles D, i.e.,

$$k_0 = 8\pi DR. \tag{5.15}$$

D is given by the Stokes–Einstein equation,

$$D = \frac{kT}{6\pi\eta R}. \tag{5.16}$$

Combining equations (5.15) and (5.16),

$$k_0 = \frac{4}{3}\frac{kT}{\eta} = 5.5 \times 10^{-18} \text{ m}^3 \text{ s}^{-1} \text{ for water at 25 °C}. \tag{5.17}$$

The half-life $t_{1/2}$ (n = (1/2)n_0) can be calculated at various n_0 or volume fraction ϕ as give in Tab. 5.2.

Tab. 5.2: Half-life of dispersion flocculation.

R / μm	ϕ			
	10^{-5}	10^{-2}	10^{-1}	5×10^{-1}
0.1	765 s	76 ms	7.6 ms	1.5 ms
1.0	21 h	76 s	7.6 s	1.5 s
10.0	4 months	21 h	2 h	25 min

Slow flocculation kinetics was treated by Fuchs [12] who related the rate constant k to the Smoluchowski rate by the stability constant W,

$$W = \frac{k_0}{k}. \tag{5.18}$$

W is related to G_{max} by the following expression,

$$W = \frac{1}{2}k_0 \exp\left(\frac{G_{max}}{kT}\right). \tag{5.19}$$

Since G_{max} is determined by the salt concentration C and valency Z, one can derive an expression relating W to C and Z [13],

$$\log W = -2.06 \times 10^9 \left(\frac{R\gamma^2}{Z^2}\right) \log C, \tag{5.20}$$

where y is a function that is determined by the surface potential ψ_0,

$$\gamma = \left[\frac{\exp(Ze\psi_0/kT) - 1}{\exp(Ze\psi_0/kT) + 1}\right]. \tag{5.21}$$

Plots of log W versus log C are shown in Fig. 5.9. The condition log W = 0 (W = 1) is the onset of fast flocculation. The electrolyte concentration at this point defines the critical flocculation concentration ccc. Above the ccc W < 1 (due to the contribution of van der Waals attraction which accelerates the rate above the Smoluchowski value). Below the ccc, W > 1 and it increases with decreasing electrolyte concentration. Fig. 5.9 also shows that the ccc decreases with increasing valency. At low surface potentials, ccc $\propto$ $1/Z^2$. This referred to as the Schultze–Hardy rule.

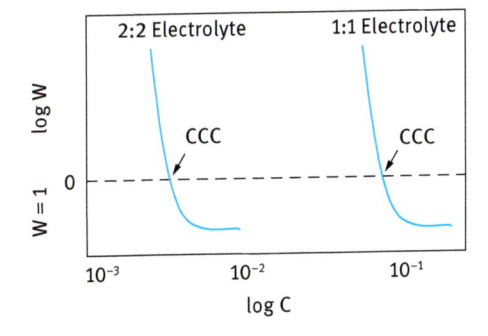

Fig. 5.9: log W-log C curves for electrostatically stabilized emulsions.

Another mechanism of flocculation is that involving the secondary minimum (G_{min}) which is few kT units. In this case flocculation is weak and reversible and hence one must consider both the rate of flocculation (forward rate k_f) and deflocculation (backward rate k_b). In this case the rate of decrease of particle number with time is given by the expression,

$$-\frac{dn}{dt} = -k_f n^2 + k_b n. \tag{5.22}$$

The backward reaction (break-up of weak flocs) reduces the overall rate of flocculation.

Another process of flocculation that occurs under shearing conditions is referred to as orthokinetic (to distinguish it from the diffusion controlled perikinetic process). In this case the rate of flocculation is related to the shear rate by the expression,

$$-\frac{dn}{dt} = \frac{16}{3}\alpha^2 \dot{\gamma} R^3 \tag{5.23}$$

where α is the collision frequency, i.e. the fraction of collisions that result in permanent aggregates.

5.4 Criteria for stabilization of dispersions with double layer interaction

The two main criteria for stabilization are [7, 8]: (i) High surface or Stern potential (zeta potential), high surface charge. As shown in equation (5.9), the repulsive energy G_{el} is proportional to ψ_0^2. In practice, ψ_0 cannot be directly measured and, therefore, one instead uses the measurable zeta potential. (ii) Low electrolyte concentration and low valency of counter- and co-ions. As shown in Fig. 5.8, the energy maximum increases with decreasing electrolyte concentration. The latter should be lower than 10^{-2} mol dm^{-3} for 1:1 electrolyte and lower than 10^{-3} mol dm^{-3} for 2:2 electrolyte. One should ensure that an energy maximum in excess of 25 kT should exist in the energy-distance curve. When $G_{max} \gg kT$, the particles in the dispersion cannot overcome the energy barrier, thus preventing coagulation. In some cases, particularly with large and asymmetric particles, flocculation into the secondary minimum may occur. This flocculation is usually weak and reversible and may be advantageous for preventing the formation of hard sediments.

5.5 Steric repulsion

This is produced by using nonionic surfactants or polymers, e.g. alcohol ethoxylates, or A–B–A block copolymers PEO–PPO–PEO (where PEO refers to polyethylene oxide and PPO refers to polypropylene oxide), as illustrated in Fig. 5.10.

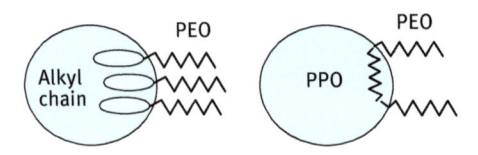

Fig. 5.10: Schematic representation of adsorbed layers.

When two particles or droplets each with a radius R and containing an adsorbed polymer layer with a hydrodynamic thickness δ_h, approach each other to a surface–surface separation distance h that is smaller than $2\delta_h$, the polymer layers interact with each other resulting in two main situations [14]: (i) The polymer chains may overlap with each other. (ii) The polymer layer may undergo some compression. In both cases, there will be an increase in the local segment density of the polymer chains in the interaction region. This is schematically illustrated in Fig. 5.11. The real situation is perhaps in between these two cases, i.e. the polymer chains may undergo some interpenetration and some compression.

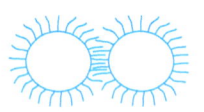

Interpenetration without compression   Compression without interpenetration

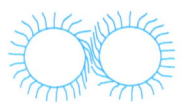

Fig. 5.11: Schematic representation of the interaction between particles or droplets containing adsorbed polymer layers.

Provided the dangling chains (the A chains in A–B, A–B–A block or BA$_n$ graft copolymers) are in a good solvent, this local increase in segment density in the interaction zone will result in strong repulsion as a result of two main effects [14]: (i) Increase in the osmotic pressure in the overlap region as a result of the unfavourable mixing of the polymer chains, when these are in good solvent conditions. This is referred to as osmotic repulsion or mixing interaction and it is described by a free energy of interaction G_{mix}. (ii) Reduction of the configurational entropy of the chains in the interaction zone; this entropy reduction results from the decrease in the volume available for the chains when these are either overlapped or compressed. This is referred to as volume restriction interaction, entropic or elastic interaction and it is described by a free energy of interaction G_{el}.

Combination of G_{mix} and G_{el} is usually referred to as the steric interaction free energy, G_s, i.e.,

$$G_s = G_{mix} + G_{el} \tag{5.24}$$

The sign of G_{mix} depends on the solvency of the medium for the chains. If in a good solvent, i.e. the Flory–Huggins interaction parameter χ is less than 0.5, then G_{mix} is positive and the mixing interaction leads to repulsion (see below). In contrast, if $\chi > 0.5$ (i.e. the chains are in a poor solvent condition), G_{mix} is negative and the mixing interaction becomes attractive. G_{el} is always positive and hence in some cases one can produce stable dispersions in a relatively poor solvent (enhanced steric stabilization).

5.5.1 Mixing interaction G_{mix}

This results from the unfavourable mixing of the polymer chains, when these are in good solvent conditions. This is schematically shown in Fig. 5.12.

Consider two spherical particles or droplets with the same radius and each containing an adsorbed polymer layer with thickness δ. Before overlap, one can define in each polymer layer a chemical potential for the solvent μ_i^α and a volume fraction for the polymer in the layer ϕ_2^α. In the overlap region (volume element dV), the chemical potential of the solvent is reduced to μ_i^β. This results from the increase in polymer segment concentration in this overlap region which is now ϕ_2^β.

In the overlap region, the chemical potential of the polymer chains is now higher than in the rest of the layer (with no overlap). This amounts to an increase in the osmotic pressure in the overlap region; as a result solvent will diffuse from the bulk to the overlap region, thus separating the particles and hence a strong repulsive energy arises from this effect. The above repulsive energy can be calculated by considering the free energy of mixing of two polymer solutions, as for example treated by Flory and Krigbaum [15]. The free energy of mixing is given by two terms: (i) an entropy term that depends on the volume fraction of polymer and solvent and (ii) an energy term that is

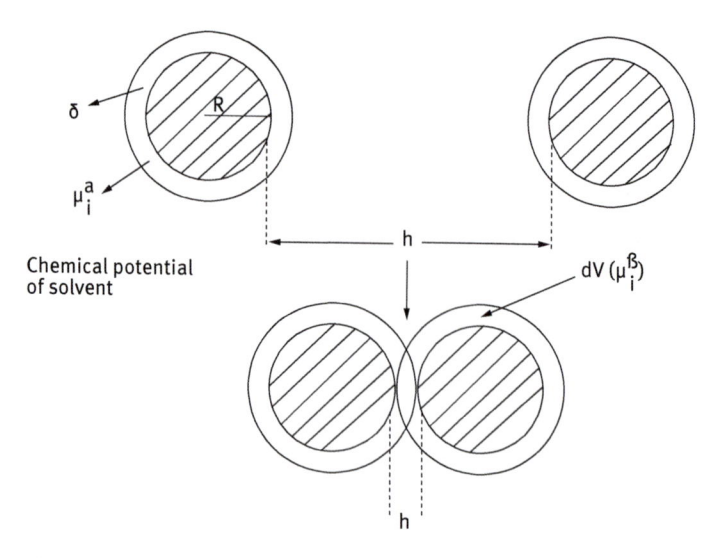

Fig. 5.12: Schematic representation of polymer layer overlap.

determined by the Flory–Huggins interaction parameter,

$$\delta(G_{mix}) = kT(n_1 \ln \phi_1 + n_2 \ln \phi_2 + \chi n_1 \phi_2), \qquad (5.25)$$

where n_1 and n_2 are the number of moles of solvent and polymer with volume fractions ϕ_1 and ϕ_2, k is the Boltzmann constant and T is the absolute temperature.

The total change in free energy of mixing for the whole interaction zone, V, is obtained by summing over all the elements in V,

$$G_{mix} = \frac{2kTV_2^2}{V_1} \nu_2 \left(\frac{1}{2} - \chi\right) R_{mix}(h), \qquad (5.26)$$

where V_1 and V_2 are the molar volumes of solvent and polymer respectively, ν_2 is the number of chains per unit area and $R_{mix}(h)$ is a geometric function which depends on the form of the segment density distribution of the chain normal to the surface, $\rho(z)$. k is the Boltzmann constant and T is the absolute temperature.

Using the above theory one can derive an expression for the free energy of mixing of two polymer layers (assuming a uniform segment density distribution in each layer) surrounding two spherical droplets as a function of the separation distance h between the particles [16].

The expression for G_{mix} is,

$$G_{mix} = \left(\frac{2V_2^2}{V_1}\right) \nu_2 \left(\frac{1}{2} - \chi\right) \left(\delta - \frac{h}{2}\right)^2 \left(3R + 2\delta + \frac{h}{2}\right) \qquad (5.27)$$

The sign of G_{mix} depends on the value of the Flory–Huggins interaction parameter χ: if $\chi < 0.5$, G_{mix} is positive and the interaction is repulsive; if $\chi > 0.5$, G_{mix} is negative and the interaction is attractive. The condition $\chi = 0.5$ and $G_{mix} = 0$ is termed the θ-condition. The latter corresponds to the case where the polymer mixing behaves as ideal, i.e. mixing of the chains does not lead to an increase or a decrease of the free energy.

5.5.2 Elastic interaction G_{el}

This arises from the loss in configurational entropy of the chains on the approach of a second drop. As a result of this approach, the volume available for the chains becomes restricted, resulting in loss of the number of configurations. This can be illustrated by considering a simple molecule, represented by a rod that rotates freely in a hemisphere across a surface (Fig. 5.13).

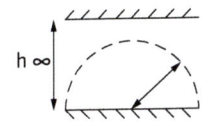

Fig. 5.13: Schematic representation of configurational entropy loss on approach of a second drop.

When the two surfaces are separated by an infinite distance ∞ the number of configurations of the rod is $\Omega(\infty)$, which is proportional to the volume of the hemisphere. When a second particle or drop approaches to a distance h such that it cuts the hemisphere (losing some volume), the volume available to the chains is reduced and the number of configurations become $\Omega(h)$ which is less than $\Omega(\infty)$. For two flat plates, G_{el} is given by the following expression [17],

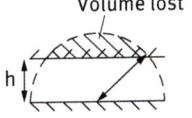

$$\frac{G_{el}}{kT} = -2\nu_2 \ln\left[\frac{\Omega(h)}{\Omega(\infty)}\right] = -2\nu_2 R_{el}(h), \quad (5.28)$$

where $R_{el}(h)$ is a geometric function whose form depends on the segment density distribution. It should be stressed that G_{el} is always positive and could play a major role in steric stabilization. It becomes very strong when the separation distance between the particles becomes comparable to the adsorbed layer thickness δ.

5.5.3 Total energy of interaction

Combining G_{mix} and G_{el} with G_A gives the total energy of interaction G_T (assuming there is no contribution from any residual electrostatic interaction) [18], i.e.,

$$G_T = G_{mix} + G_{el} + G_A \quad (5.29)$$

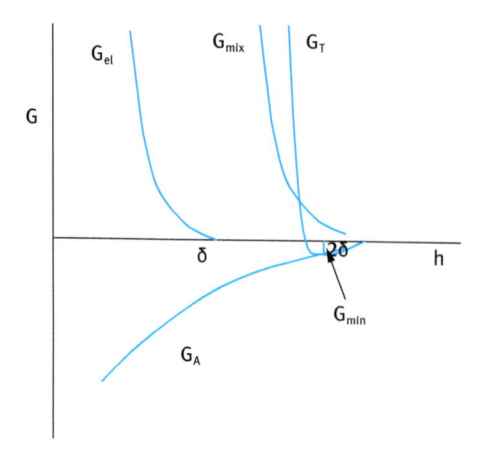

Fig. 5.14: Schematic representation of the energy-distance curve for a sterically stabilized emulsion.

A schematic representation of the variation of G_{mix}, G_{el}, G_A and G_T with surface–surface separation distance h is shown in Fig. 5.14.

G_{mix} increases very sharply with decreasing h, when $h < 2\delta$. G_{el} increases very sharply with decreasing h, when $h < \delta$. G_T versus h shows a minimum, G_{min}, at separation distances comparable to 2δ. When $h < 2\delta$, G_T shows a rapid increase with decreasing h. The depth of the minimum depends on the Hamaker constant A, the particle radius R and adsorbed layer thickness δ. G_{min} increases with increasing A and R. At a given A and R, G_{min} decreases with increasing δ (i.e. with an increase in the molecular weight, M_w, of the stabilizer). This is illustrated in Fig. 5.15 which shows the energy–distance curves as a function of δ/R. The larger the value of δ/R, the smaller the value of G_{min}. In this case the system may approach thermodynamic stability as is the case with nanodispersions.

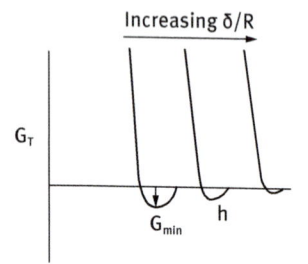

Fig. 5.15: Variation of G_T with h at various δ/R values.

5.5.4 Criteria for effective steric stabilization

(i) The droplets should be completely covered by the polymer (the amount of polymer should correspond to the plateau value). Any bare patches may cause flocculation either by van der Waals attraction (between the bare patches) or by bridging floc-

culation (whereby a polymer molecule will become simultaneously adsorbed on two or more drops).
(ii) The polymer should be strongly "anchored" to the droplet surfaces, to prevent any displacement during drop approach. This is particularly important for concentrated emulsions. For this purpose A–B, A–B–A block and BA$_n$ graft copolymers are the most suitable where the chain B is chosen to be highly insoluble in the medium and has a strong affinity to the surface or soluble in the oil. Examples of B groups for nonpolar oils in aqueous media are polystyrene, polypropylene oxide and polymethylmethacrylate.
(iii) The stabilizing chain A should be highly soluble in the medium and strongly solvated by its molecules. Examples of A chains in aqueous media are poly(ethylene oxide) and poly(vinyl alcohol).
(iv) δ should be sufficiently large (> 5 nm) to prevent weak flocculation.

5.5.5 Flocculation of sterically stabilized dispersions

Four main types of flocculation may be distinguished.

5.5.5.1 Weak flocculation

This occurs when the thickness of the adsorbed layer is small (usually < 5 nm), particularly when the particle radius and Hamaker constant are large. The minimum depth required for causing weak flocculation depends on the volume fraction of the suspension. The higher the volume fraction, the lower the minimum depth required for weak flocculation. This can be understood if one considers the free energy of flocculation that consists of two terms, an energy term determined by the depth of the minimum (G_{min}) and an entropy term that is determined by reduction in configurational entropy on aggregation of particles,

$$\Delta G_{flocc} = \Delta H_{flocc} - T\Delta S_{flocc}. \qquad (5.30)$$

With dilute suspension, the entropy loss on flocculation is larger than with concentrated suspensions. Hence for flocculation of a dilute suspension, a higher energy minimum is required when compared to the case with concentrated suspensions.

The above flocculation is weak and reversible, i.e. on shaking the container redispersion of the suspension occurs. On standing, the dispersed particles aggregate to form a weak "gel". This process (referred to as sol ↔ gel transformation) leads to reversible time dependence of viscosity (thixotropy). On shearing the dispersion, the viscosity decreases and when the shear is removed, the viscosity is recovered. This phenomenon is applied in paints. On application of the formulation, the gel is fluidized, allowing uniform coating of the product. When shearing is stopped, the film recovers its viscosity and this avoids any dripping.

5.5.5.2 Incipient flocculation

This occurs when the solvency of the medium is reduced to become worse than θ-solvent (i.e. $\chi > 0.5$). This is illustrated in Fig. 5.16 where χ was increased from < 0.5 (good solvent) to > 0.5 (poor solvent).

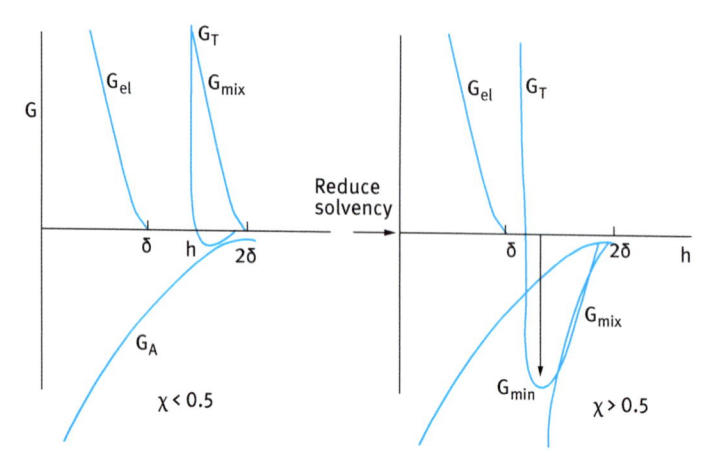

Fig. 5.16: Influence of reduction in solvency on the energy–distance curve.

When $\chi > 0.5$, G_{mix} becomes negative (attractive) which when combined with the van der Waals attraction at this separation distance gives a deep minimum causing flocculation. In most cases, there is a correlation between the critical flocculation point and the θ-condition of the medium. Good correlation is found in many cases between the critical flocculation temperature (CFT) and θ-temperature of the polymer in solution (with block and graft copolymers one should consider the θ-temperature of the stabilizing chains A). Good correlation is also found between the critical volume fraction (CFV) of a nonsolvent for the polymer chains and their θ-point under these conditions. However, in some cases such correlation may break down, particularly the case for polymers which adsorb by multipoint attachment. This situation has been described by Napper [14] who referred to it as "enhanced" steric stabilization.

Thus by measuring the θ-point (CFT or CFV) for the polymer chains (A) in the medium under investigation (which could be obtained from viscosity measurements) one can establish the stability conditions for a dispersion, before its preparation. This procedure helps also in designing effective steric stabilizers such as block and graft copolymers.

5.5.5.3 Depletion flocculation

Depletion flocculation is produced by addition of "free" non-adsorbing polymer [19]. In this case, the polymer coils cannot approach the particles to a distance Δ (that is determined by the radius of gyration of free polymer R_G), since the reduction of entropy on close approach of the polymer coils is not compensated by an adsorption energy. The dispersion particles or droplets will be surrounded by a depletion zone with thickness Δ. Above a critical volume fraction of the free polymer, ϕ_p^+, the polymer coils are "squeezed out" from between the particles and the depletion zones begin to interact. The interstices between the particles are now free from polymer coils and hence an osmotic pressure is exerted outside the particle surface (the osmotic pressure outside is higher than in between the particles) resulting in weak flocculation [19]. A schematic representation of depletion flocculation is shown in Fig. 5.17.

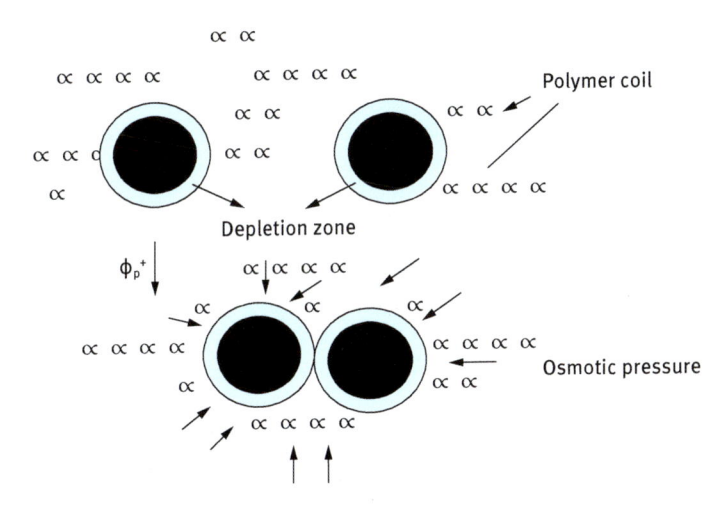

Fig. 5.17: Schematic representation of depletion flocculation.

The magnitude of the depletion attraction free energy, G_{dep}, is proportional to the osmotic pressure of the polymer solution, which in turn is determined by ϕ_p and molecular weight M. The range of depletion attraction is proportional to the thickness of the depletion zone, Δ, which is roughly equal to the radius of gyration, R_G, of the free polymer. A simple expression for G_{dep} is [19],

$$G_{dep} = \frac{2\pi R \Delta^2}{V_1}(\mu_1 - \mu_1^0)\left(1 + \frac{2\Delta}{R}\right), \qquad (5.31)$$

where V_1 is the molar volume of the solvent, μ_1 is the chemical potential of the solvent in the presence of free polymer with volume fraction ϕ_p and μ_1^0 is the chemical potential of the solvent in the absence of free polymer. $(\mu_1 - \mu_1^0)$ is proportional to the osmotic pressure of the polymer solution.

5.5.5.4 Bridging flocculation by polymers and polyelectrolytes

Certain long-chain polymers may adsorb in such a way that different segments of the same polymer chain are adsorbed on different particles, thus binding or "bridging" the particles together, despite the electrical repulsion [20]. With polyelectrolytes of opposite charge to the particles, another possibility exists; the particle charge may be partly or completely neutralized by the adsorbed polyelectrolyte, thus reducing or eliminating the electrical repulsion and destabilizing the particles.

Effective flocculants are usually linear polymers, often of high molecular weight, which may be nonionic, anionic or cationic in character. Ionic polymers should be strictly referred to as polyelectrolytes. The most important properties are molecular weight and charge density. There are several polymeric flocculants that are based on natural products, e.g. starch and alginates, but the most commonly used flocculants are synthetic polymers and polyelectrolytes, e.g. polyacrylamide and copolymers of acrylamide and a suitable cationic monomer such as dimethylaminoethyl acrylate or methacrylate. Other synthetic polymeric flocculants are poly(vinyl alcohol), poly(ethylene oxide) (nonionic), sodium polystyrene sulphonate (anionic) and polyethyleneimine (cationic).

As mentioned above, bridging flocculation occurs because segments of a polymer chain adsorb simultaneously on different particles thus linking them together. Adsorption is an essential step and this requires favourable interaction between the polymer segments and the particles. Several types of interactions are responsible for adsorption that is irreversible in nature: (i) Electrostatic interaction when a polyelectrolyte adsorbs on a surface bearing oppositely charged ionic groups, e.g. adsorption of a cationic polyelectrolyte on a negative oxide surface such as silica. (ii) Hydrophobic bonding that is responsible for adsorption of nonpolar segments on a hydrophobic surface, e.g. partially hydrolysed poly(vinyl acetate) (PVA) on a hydrophobic surface such as polystyrene. (iii) Hydrogen bonding as for example interaction of the amide group of polyacrylamide with hydroxyl groups on an oxide surface. (iv) Ion binding as is the case of adsorption of anionic polyacrylamide on a negatively charged surface in the presence of Ca^{2+}.

Effective bridging flocculation requires that the adsorbed polymer extends far enough from the particle surface to attach to other particles and that there is sufficient free surface available for adsorption of these segments of extended chains. When excess polymer is adsorbed, the particles can be restabilized, either because of surface saturation or by steric stabilization as discussed before. This is one explanation of the fact that an "optimum dosage" of flocculant is often found; at low concentration there is insufficient polymer to provide adequate links and with larger amounts restabilization may occur. A schematic picture of bridging flocculation and restabilization by adsorbed polymer is given in Fig. 5.18.

If the fraction of particle surface covered by polymer is θ then the fraction of uncovered surface is $(1 - \theta)$ and the successful bridging encounters between the particles

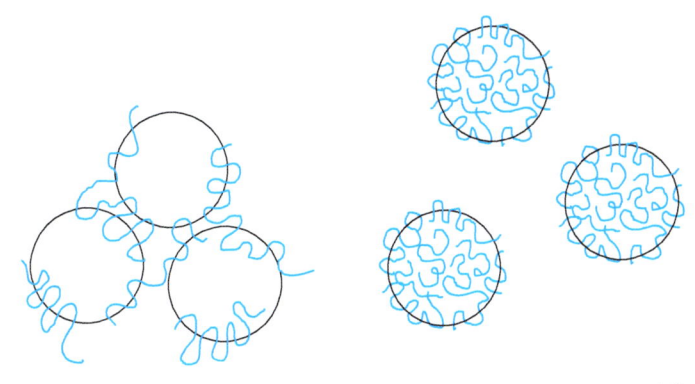

Fig. 5.18: Schematic illustration of bridging flocculation (left) and restabilization (right) by adsorbed polymer.

should be proportional to $\theta(1-\theta)$, which has its maximum when $\theta = 0.5$. This is the well-known condition of "half-surface-coverage" that has been suggested as giving the optimum flocculation.

An important condition for bridging flocculation with charged particles is the role of electrolyte concentration. The latter determines the extension ("thickness") of the double layer which can reach values as high as 100 nm (in 10^{-5} mol dm^{-5} 1:1 electrolyte such as NaCl). For bridging flocculation to occur, the adsorbed polymer must extend far enough from the surface to a distance over which electrostatic repulsion occurs (> 100 nm in the above example). This means that at low electrolyte concentrations quite high molecular weight polymers are needed for bridging to occur. As the ionic strength is increased, the range of electrical repulsion is reduced and lower molecular weight polymers should be effective.

In many cosmetic formulations, it has been found that the most effective flocculants are polyelectrolytes with a charge opposite to that of the particle or droplet. In aqueous media most particles or droplets are negatively charged, and cationic polyelectrolytes such as polyethyleneimine are often necessary. With oppositely charged polyelectrolytes it is likely that adsorption occurs to give a rather flat configuration of the adsorbed chain, due to the strong electrostatic attraction between the positive ionic groups on the polymer and the negative charged sites on the particle surface.

This would probably reduce the probability of bridging contacts with other particles, especially with fairly low molecular weight polyelectrolytes with high charge density. However, the adsorption of a cationic polyelectrolyte on a negatively charged particle will reduce the surface charge of the latter, and this charge neutralization could be an important factor in destabilizing the particles. Another mechanism for destabilization has been suggested by Gregory [18] who proposed an "electrostatic-patch" model. This applied to cases where the particles have a fairly low density of immobile charges and the polyelectrolyte has a fairly high charge density. Under these conditions, it is not physically possible for each surface site to be neutralized by a charged segment on the polymer chain, even though the particle may have sufficient adsorbed polyelectrolyte to achieve overall neutrality. There are then "patches" of excess positive charge, corresponding to the adsorbed polyelectrolyte chains (probably in a rather flat configuration), surrounded by areas of negative charge, representing the original particle surface. Particles which have this "patchy" or "mosaic" type of surface charge distribution may interact in such a way that the positive and negative "patches" come into contact, giving quite strong attraction (although not as strong as in the case of bridging flocculation). A schematic illustration of this type of interaction is given in Fig. 5.19. The electrostatic-patch concept (which can be regarded as another form of "bridging") can explain a number of features of flocculation of negatively charged particles with positive polyelectrolytes. These include the rather small effect of increasing the molecular weight and the effect of ionic strength on the breadth of the flocculation dosage range and the rate of flocculation at optimum dosage.

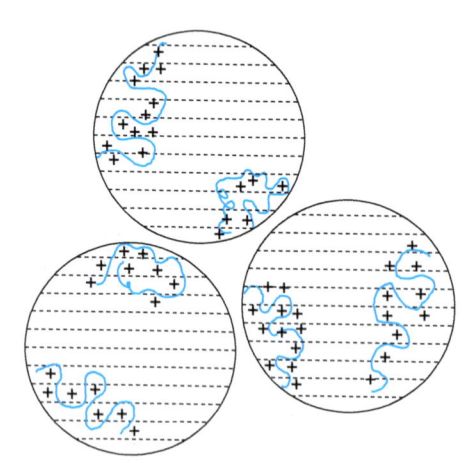

Fig. 5.19: "Electrostatic-patch" model for the interaction of negatively charged particles with adsorbed cationic polyelectrolytes.

References

[1] Hamaker, H. C., Physica (Utrecht), **4**, 1058 (1937).
[2] Gouy, G., J. Phys., **9**, 457 (1910); Ann. Phys., **7**, 129 (1917).
[3] Chapman, D. L., Phil. Mag., **25**, 475 (1913).
[4] Stern, O., Z. Elektrochem., **30**, 508 (1924).
[5] Grahame, D. C., Chem. Rev., **41**, 44 (1947).
[6] Bijesterbosch, B. H., "Stability of Solid/Liquid Dispersions", in "Solid/Liquid Dispersions", Th. F. Tadros (ed.), Academic Press, London (1987).
[7] Tadros, Th. F., "Applied Surfactants", Wiley-VCH, Germany (2005).
[8] Tadros, Th. F., "Dispersion of Powders in Liquids and Stabilisation of Suspensions", Wiley-VCH, Germany (2012).
[9] Deryaguin, B. V. and Landau, L., Acta Physicochem. USSR, **14**, 633 (1941).

[10] Verwey, E. J. W. and Overbeek, J. Th. G., "Theory of Stability of Lyophobic Colloids", Elsevier, Amsterdam (1948).
[11] Smoluchowski, M. V., Z. Phys. Chem., **92**, 129 (1927).
[12] Fuchs, N., Z. Physik, **89**, 736 (1936).
[13] Reerink, H. and Overbeek, J. Th. G., Discussion Faraday Soc., **18**, 74 (1954).
[14] Napper, D. H., "Polymeric Stabilization of Colloidal Dispersions", Academic Press, London (1983).
[15] Flory, P. J. and Krigbaum, W. R., J. Chem. Phys. **18**, 1086 (1950).
[16] Fischer, E. W., Kolloid Z. **160**, 120 (1958).
[17] Mackor, E. L. and van der Waals, J. H., J. Colloid Sci., **7**, 535 (1951).
[18] Hesselink, F. Th., Vrij, A. and Overbeek, J. Th. G., J. Phys. Chem. **75**, 2094 (1971).
[19] Asakura, S. and Oosawa, F., J. Phys. Chem., **22**, 1255 (1954); Asakura, S. and Oosawa, F., J. Polym. Sci., **33**, 183 (1958).
[20] Gregory, J., in "Solid/Liquid Dispersions", Th. F. Tadros (ed.), Academic Press, London (1987).

6 Formulation of cosmetic emulsions

6.1 Introduction

Cosmetic emulsions need to satisfy a number of benefits. For example, such systems should deliver a functional benefit such as cleaning (e.g. hair, skin, etc.), provide a protective barrier against water loss from the skin and in some cases they should screen out damaging UV light (in which case a sunscreen agent such as titania is incorporated in the emulsion). These systems should also impart a pleasant odour and make the skin feel smooth. Both oil-in-water (O/W) and water-in-oil (W/O) emulsions are used in cosmetic applications [1]. As will be discussed in Chapter 8, more complex systems such as multiple emulsions have been applied in recent years [1, 2].

The main physicochemical characteristics that need to be controlled in cosmetic emulsions are their formation and stability on storage as well as their rheology, which controls spreadability and skin feel. The life span of most cosmetic and toiletry brands is relatively short (3–5 years) and hence development of the product should be fast. For this reason, accelerated storage testing is needed for prediction of stability and change of rheology with time. These accelerated tests represent a challenge to the formulation chemist [1, 2].

As mentioned in Chapter 1, the main criterion for any cosmetic ingredient should be medical safety (free of allergens, sensitizers and irritants and impurities that have systemic toxic effects). These ingredients should be suitable for producing stable emulsions that can deliver the functional benefit and the aesthetic characteristics. The main components of an emulsion are the water and oil phases and the emulsifier. Several water soluble ingredients may be incorporated in the aqueous phase and oil soluble ingredients in the oil phase. Thus, the water phase may contain functional materials such as proteins, vitamins, minerals and many natural or synthetic water soluble polymers. The oil phase may contain perfumes and/or pigments (e.g. in make-up). The oil phase may be a mixture of several mineral or vegetable oils. Examples of oils used in cosmetic emulsions are lanolin and its derivatives, paraffin and silicone oils. The oil phase provides a barrier against water loss from the skin.

6.2 Thermodynamics of emulsion formation

The process of emulsion formation is determined by the property of the interface, in particular the interfacial tension which is determined by the concentration and type of the emulsifier [3–7]. This is illustrated as follows. Consider a system in which an oil is represented by a large drop 2 of area A_1 immersed in a liquid 2, which is now subdivided into a large number of smaller droplets with total area A_2 ($A_2 \gg A_1$) as shown in

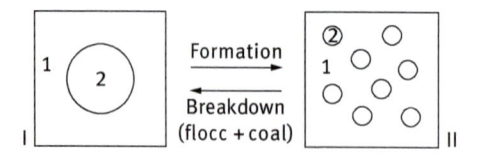

Fig. 6.1: Schematic representation of emulsion formation and breakdown.

Fig. 6.1. The interfacial tension γ_{12} is the same for the large and smaller droplets since the latter are generally in the region of 0.1 to few μm.

The change in free energy in going from state I to state II is made from two contributions: a surface energy term (that is positive) that is equal to $\Delta A \gamma_{12}$ (where $\Delta A = A_2 - A_1$) and an entropy of dispersions term which is also positive (since producing a large number of droplets is accompanied by an increase in configurational entropy) which is equal to $T\Delta S^{conf}$.

From the second law of thermodynamics,

$$\Delta G^{form} = \Delta A \gamma_{12} - T\Delta S^{conf}. \tag{6.1}$$

In most cases $\Delta A \gamma_{12} \gg T\Delta S^{conf}$, which means that ΔG^{form} is positive, i.e. the formation of emulsions is non-spontaneous and the system is thermodynamically unstable. In the absence of any stabilization mechanism, the emulsion will break by flocculation, coalescence, Ostwald ripening or combination of all these processes. This is illustrated in Fig. 6.2 which shows several paths for emulsion breakdown processes.

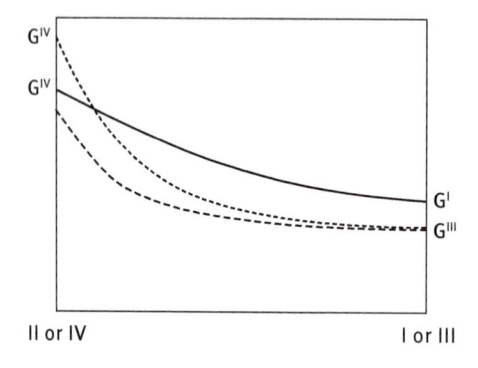

Fig. 6.2: Free energy path in emulsion breakdown: ———, flocc. + coal., – – –, flocc. + coal. + sed., ·······, flocc. + coal. + sed. + Ostwald ripening.

In the presence of a stabilizer (surfactant and/or polymer), an energy barrier is created between the droplets and therefore the reversal from state II to state I becomes non-continuous as a result of the presence of these energy barriers. This is illustrated in Fig. 6.3. In the presence of the above energy barriers, the system becomes kinetically stable.

Several emulsifiers, mostly nonionic or polymeric, are used for preparation of O/W or W/O emulsions and their subsequent stabilization. For W/O emulsion, the

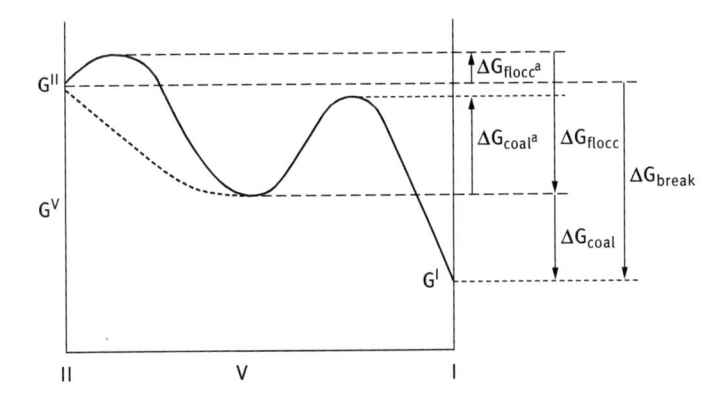

[Figure 6.3: Schematic free energy diagram with labels G^{II}, G^{V}, G^{I} on the axes and energy differences ΔG_{flocc^a}, ΔG_{coal^a}, ΔG_{flocc}, ΔG_{break}, ΔG_{coal} marked. Horizontal axis labeled II, V, I.]

Fig. 6.3: Schematic representation of free energy path for breakdown (flocculation and coalescence) for systems containing an energy barrier.

hydrophilic–lipophilic balance (HLB) range (see below) of the emulsifier is in the range 3–6, whereas for O/W emulsions this range is 8–18.

The HLB number is based on the relative percentage of hydrophilic to lipophilic (hydrophobic) groups in the surfactant molecule(s) as will be discussed below. For an O/W emulsion droplet the hydrophobic chain resides in the oil phase whereas the hydrophilic head group resides in the aqueous phase. For a W/O emulsion droplet, the hydrophilic group(s) reside in the water droplet, whereas the lipophilic groups reside in the hydrocarbon phase.

6.3 Emulsion breakdown processes and their prevention

Several breakdown processes may occur on storage depending on: particle size distribution and density difference between the droplets and the medium; magnitude of the attractive versus repulsive forces which determines flocculation; solubility of the disperse droplets and the particle size distribution which determines Ostwald ripening; stability of the liquid film between the droplets that determines coalescence; phase inversion, where the two phases exchange, e.g. an O/W emulsion inverting to W/O and vice versa. Phase inversion can be catastrophic as is the case when the oil phase in an O/W emulsion exceeds a critical value. The inversion can be transient when for example the emulsion is subjected to temperature increase.

The various breakdown processes are illustrated in the Fig. 6.4. The physical phenomena involved in each breakdown process are not simple and it requires analysis of the various surface forces involved. In addition, the above processes may take place simultaneously rather than consecutively and this complicates the analysis. Model emulsions, with monodisperse droplets cannot be easily produced and hence any theoretical treatment must take into account the effect of droplet size distribution.

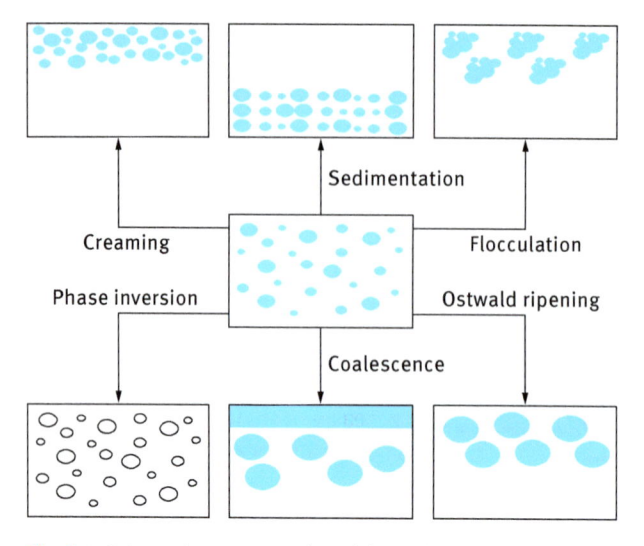

Fig. 6.4: Schematic representation of the various breakdown processes in emulsions.

Theories that take into account the polydispersity of the system are complex and in many cases only numerical solutions are possible. In addition, measurements of surfactant and polymer adsorption in an emulsion are not easy and one has to extract such information from measurements at a planer interface.

A summary of each of the above breakdown processes and details of each process and methods for its prevention are given below.

6.3.1 Creaming and sedimentation

This process, with no change in droplet size, results from external forces usually gravitational or centrifugal. When such forces exceed the thermal motion of the droplets (Brownian motion), a concentration gradient builds up in the system with the larger droplets moving faster to the top (if their density is lower than that of the medium) or to the bottom (if their density is larger than that of the medium) of the container. In the limiting cases, the droplets may form a close-packed (random or ordered) array at the top or bottom of the system with the remainder of the volume occupied by the continuous liquid phase.

The most commonly used method for prevention of creaming or sedimentation is to add a rheology modifier, sometimes referred to as a thickener, usually a high molecular weight polymer such as hydroxyethyl cellulose, associative thickener (hydrophobically modified polymer) or a microgel such as Carbopol (crosslinked polyacrylate which on neutralization with alkali forms a microgel). All these systems give a non-Newtonian system (see section on rheology of cosmetic emulsions) with a very high viscosity at low shear rate (referred to as residual or "zero-shear" viscosity $\eta(0)$). This high viscosity (usually above 10 Pas) prevents any creaming or sedimentation of the emulsion.

6.3.2 Flocculation

This process refers to aggregation of the droplets (without any change in primary droplet size) into larger units. It is the result of van der Waals attraction which is universal with all disperse systems. The main force of attraction arises from the London dispersion force that results from charge fluctuations of the atoms or molecules in the disperse droplets. The van der Waals attraction increases with decreasing separation distance between the droplets and at small separation distances the attraction becomes very strong resulting in droplet aggregation or flocculation. The latter occurs when there is not sufficient repulsion to keep the droplets apart to distances where the van der Waals attraction is weak. Flocculation may be "strong" or "weak", depending on the magnitude of the attractive energy involved. In cases where the net attractive forces are relatively weak, an equilibrium degree of flocculation may be achieved (so-called weak flocculation), associated with the reversible nature of the aggregation process. The exact nature of the equilibrium state depends on the characteristics of the system. One can envisage the build-up of aggregate size distribution and an equilibrium may be established between single droplets and large aggregates. With a strongly flocculated system, one refers to a system in which all the droplets are present in aggregates due to the strong van der Waals attraction between the droplets.

Two main rules can be applied for reducing (eliminating) flocculation depending on the stabilization mechanism: (i) With charge stabilized emulsions, e.g. using ionic surfactants, the most important criterion is to make the energy barrier in the energy-distance curve (see Chapter 5), Gmax, as high as possible. This is achieved by three main conditions: high surface or zeta potential, low electrolyte concentration and low valency of ions. (ii) For sterically stabilized emulsions four main criteria are necessary: (a) Complete coverage of the droplets by the stabilizing chains. (b) Firm attachment (strong anchoring) of the chains to the droplets. This requires the chains to be insoluble in the medium and soluble in the oil. However, this is incompatible with stabilization which requires a chain that is soluble in the medium and strongly solvated by its molecules. These conflicting requirements are solved by the use of A–B, A–B–A block or BA_n graft copolymers (B is the "anchor" chain and A is the stabilizing chain(s)). Examples for the B chains for O/W emulsions are polystyrene, polymethylmethacrylate, polypropylene oxide and alkyl polypropylene oxide. For the A chain(s), polyethylene oxide (PEO) or polyvinyl alcohol are good examples. For W/O emulsions, PEO can form the B chain, whereas the A chain(s) could be polyhydroxy stearic acid (PHS) which is strongly solvated by most oils. (c) Thick adsorbed layers; the adsorbed layer thickness should be in the region of 5–10 nm. This means that the molecular weight of the stabilizing chains could be in the region of 1000–5000. (d) The stabilizing chain should be maintained in good solvent conditions (the Flory–Huggins interaction parameter $\chi < 0.5$) under all conditions of temperature changes on storage.

6.3.3 Ostwald ripening (disproportionation)

This results from the finite solubility of the liquid phases. Liquids which are referred to as being immiscible often have mutual solubilities which are not negligible. With emulsions which are usually polydisperse, the smaller droplets will have larger solubility when compared with the larger ones (due to curvature effects). With time, the smaller droplets disappear and their molecules diffuse to the bulk and become deposited on the larger droplets. With time the droplet size distribution shifts to larger values.

Two general methods may be applied to reduce Ostwald ripening [1–3]: (i) Addition of a second disperse phase component which is insoluble in the continuous medium (e.g. squalane). In this case partitioning between different droplet sizes occurs, with the component having low solubility expected to be concentrated in the smaller droplets. During Ostwald ripening in a two component system, equilibrium is established when the difference in chemical potential between different sized droplets (which results from curvature effects) is balanced by the difference in chemical potential resulting from partitioning of the two components. This effect reduces further growth of droplets. (ii) Modification of the interfacial film at the O/W interface: reduction in γ results in reduction of Ostwald ripening rate. By using surfactants that are strongly adsorbed at the O/W interface (i.e. polymeric surfactants) and which do not desorb during ripening (by choosing a molecule that is insoluble in the continuous phase) the rate could be significantly reduced. An increase in the surface dilational modulus ε (= $d\gamma/d\ln A$) and decrease in γ would be observed for the shrinking drop and this tends to reduce further growth [1–3].

A–B–A block copolymers such as PHS–PEO–PHS (which is soluble in the oil droplets but insoluble in water) can be used to achieve the above effect. Similar effects can also be obtained using a graft copolymer of hydrophobically modified inulin, namely INUTEC®SP1 (ORAFTI, Belgium). This polymeric surfactant adsorbs with several alkyl chains (which may dissolve in the oil phase) leaving loops and tails of strongly hydrated inulin (polyfructose) chains. The molecule has limited solubility in water and hence it resides at the O/W interface. These polymeric emulsifiers enhance the Gibbs elasticity thus significantly reducing the Ostwald ripening rate.

6.3.4 Coalescence

This refers to the process of thinning and disruption of the liquid film between the droplets which may be present in a creamed or sedimented layer, in a floc or simply during droplet collision, with the result of fusion of two or more droplets into larger ones. This process of coalescence results in a considerable change in the droplet size distribution, which shifts to larger sizes. The limiting case for coalescence is the complete separation of the emulsion into two distinct liquid phases. The thinning and disruption of the liquid film between the droplets is determined by the relative magnitudes of the attractive versus repulsive forces. To prevent coalescence, the repulsive forces must exceed the van der Waals attraction, thus preventing film rupture.

Several methods may be applied to achieve the above effects: (i) Use of mixed surfactant films. In many cases using mixed surfactants, say anionic and nonionic or long chain alcohols can reduce coalescence as a result of several effects: high Gibbs elasticity; high surface viscosity; hindered diffusion of surfactant molecules from the film. (ii) Formation of lamellar liquid crystalline phases at the O/W interface. Surfactant or mixed surfactant film can produce several bilayers that "wrap" the droplets as discussed in Chapter 4. As a result of these multilayer structures, the potential drop is shifted to longer distances thus reducing the van der Waals attraction. For coalescence to occur, these multilayers have to be removed "two-by-two" and this forms an energy barrier preventing coalescence.

6.3.5 Phase Inversion

This refers to the process in which there is an exchange between the disperse phase and the medium. For example an O/W emulsion may with time or change of conditions invert to a W/O emulsion. In many cases, phase inversion passes through a transition state whereby multiple emulsions are produced. For example with an O/W emulsion, the aqueous continuous phase may become emulsified in the oil droplets forming a W/O/W multiple emulsion. This process may continue until all the continuous phase is emulsified into the oil phase thus producing a W/O emulsion.

6.4 Selection of emulsifiers

6.4.1 The Hydrophilic-Lipophilic Balance (HLB) concept

The hydrophilic-lipophilic balance (HLB number) is a semi-empirical scale for selecting surfactants developed by Griffin [8]. This scale is based on the relative percentage of hydrophilic to lipophilic (hydrophobic) groups in the surfactant molecule(s). For an O/W emulsion droplet the hydrophobic chain resides in the oil phase whereas the hydrophilic head group resides in the aqueous phase. For a W/O emulsion droplet, the hydrophilic group(s) reside in the water droplet, whereas the lipophilic groups reside in the hydrocarbon phase.

Tab. 6.1 gives a guide to the selection of surfactants for a particular application. The HLB number depends on the nature of the oil. As an illustration, Tab. 6.2 gives the required HLB numbers to emulsify various oils. Examples of HLB numbers of a list of surfactants are given in Tab. 6.3.

Tab. 6.1: Summary of HLB ranges and their applications.

HLB Range	Application
3–6	W/O Emulsifier
7–9	Wetting agent
8–18	O/W Emulsifier
13–15	Detergent
15–18	Solubilizer

Tab. 6.2: Required HLB numbers to emulsify various oils.

Oil	W/O Emulsion	O/W Emulsion
Paraffin oil	4	10
Beeswax	5	9
Lanolin, anhydrous	8	12
Cyclohexane	—	15
Toluene	—	15
Silicone oil (volatile)	—	7–8
Isopropyl myristate	—	11–12
Isohexadecyl alcohol	—	11–12
Castor oil		14

Tab. 6.3: HLB numbers of some surfactants.

Surfactant	Chemical Name	HLB
Span 85	Sorbitan trioleate	1.8
Span 80	Sorbitan mono-oleate	4.3
Brij 72	Ethoxylated (2 mol ethylene oxide) stearyl alcohol	4.9
Triton X-35	Ethoxylated octylphenol	7.8
Tween 85	Ethoxylated (20 mol ethylene oxide) sorbitan trioleate	11.0
Tween 80	Ethoxylated (20 mol ethylene oxide) sorbitan mono-oleate	15.0

The relative importance of the hydrophilic and lipophilic groups was first recognized when using mixtures of surfactants containing varying proportions of a low and high HLB number. The efficiency of any combination (as judged by phase separation) was found to pass a maximum when the blend contained a particular proportion of the surfactant with the higher HLB number. This is illustrated in Fig. 6.5 which shows the variation of emulsion stability, droplet size and interfacial tension with % surfactant with high HLB number.

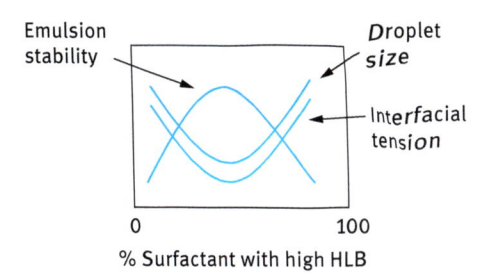

Fig. 6.5: Variation of emulsion stability, droplet size and interfacial tension with % surfactant with high HLB number.

The average HLB number may be calculated from additivity,

$$\text{HLB} = x_1 \text{HLB}_1 + x_2 \text{HLB}_2, \quad (6.2)$$

x_1 and x_2 are the weight fractions of the two surfactants with HLB_1 and HLB_2.

Griffin [8] developed simple equations for calculating the HLB number of relatively simple nonionic surfactants. For a polyhydroxy fatty acid ester,

$$\text{HLB} = 20\left(1 - \frac{S}{A}\right). \quad (6.3)$$

S is the saponification number of the ester and A is the acid number. For a glyceryl monostearate, $S = 161$ and $A = 198$; the HLB is 3.8 (suitable for W/O emulsion).

For a simple alcohol ethoxylate, the HLB number can be calculated from the weight percent of ethylene oxide (E) and polyhydric alcohol (P),

$$\text{HLB} = \frac{E + P}{5}. \quad (6.4)$$

If the surfactant contains PEO as the only hydrophilic group, the contribution from one OH group can be neglected,

$$\text{HLB} = \frac{E}{5}. \quad (6.5)$$

For a nonionic surfactant $C_{12}H_{25}-O-(CH_2-CH_2-O)_6$, the HLB is 12 (suitable for O/W emulsion).

The above simple equations cannot be used for surfactants containing propylene oxide or butylene oxide. They also cannot be applied for ionic surfactants. Davies [9, 10] devised a method for calculating the HLB number for surfactants from their

Tab. 6.4: HLB group numbers.

	Group Number
Hydrophilic	
$-SO_4Na^+$	38.7
$-COOK$	21.2
$-COONa$	19.1
N(tertiary amine)	9.4
Ester (sorbitan ring)	6.8
$-O-$	1.3
CH–(sorbitan ring)	0.5
Lipophilic	
$(-CH-), (-CH_2-), CH_3$	0.475
Derived	
$-CH_2-CH_2-O$	0.33
$-CH_2-CHCH_3-O$	-0.11

chemical formulae, using empirically determined group numbers. A group number is assigned to various component groups. A summary of the group numbers for some surfactants is given in Tab. 6.4.

The HLB is given by the empirical equation (6.6),

$$\text{HLB} = 7 + \sum(\text{hydrophilic group numbers}) - \sum(\text{lipohilic group numbers}). \quad (6.6)$$

Davies has shown that the agreement between HLB numbers calculated from equation (6.6) and those determined experimentally is quite satisfactory.

Various other procedures have been developed to obtain a rough estimate of the HLB number. Griffin found good correlation between the cloud point of 5 % solution of various ethoxylated surfactants and their HLB number.

Davies [9, 10] attempted to relate the HLB values to the selective coalescence rates of emulsions. Such correlations were not realized since it was found that the emulsion stability and even its type depend to a large extent on the method of dispersing the oil into the water and vice versa. At best the HLB number can only be used as a guide for selecting optimum compositions of emulsifying agents.

One may take any pair of emulsifying agents which fall at opposite ends of the HLB scale, e.g. Tween 80 (sorbitan mono-oleate with 20 mol EO, HLB = 15) and Span 80 (sorbitan mono-oleate, HLB = 5) using them in various proportions to cover a wide range of HLB numbers. The emulsions should be prepared in the same way, with a few percent of the emulsifying blend. For example, a 20 % O/W emulsion is prepared by using 4 % emulsifier blend (20 % with respect to oil) and 76 % water. The stability of the emulsions is then assessed at each HLB number from the rate of coalescence or qualitatively by measuring the rate of oil separation. In this way one may be able to find the optimum HLB number for a given oil. For example with a given oil, the opti-

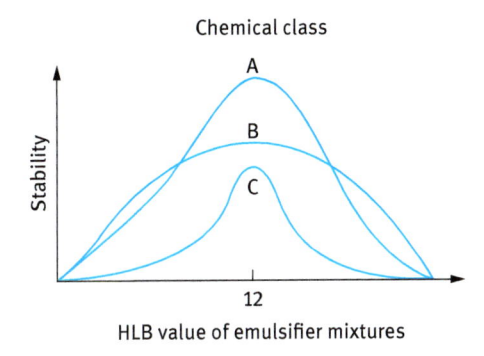

Fig. 6.6: Stabilization of emulsion by different classes of surfactants as a function of HLB.

mum HLB number is found to be 10.3. The latter can be determined more exactly by using mixtures of surfactants with narrower HLB range, say between 9.5 and 11. Having found the most effective HLB value, various other surfactant pairs are compared at this HLB value, to find the most effective pair. This is illustrated in Fig. 6.6 which shows schematically the difference between three chemical classes of surfactants. Although the different classes give a stable emulsion at HLB = 12, mixture A gives the best emulsion stability.

The HLB value of a given magnitude can be obtained by mixing emulsifiers of different chemical types. The "correct" chemical type is as important as the "correct" HLB number. This is illustrated in Fig. 6.7, which shows that an emulsifier with unsaturated alkyl chain such as oleate (ethoxylated sorbitan mono-oleate, Tween 80) is more suitable for emulsifying an unsaturated oil [6]. An emulsifier with saturated alkyl chain (stearate in Tween 60) is better for emulsifying a saturated oil).

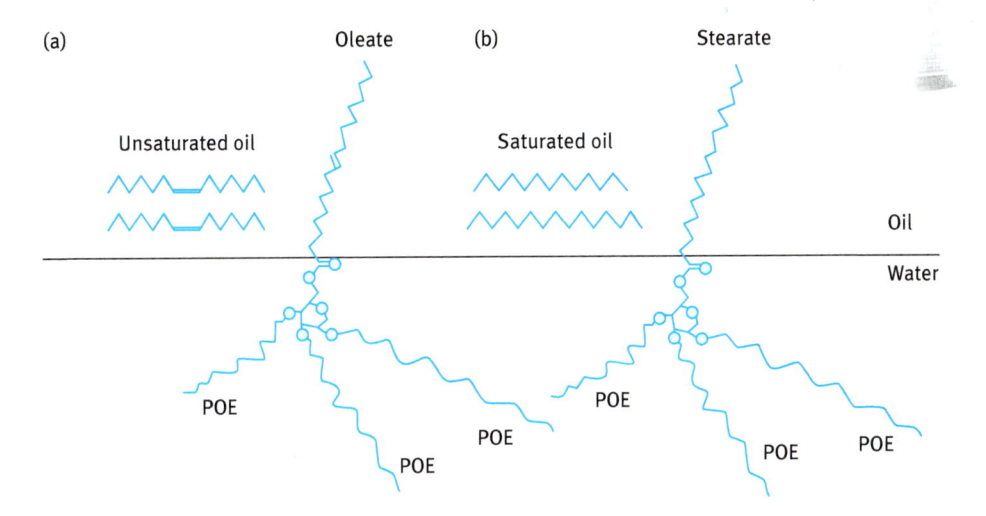

Fig. 6.7: Selection of Tween type to correspond to the type of the oil to be emulsified.

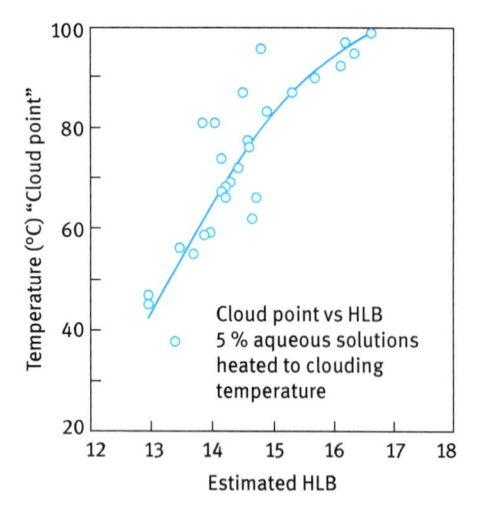

Fig. 6.8: Relationship between cloud point and HLB number.

Various procedures have been developed to determine the HLB of different surfactants. Griffin [8] found a correlation between the HLB and the cloud points of 5 % aqueous solution of ethoxylated surfactants as illustrated in Fig. 6.8.

A titration procedure was developed [7] for estimating the HLB number. In this method, a 1 % solution of surfactant in benzene plus dioxane is titrated with distilled water at constant temperature until a permanent turbidity appears. The authors found a good linear relationship between the HLB number and the water titration value for polyhydric alcohol esters as shown in Fig. 6.9. However, the slope of the line depends on the class of material used.

Gas liquid chromatography (GLC) could also be used to determine the HLB number [7]. Since in GLC the efficiency of separation depends on the polarity of the substrate with respect to the components of the mixture, it should be possible to determine the HLB directly by using the surfactant as the substrate and passing an oil phase down the column. Thus, when a 50 : 50 mixture of ethanol and hexane is passed down a column of a simple nonionic surfactant, such as sorbitan fatty acid esters and polyoxyethylated sorbitan fatty acid esters, two well-defined peaks, corresponding to hexane (which appears first) and ethanol appears on the chromatograms. A good correlation was found between the retention time ratio R_t (ethanol/hexane) and the HLB value. This is illustrated in Fig. 6.10. Statistical analysis of the data gave the following empirical relationship between R_t and HLB,

$$\text{HLB} = 8.55 R_t - 6.36, \tag{6.7}$$

where

$$R_t = \frac{R_t^{\text{ETOH}}}{R_t^{\text{hexane}}}. \tag{6.8}$$

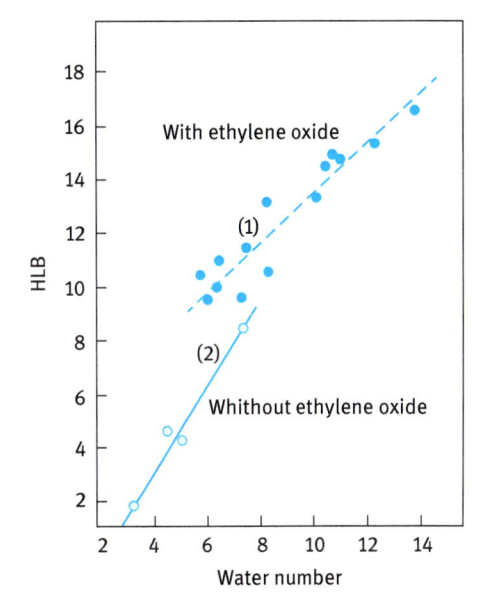

Fig. 6.9: Correlation of HLB with water number.

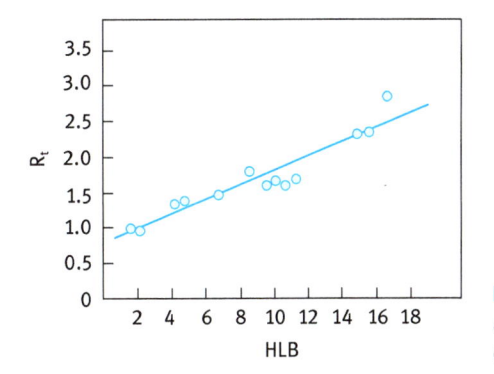

Fig. 6.10: Correlation between retention time and HLB of sorbitan fatty acid esters and polyoxyethylated fatty acid esters.

6.4.2 The Phase Inversion Temperature (PIT) concept

Shinoda and co-workers [11, 12] found that many O/W emulsions stabilized with nonionic surfactants undergo a process of inversion at a critical temperature (PIT). The PIT can be determined by following the emulsion conductivity (small amount of electrolyte is added to increase the sensitivity) as a function of temperature as illustrated in Fig. 6.11. The conductivity of the O/W emulsion increases with increasing temperature until the PIT is reached, above which there will be a rapid reduction in conductivity (W/O emulsion is formed). Shinoda and co-workers [11, 12] found that the PIT is influenced by the HLB number of the surfactant as shown in Fig. 6.12. For any given oil, the PIT increases with increasing HLB number. The size of the emulsion droplets was found to depend on the temperature and HLB number of the emulsifiers. The droplets

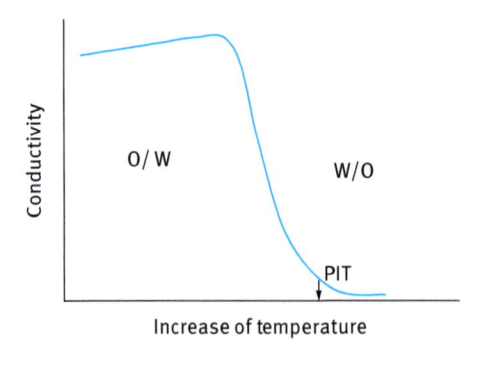

Fig. 6.11: Variation of conductivity with temperature for an O/W emulsion.

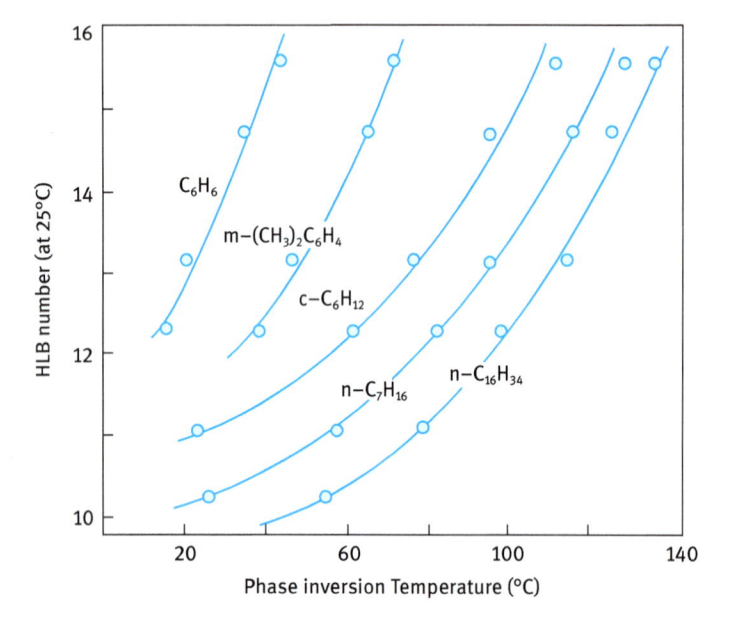

Fig. 6.12: Correlation between HLB number and PIT for various O/W (1 : 1) emulsions stabilized with nonionic surfactants (1.5 wt%).

are less stable towards coalescence close to the PIT. However, by rapid cooling of the emulsion a stable system may be produced. Relatively stable O/W emulsions were obtained when the PIT of the system was 20–65 °C higher than the storage temperature. Emulsions prepared at a temperature just below the PIT followed by rapid cooling generally have smaller droplet sizes. This can be understood if one considers the change of interfacial tension with temperature as is illustrated in Fig. 6.13. The interfacial tension decreases with increasing temperature reaching a minimum close to the PIT, after which it increases.

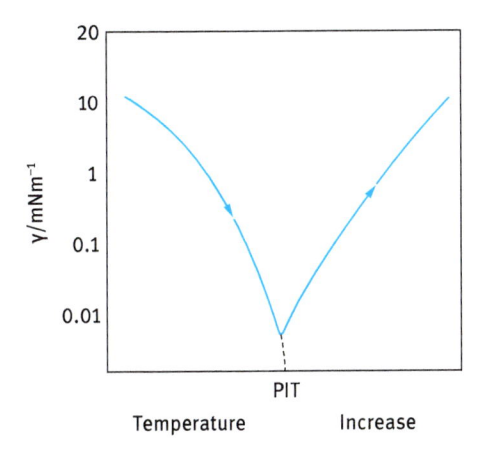

Fig. 6.13: Variation of interfacial tension with temperature increase for an O/W emulsion.

Thus, the droplets prepared close to the PIT are smaller than those prepared at lower temperatures. These droplets are relatively unstable towards coalescence near the PIT, but by rapid cooling of the emulsion one can retain the smaller size. This procedure may be applied to prepare mini- (nano)emulsions.

The optimum stability of the emulsion was found to be relatively insensitive to changes in the HLB value or the PIT of the emulsifier, but instability was very sensitive to the PIT of the system.

It is essential, therefore to measure the PIT of the emulsion as a whole (with all other ingredients).

At a given HLB value, stability of the emulsions against coalescence increases markedly as the molar mass of both the hydrophilic and lipophilic components increases. The enhanced stability using high molecular weight surfactants (polymeric surfactants) can be understood from a consideration of steric repulsion which produces more stable films. Films produced using macromolecular surfactants resist thinning and disruption, thus reducing the possibility of coalescence. The emulsions showed maximum stability when the distribution of the PEO chains was broad. The cloud point is lower but the PIT is higher than in the corresponding case for narrow size distributions. The PIT and HLB number are directly related parameters.

Addition of electrolytes reduces the PIT and hence an emulsifier with a higher PIT value is required when preparing emulsions in the presence of electrolytes. Electrolytes cause dehydration of the PEO chains and in effect this reduces the cloud point of the nonionic surfactant. One needs to compensate for this effect by using a surfactant with higher HLB. The optimum PIT of the emulsifier is fixed if the storage temperature is fixed.

In view of the above correlation between PIT and HLB and the possible dependence of the kinetics of droplet coalescence on the HLB number, Sherman and coworkers suggested the use of PIT measurements as a rapid method for assessing emulsion stability. However, one should be careful in using such methods for assessment of the long-term stability since the correlations were based on a very limited number of surfactants and oils.

Measurement of the PIT can at best be used as a guide for preparation of stable emulsions. Assessment of the stability should be evaluated by following the droplet size distribution as a function of time using a Coulter counter or light diffraction techniques. Following the rheology of the emulsion as a function of time and temperature may also be used for assessment of the stability against coalescence. Care should be taken in analysing the rheological results. Coalescence results in an increase in droplet size and this is usually followed by a reduction in the viscosity of the emulsion. This trend is only observed if the coalescence is not accompanied by flocculation of the emulsion droplets (which results in an increase in the viscosity). Ostwald ripening can also complicate the analysis of the rheological data.

6.4.3 The Cohesive Energy Ratio (CER) concept

Beerbower and Hills [13] considered the dispersing tendency on the oil and water interfaces of the surfactant or emulsifier in terms of the ratio of the cohesive energies of the mixtures of oil with the lipophilic portion of the surfactant, and the water with the hydrophilic portion. They used the Winsor R_0 concept which is the ratio of the intermolecular attraction of oil molecules (O) and lipophilic portion of surfactant (L), C_{LO}, to that of water (W) and hydrophilic portion (H), C_{HW},

$$R_0 = \frac{C_{LO}}{C_{HW}} \qquad (6.9)$$

Several interaction parameters may be identified at the oil and water sides of the interface. One can identify at least nine interaction parameters as schematically represented in Fig. 6.14.

In the absence of emulsifier, there will be only three interaction parameters: C_{OO}, C_{WW}, C_{OW}; if $C_{OW} \ll C_{WW}$, the emulsion breaks.

C_{LL}, C_{OO}, C_{LO} (at oil side)

C_{HH}, C_{WW}, C_{HW} (at water side)

C_{LW}, C_{HO}, C_{LH} (at the interface)

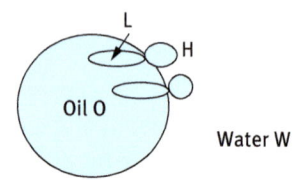

Fig. 6.14: The cohesive energy ratio concept.

The above interaction parameters may be related to the Hildebrand solubility parameter [14] δ (at the oil side of the interface) and the Hansen [15] nonpolar, hydrogen bonding and polar contributions to δ at the water side of the interface.

The solubility parameter of any component is related to its heat of vaporization ΔH by the expression,

$$\delta^2 = \frac{\Delta H - RT}{V_m}, \tag{6.10}$$

where V_m is the molar volume.

Hansen [15] considered δ (at the water side of the interface) to consist of three main contributions, a dispersion contribution, δ_d, a polar contribution, δ_p and a hydrogen bonding contribution, δ_h. These contributions have different weighting factors,

$$\delta^2 = \delta_d^2 + 0.25\delta_p^2 + 0.25\delta_h^2. \tag{6.11}$$

Beerbower and Hills [13] used the following expression for the HLB number,

$$HLB = 20\left(\frac{M_H}{M_L + M_H}\right) = 20\left(\frac{V_H \rho_H}{V_L \rho_L + V_H \rho_H}\right) \tag{6.12}$$

where M_H and M_L are the molecular weights of the hydrophilic and lipophilic portions of the surfactants. V_L and V_H are their corresponding molar volumes whereas ρ_H and ρ_L are the densities respectively.

The cohesive energy ratio was originally defined by Winsor, equation (6.8).

When $C_{LO} > C_{HW}$, R > 1 and a W/O emulsion forms. If $C_{LO} < C_{HW}$, R < 1 and an O/W emulsion forms. If $C_{LO} = C_{HW}$, R = 1 and a planer system results; this denotes the inversion point.

R_0 can be related to V_L, δ_L and V_H, δ_H by the expression,

$$R_0 = \frac{V_L \delta_L^2}{V_H \delta_H^2}. \tag{6.13}$$

Using equation (6.12),

$$R_0 = \frac{V_L(\delta_d^2 + 0.25\delta_p^2 + 0.25\delta_h^2)_L}{V_h(\delta_d^2 + 0.25\delta_p^2 + 0.25\delta_h^2)_H} \tag{6.14}$$

Combining equations (6.11) and (6.13), one obtains the following general expression for the cohesive energy ratio,

For an O/W system, HLB = 12–15 and R_0 = 0.58–0.29 (R_0 < 1). For a W/O system, HLB = 5–6 and R_0 = 2.3–1.9 (R_0 > 1). For a planer system, HLB = 8–10 and R_0 = 1.25–0.85 (R_0 ≈ 1).

$$R_0 = \left(\frac{20}{HLB} - 1\right)\frac{\rho_h(\delta_d^2 + 025\delta_p^2 + 0.25\delta_h^2)_L}{\rho_L(\delta_d^2 + 0.25\delta_p^2 + 0.25\delta_p^2)_L}. \tag{6.15}$$

The R_0 equation combines both the HLB and cohesive energy densities; it gives a more quantitative estimate of emulsifier selection. R_0 considers HLB, molar volume and chemical match. The success of this approach depends on the availability of data on the solubility parameters of the various surfactant portions. Some values are tabulated in the book by Barton [16].

6.4.4 The Critical Packing Parameter (CPP) for emulsion selection

The critical packing parameter (CPP) is a geometric expression relating the hydrocarbon chain volume (v) and length (l) and the interfacial area occupied by the head group (a_0) [17],

$$\text{CPP} = \frac{v}{l_c a_0}. \tag{6.16}$$

a_0 is the optimal surface area per head group and l_c is the critical chain length.

Regardless of the shape of any aggregated structure (spherical or cylindrical micelle or a bilayer), no point within the structure can be farther from the hydrocarbon-water surface than l_c. The critical chain length, l_c, is roughly equal to but less than the fully extended length of the alkyl chain.

The above concept can be applied to predict the shape of an aggregated structure. Consider a spherical micelle with radius r and aggregation number n; the volume of the micelle is given by,

$$\frac{4}{3}\pi r^3 = nv, \tag{6.17}$$

where v is the volume of a surfactant molecule.

The area of the micelle is given by,

$$4\pi r^2 = na_0, \tag{6.18}$$

where a_0 is the area per surfactant head group.

Combining equations (6.17) and (6.18),

$$a_0 = \frac{3v}{r}. \tag{6.19}$$

The cross-sectional area of the hydrocarbon chain a is given by the ratio of its volume to its extended length l_c,

$$a = \frac{v}{l_c}. \tag{6.20}$$

From equations (6.18) and (6.19),

$$\text{CPP} = \frac{a}{a_0} = \frac{1}{3}\frac{r}{l_c}. \tag{6.21}$$

Since $r < l_c$, then CPP $\leq (1/3)$.

For a cylindrical micelle with length d and radius r,

$$\text{Volume of the micelle} = \pi r^2 d = nv. \tag{6.22}$$

Tab. 6.5: Shape of micelles and CPP.

Lipid	Critical packing parameter $v/a_0 l_c$	Critical packing shape	Structures formed
Single-chained lipids (surfactants) with large head-group areas: – SDS in low salt	< 1/3	Cone	Spherical micelles
Single-chained lipids with small head-group areas: – SDS and CTAB in high salt – nonionic lipids	1/3–1/2	Truncated cone	Cylindrical micelles
Double-chained lipids with large head-group areas, fluid chains: – Phosphatidyl choline (lecithin) – phosphatidyl serine – phosphatidyl glycerol – phosphatidyl inositol – phosphatidic acid – sphingomyelin, DGDG[a] – dihexadecyl phosphate – dialkyl dimethyl ammonium – salts	1/2–1	Truncated one	Flexible bilayers, vesicles
Double-chained lipids with small head-group areas, anionic lipids in high salt, saturated frozen chains: – phosphatidyl ethanaiamine – phosphatidyl serine + Ca^{2+}	≈ 1	Cylinder	Planar bilayers
Double-chained lipids with small head-group areas, nonionic lipids, poly(cis) unsaturated chains, high T: – unsat. phosphatidyl ethanolamine – cardiolipin + Ca^{2+} – phosphatidic acid + Ca^{2+} – cholesterol, MGDG[b]	> 1	Inverted truncated cone or wedge	Inverted micelles

a DGDG: digalactosyl diglyceride, diglucosyldiglyceride
b MGDG: monogalactosyl diglyceride, monoglucosyl diglyceride

Combining equations (6.21) and (6.22),

$$\text{Area of the micelle} = 2\pi r d = n a_0 \tag{6.23}$$

$$a_0 = \frac{2v}{r} \tag{6.24}$$

$$a = \frac{v}{l_c} \tag{6.25}$$

$$CPP = \frac{a}{a_0} = \frac{1}{2}\frac{r}{l_c}. \tag{6.26}$$

Since $r < l_c$, then $(1/3) < CPP \le (1/2)$.

For vesicles (liposomes) $1 > CPP \ge (2/3)$ and for lamellar micelles $P \approx 1$. For inverse micelles $CPP > 1$. A summary of the various shapes of micelles and their CPP is given in Tab. 6.5.

Surfactants that make spherical micelles with the above packing constraints, i.e. $CPP \le (1/3)$, are more suitable for O/W emulsions. Surfactants with $CPP > 1$, i.e. forming inverted micelles, are suitable for formation of W/O emulsions.

6.5 Manufacture of cosmetic emulsions

For the manufacture of cosmetic emulsions (which are sometimes referred to as cosmetic creams), it is necessary to control the process that determines the droplet size distribution, since this controls the rheology of the resulting emulsion. Usually, one starts to make the emulsion on a lab scale (of the order of 1–2 liters), which has to be scaled-up to a pilot plant and manufacturing scale. At each stage, it is necessary to control the various process parameters which need to be optimized to produce the desirable effect. It is necessary to relate the process variable from the lab to the pilot plant to the manufacturing scale and this requires a great deal of understanding of emulsion formation that is controlled by the interfacial properties of the surfactant film. Two main factors should be considered, namely the mixing conditions and selection of production equipment. For proper mixing, sufficient agitation that produces turbulent flow is necessary in order to break up the liquid (disperse phase) into small droplets. Various parameters should be controlled such as flow rate and turbulence, type of impellers, viscosity of the internal and external phases and the interfacial properties such as surface tension, surface elasticity and viscosity. The selection of production equipment depends on the characteristics of the emulsion to be produced. Propeller and turbine agitators are normally used for low and medium viscosity emulsions. Agitators that are capable of scrapping the walls of the vessel are essential for high viscosity emulsions. Very high shear rates can be produced by using ultrasonics, colloid mills and homogenizers. It is essential to avoid too much heating in the emulsion during preparation, which may produce undesirable effects such as flocculation and coalescence.

6.5.1 Mechanism of emulsification

This can be considered from a consideration of the energy required to expand the interface, $\Delta A \gamma$ (where ΔA is the increase in interfacial area when the bulk oil with area A_1 produces a large number of droplets with area A_2; $A_2 \gg A_1$ and γ is the interfacial tension). Since γ is positive, the energy to expand the interface is large and positive; this energy term cannot be compensated by the small entropy of dispersion $T\Delta S^{conf}$ (which is also positive) and the total free energy of formation of an emulsion, ΔG^{form}, is positive as given by equation (6.1). Thus, emulsion formation is non-spontaneous and energy is required to produce the droplets.

The formation of large droplets (few μm) as is the case for macroemulsions is fairly easy and hence high speed stirrers such as the Ultra-Turrax or Silverson Mixer are sufficient to produce the emulsion. In contrast, the formation of small drops (submicron as is the case with nanoemulsions) is difficult and this requires a large amount of surfactant and/or energy. The high energy required for formation of nanoemulsions can be understood from a consideration of the Laplace pressure Δp (the difference in pressure between inside and outside the droplet) as given by equations (6.27) and (6.28),

$$\Delta p = \gamma \left(\frac{1}{r_1} + \frac{1}{r_2} \right), \qquad (6.27)$$

where r_1 and r_2 are the two principal radii of curvature.
For a perfectly spherical droplet $r_1 = r_2 = r$ and

$$\Delta p = \frac{2\gamma}{r}. \qquad (6.28)$$

To break up a drop into smaller ones, it must be strongly deformed and this deformation increases Δp. Consequently, the stress needed to deform the drop is higher for a smaller drop. Since the stress is generally transmitted by the surrounding liquid via agitation, higher stresses need more vigorous agitation, and hence more energy is needed to produce smaller drops.

Surfactants play major roles in the formation of emulsions [3–7]. By lowering the interfacial tension, Δp is reduced and hence the stress needed to break up a drop is reduced. Surfactants also prevent coalescence of newly formed drops (see below).

To describe emulsion formation one has to consider two main factors: hydrodynamics and interfacial science. In hydrodynamics one has to consider the type of flow: laminar flow and turbulent flow. This depends on the Reynolds number as will be discussed below.

To assess emulsion formation, one usually measures the droplet size distribution using for example laser diffraction techniques. If the number frequency of droplets as a function of droplet diameter d is given by f(d), the n-th moment of the distribution is,

$$S_n = \int_0^\infty d^n f(d)\, \partial d. \qquad (6.29)$$

6 Formulation of cosmetic emulsions

The mean droplet size is defined as the ratio of selected moments of the size distribution,

$$d_{nm} = \left[\frac{\int_0^\infty d^n f(d)\, \partial d}{\int_0^\infty d^m f(d)\, \partial d} \right]^{1/(n-m)}, \quad (6.30)$$

where n and m are integers and n > m and typically n does not exceed 4.
Using equation (6.30) one can define several mean average diameters:

– The Sauter mean diameter with n = 3 and m = 2.

$$d_{32} = \left[\frac{\int_0^\infty d^3 f(d)\, \partial d}{\int_0^\infty d^2 f(d)\, \partial d} \right]. \quad (6.31)$$

– The mass mean diameter,

$$d_{43} = \left[\frac{\int_0^\infty d^4 f(d)\, \partial d}{\int_0^\infty d^3 f(d)\, \partial d} \right]. \quad (6.32)$$

– The number mean diameter,

$$d_{10} = \left[\frac{\int_0^\infty d^1 f(d)\, \partial d}{\int_0^\infty f(d)\, \partial d} \right]. \quad (6.33)$$

In most cases d_{32} (the volume/surface average or Sauter mean) is used. The width of the size distribution can be given as the variation coefficient c_m which is the standard deviation of the distribution weighted with d_m divided by the corresponding average d. Generally C_2 will be used which corresponds to d_{32}.

Another is the specific surface area A (surface area of all emulsion droplets per unit volume of emulsion),

$$A = \pi S_2 = \frac{6\phi}{d_{32}}. \quad (6.34)$$

Surfactants lower the interfacial tension γ and this causes a reduction in droplet size. The width of the size distribution decreases with decreasing γ. For laminar flow the droplet diameter is proportional to γ; for turbulent inertial regime, the droplet diameter is proportional to $\gamma^{3/5}$.

The surfactant can lower the interfacial tension γ_0 of a clean oil-water interface to a value γ and,

$$\pi = \gamma_0 - \gamma, \quad (6.35)$$

where π is the surface pressure. The dependence of π on the surfactant activity a or concentration C is given by the Gibbs equation,

$$d\pi = -d\gamma = RT\Gamma\, d\ln a = RT\Gamma\, d\ln C, \quad (6.36)$$

where R is the gas constant, T is the absolute temperature and Γ is the surface excess (number of moles adsorbed per unit area of the interface).

At high a, the surface excess Γ reaches a plateau value; for many surfactants it is of the order of 3 mg m^{-2}. Γ increases with increasing surfactant concentration and eventually it reaches a plateau value (saturation adsorption). The value of C needed to obtain the same Γ is much smaller for the polymer when compared with the surfactant. In contrast, the value of γ reached at full saturation of the interface is lower for a surfactant (mostly in the region of 1–3 mN m^{-1} depending on the nature of surfactant and oil) when compared with a polymer (with γ values in the region of 10–20 mN m^{-1} depending on the nature of polymer and oil). This is due to the much closer packing of the small surfactant molecules at the interface when compared with the much larger polymer molecule that adopts tail-train-loop-tail conformation.

Another important role of the surfactant is its effect on the interfacial dilational modulus ε,

$$\varepsilon = \frac{d\gamma}{d\ln A}. \qquad (6.37)$$

ε is the absolute value of a complex quantity, composed of an elastic and a viscous term.

During emulsification an increase in the interfacial area A takes place and this causes a reduction in Γ. The equilibrium is restored by adsorption of surfactant from the bulk, but this takes time (shorter times occur at higher surfactant activity). Thus ε is small at small a and also at large a. Because of the lack or slowness of equilibrium with polymeric surfactants, ε will not be the same for expansion and compression of the interface.

In practice emulsifiers are generally made of surfactant mixtures, often containing different components and these have pronounced effects on γ and ε. Some specific surfactant mixtures give lower γ values than either of the two individual components. The presence of more than one surfactant molecule at the interface tends to increase ε at high surfactant concentrations. The various components vary in surface activity. Those with the lowest γ tend to predominate at the interface, but if present at low concentrations, it may take a long time before reaching the lowest value. Polymer-surfactant mixtures may show some synergetic surface activity.

During emulsification, surfactant molecules are transferred from the solution to the interface and this leaves an ever lower surfactant activity [3–7]. Consider for example an O/W emulsion with a volume fraction ϕ = 0.4 and a Sauter diameter d_{32} = 1 μm. According to equation (6.9), the specific surface area is 2.4 m^2 ml^{-1} and for a surface excess Γ of 3 mg m^{-2}, the amount of surfactant at the interface is 7.2 mg ml^{-1} emulsion, corresponding to 12 mg ml^{-1} aqueous phase (or 1.2 %). Assuming that the concentration of surfactant, C_{eq} (the concentration left after emulsification), leading to a plateau value of Γ equals 0.3 mg ml^{-1} then the surfactant concentration decreases from 12.3 to 0.3 mg ml^{-1} during emulsification. This implies that the effective γ value increases during the process. If insufficient surfactant is present to leave a concentration C_{eq} after emulsification, even the equilibrium γ value would increase.

Another aspect is that the composition of surfactant mixture in solution may alter during emulsification. If some minor components are present that give a relatively small γ value, this will predominate at a macroscopic interface, but during emulsification, as the interfacial area increases, the solution will soon become depleted of these components. Consequently, the equilibrium value of γ will increase during the process and the final value may be markedly larger than what is expected on the basis of the macroscopic measurement.

During droplet deformation, its interfacial area is increased. The drop will commonly have acquired some surfactant, and it may even have a Γ value close to the equilibrium at the prevailing (local) surface activity. The surfactant molecules may distribute themselves evenly over the enlarged interface by surface diffusion or by spreading. The rate of surface diffusion is determined by the surface diffusion coefficient D_s that is inversely proportional to the molar mass of the surfactant molecule and also inversely proportional to the effective viscosity felt. D_s also decreases with increasing Γ. Sudden extension of the interface or sudden application of a surfactant to an interface can produce a large interfacial tension gradient and in such a case spreading of the surfactant can occur.

Surfactants allow the existence of interfacial tension gradients which is crucial for formation of stable droplets. In the absence of surfactants (clean interface), the interface cannot withstand a tangential stress; the liquid motion will be continuous across a liquid interface.

If the γ-gradient can become large enough, it will arrest the interface [3–7]. The largest value attainable for dγ equals about π_{eq}, i.e. $\gamma_0 - \gamma_{eq}$. If it acts over a small distance, a considerable stress can develop, of the order of 10 kPa.

Interfacial tension gradients are very important in stabilizing the thin liquid film between the droplets which is very important during the beginning of emulsification, when films of the continuous phase may be drawn through the disperse phase or when collision of the still large deformable drops causes the film to form between them. The magnitude of the γ-gradients and of the Marangoni effect depends on the surface dilational modulus ε, which for a plane interface with one surfactant-containing phase is given by the expressions,

$$\varepsilon = \frac{-d\gamma/d\ln \Gamma}{(1 + 2\xi + 2\xi^2)^{1/2}} \tag{6.38}$$

$$\xi = \frac{dm_C}{d\Gamma}\left(\frac{D}{2\omega}\right)^{1/2} \tag{6.39}$$

$$\omega = \frac{d\ln A}{dt}, \tag{6.40}$$

where D is the diffusion coefficient of the surfactant and ω represents a timescale (time needed for doubling the surface area) that is roughly equal to τ_{def}.

During emulsification, ε is dominated by the magnitude of the numerator in equation (6.38) because ξ remains small. The value of $dm_C/d\Gamma$ tends to go to very high values when Γ reaches its plateau value; ε goes to a maximum when m_C is increased. However, during droplet deformation, Γ will always remain smaller. Taking reasonable values for the variables; $dm_C/d\Gamma = 10^2-10^4$ m^{-1}, $D = 10^{-9}-10^{-11}$ m^2 s^{-1} and $\tau_{def} = 10^{-2}-10^{-6}$ s, $\xi < 0.1$ at all conditions. The same conclusion can be drawn for values of ε in thin films, e.g. between closely approaching drops. It may be concluded that for conditions that prevail during emulsification, ε increases with m_C and follows the relation,

$$\varepsilon \approx \frac{d\pi}{d\ln\Gamma}, \qquad (6.41)$$

except for very high surfactant concentration, where π is the surface pressure (π = γ₀ − γ).

The presence of a surfactant means that during emulsification the interfacial tension need not to be the same everywhere. This has two consequences: (i) the equilibrium shape of the drop is affected; (ii) any γ-gradient formed will slow down the motion of the liquid inside the drop (this diminishes the amount of energy needed to deform and break-up the drop).

Another important role of the emulsifier is to prevent coalescence during emulsification. This is certainly not due to the strong repulsion between the droplets, since the pressure at which two drops are pressed together is much greater than the repulsive stresses. The counteracting stress must be due to the formation of γ-gradients. When two drops are pushed together, liquid will flow out from the thin layer between them, and the flow will induce a γ-gradient.

$$\tau_{\Delta\gamma} \approx \frac{2|\Delta\gamma|}{(1/2)d}. \qquad (6.42)$$

The factor 2 follows from the fact that two interfaces are involved. Taking a value of Δγ = 10 mN m^{-1}, the stress amounts to 40 kPa (which is of the same order of magnitude as the external stress). The stress due to the γ-gradient cannot as such prevent coalescence, since it only acts for a short time, but it will greatly slow down the mutual approach of the droplets. The external stress will also act for a short time, and it may well be that the drops move apart before coalescence can occur. The effective γ-gradient will depend on the value of ε as given by equation (6.41).

Closely related to the above mechanism, is the Gibbs–Marangoni effect [18–20], schematically represented in Fig. 6.15. The depletion of surfactant in the thin film between approaching drops results in γ-gradient without liquid flow being involved. This results in an inward flow of liquid that tends to drive the drops apart. Such a mechanism would only act if the drops were insufficiently covered with surfactant (Γ below the plateau value) as occurs during emulsification.

The Gibbs–Marangoni effect also explains the Bancroft rule which states that the phase in which the surfactant is most soluble forms the continuous phase. If the surfactant is in the droplets, a γ-gradient cannot develop and the drops would be prone to coalescence. Thus, surfactants with HLB > 7 tend to form O/W emulsions and those with HLB < 7 tend to form W/O emulsions.

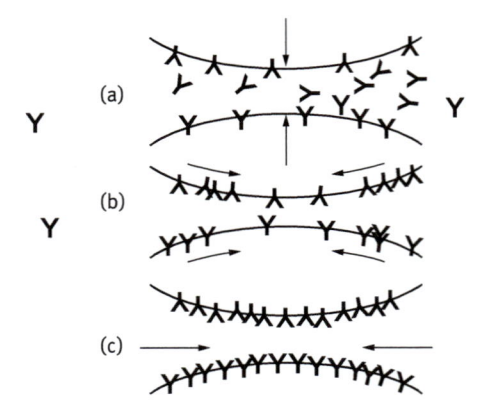

Fig. 6.15: Schematic representation of the Gibbs–Marangoni effect for two approaching drops.

The Gibbs–Marangoni effect also explains the difference between surfactants and polymers for emulsification. Polymers give larger drops when compared with surfactants. Polymers give a smaller value of ε at small concentrations when compared to surfactants.

Various other factors should also be considered for emulsification, as will be discussed below: The disperse phase volume fraction ϕ. An increase in ϕ leads to an increase in droplet collision and hence coalescence during emulsification. With increasing ϕ, the viscosity of the emulsion increases and could change the flow from being turbulent to being laminar. The presence of many particles results in a local increase in velocity gradients. In turbulent flow, increasing ϕ will induce turbulence depression (see below). This will result in larger droplets. Turbulence depression by added polymers tends to remove the small eddies, resulting in the formation of larger droplets.

If the mass ratio of surfactant to continuous phase is kept constant, increasing ϕ results in decreasing surfactant concentration and hence an increase in γ_{eq}, resulting in larger droplets. If the mass ratio of surfactant to disperse phase is kept constant, the above changes are reversed.

6.5.2 Methods of emulsification

Several procedures may be applied for emulsion preparation, these range from simple pipe flow (low agitation energy L), static mixers, rotor-stator (toothed devices such as the Ultra-Turrax and batch radial discharge mixers such as the Silverson mixers) and general stirrers (low to medium energy, L–M), colloid mills and high pressure homogenizers (high energy, H), ultrasound generators (M–H) and membrane emulsification methods. The method of preparation can be continuous (C) or batchwise (B): pipe flow – C; static mixers and general stirrers – B, C; colloid mill and high pressure homogenizers – C; ultrasound – B, C.

In all methods, there is liquid flow; unbounded and strongly confined flow. In the unbounded flow any droplets are surrounded by a large amount of flowing liquid (the confining walls of the apparatus are far away from most of the droplets). The forces can be frictional (mostly viscous) or inertial. Viscous forces cause shear stresses to act on the interface between the droplets and the continuous phase (primarily in the direction of the interface). The shear stresses can be generated by laminar flow (LV) or turbulent flow (TV); this depends on the dimensionless Reynolds numbers Re,

$$\mathrm{Re} = \frac{vl\rho}{\eta}, \qquad (6.43)$$

where v is the linear liquid velocity, ρ is the liquid density and η is its viscosity. l is a characteristic length that is given by the diameter of flow through a cylindrical tube and by twice the slit width in a narrow slit.

For laminar flow Re $\leq$ 1000, whereas for turbulent flow Re $\geq$ 2000. Thus whether the regime is linear or turbulent depends on the scale of the apparatus, the flow rate and the liquid viscosity [3–7].

Rotor-stator mixers are the most commonly used mixers for emulsification. Two main types are available. The most commonly used toothed device (schematically illustrated in Fig. 6.16) is the Ultra-Turrax (IKA works, Germany).

Fig. 6.16: Schematic representation of a toothed mixer (Ultra-Turrax).

Toothed devices are available both as in-line as well as batch mixers, and because of their open structure they have a relatively good pumping capacity; therefore in batch applications they frequently do not need an additional impeller to induce bulk flow even in relatively large mixing vessels. These mixers are used in cosmetic emulsions to manufacture creams and lotions that can be highly viscous and non-Newtonian.

Fig. 6.17: Schematic representation of batch radial discharge mixer (Silverson mixer).

Batch radial discharge mixers such as Silverson mixers (Fig. 6.17) have a relatively simple design with a rotor equipped with four blades pumping the fluid through a stationary stator perforated with differently shaped/sized holes or slots.

They are frequently supplied with a set of easily interchangeable stators enabling the same machine to be used for a range of operations. Changing from one screen to another is quick and simple. Different stators/screens used in batch Silverson mixers are shown in Fig. 6.18. The general purpose disintegrating stator (Fig. 6.18 (a)) is recommended for preparation of thick emulsions (gels) whilst the slotted disintegrating stator (Fig. 6.18 (b)) is designed for emulsions containing elastic materials such as polymers. Square hole screens (Fig. 6.18 (c)) are recommended for the preparation of emulsions whereas the standard emulsor screen (Fig. 6.18 (d)) is used for liquid/liquid emulsification.

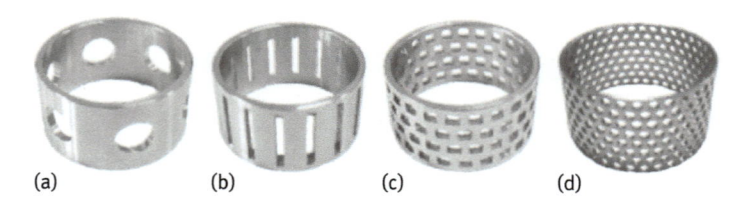

(a)   (b)   (c)   (d)

Fig. 6.18: Stators used in batch Silverson radial discharge mixers.

Batch toothed and radial discharge rotor-stator mixers are manufactured in different sizes ranging from the laboratory to the industrial scale. In lab applications mixing heads (assembly of rotor and stator) can be as small as 0.01 m (Turrax, Silverson) and the volume of processed fluid can vary from several milliliters to few liters. In models used in industrial applications mixing heads might have up to 0.5 m diameter enabling processing of several cubic meters of fluids in one batch.

In practical applications the selection of the rotor-stator mixer for a specific emulsification process depends on the required morphology of the product, frequently quantified in terms of average drop size or in terms of drop size distributions, and by the scale of the process. There is very little information enabling calculation of average drop size in rotor-stator mixers and there are no methods enabling estimation of drop size distributions. Therefore the selection of an appropriate mixer and processing conditions for a required formulation is frequently carried out by trial and error. Initially, one can carry out lab scale emulsification of given formulations testing different type/geometries of mixers they manufacture. Once the type of mixer and its operating parameters are determined at the lab scale the process needs to be scaled up. The majority of lab tests of emulsification is carried out in small batch vessels as it is easier and cheaper than running continuous processes. Therefore, prior to scaling up of the rotor-stator mixer it has to be decided whether industrial emulsification should be run as a batch or as a continuous process. Batch mixers are recommended for processes where formulation of a product requires long processing times typically associated with slow chemical reactions. They require simple control systems, but spatial homogeneity may be an issue in large vessels which could lead to a longer processing time. In processes where quality of the product is controlled by mechanical/hydrodynamic interactions between continuous and dispersed phases or by fast chemical reactions, but large amounts of energy are necessary to ensure adequate mixing, in-line rotor-stator mixers are recommended. In-line mixers are also recommended to efficiently process large volumes of fluid.

In the case of batch processing, rotor-stator devices immersed as top entry mixers is mechanically the simplest arrangement, but in some processes bottom entry mixers ensure better bulk mixing; however in this case sealing is more complex. In general, the efficiency of batch rotor-stator mixers decreases as the vessel size increases and as the viscosity of the processed fluid increases because of limited bulk mixing by rotor-stator mixers. Whilst the open structure of Ultra-Turrax mixers frequently enables sufficient bulk mixing even in relatively large vessels, if the liquid/emulsion has a low apparent viscosity, processing of very viscous emulsions requires an additional impeller (typically anchor type) to induce bulk flow and to circulate the emulsion through the rotor-stator mixer. On the other hand, batch Silverson rotor-stator mixers have a very limited pumping capacity and even at the lab scale they are mounted off the centre of the vessel to improve bulk mixing. At the large scale there is always need for at least one additional impeller and in the case of very large units more than one impeller is mounted on the same shaft.

Problems associated with the application of batch rotor-stator mixers for processing large volumes of fluid discussed above can be avoided by replacing batch mixers with in-line (continuous) mixers. There are many designs offered by different suppliers (Silverson, IKA, etc.) and the main differences are related to the geometry of the rotors and stators with stators and rotors designed for different applications. The main difference between batch and in-line rotor-stator mixers is that the latter have a strong pumping capacity, therefore they are mounted directly in the pipeline. One of the main advantages of in-line over batch mixers is that for the same power duty, a much smaller mixer is required, therefore they are better suited for processing large volumes of fluid.

When the scale of the processing vessel increases, a point is reached where it is more efficient to use an in-line rotor-stator mixer rather than a batch mixer of a large diameter. Because power consumption increases sharply with rotor diameter (to the fifth power) an excessively large motor is necessary at large scales. This transition point depends on the fluid rheology, but for a fluid with a viscosity similar to water, it is recommended to change from a batch to an in-line rotor-stator process at a volume of approximately 1 to 1.5 tons. The majority of manufacturers supply both single and multistage mixers for the emulsification of highly viscous liquids.

As mentioned above, there is liquid flow in all methods: unbounded and strongly confined flow. In unbounded flow any droplet is surrounded by a large amount of flowing liquid (the confining walls of the apparatus are far away from most of the droplets); the forces can be frictional (mostly viscous) or inertial. Viscous forces cause shear stresses to act on the interface between the droplets and the continuous phase (primarily in the direction of the interface). The shear stresses can be generated by laminar flow (LV) or turbulent flow (TV); This depends on the Reynolds number Re as given by equation (6.43). For laminar flow Re ≲ 1000, whereas for turbulent flow Re ≳ 2000. Thus whether the regime is linear or turbulent depends on the scale of the apparatus, the flow rate and the liquid viscosity. If the turbulent eddies are much larger than the droplets, they exert shear stresses on the droplets. If the turbulent eddies are much smaller than the droplets, inertial forces will cause disruption (TI). In bounded flow other relations hold; if the smallest dimension of the part of the apparatus in which the droplets are disrupted (say a slit) is comparable to droplet size, other relations hold (the flow is always laminar). A different regime prevails if the droplets are directly injected through a narrow capillary into the continuous phase (injection regime), i.e. membrane emulsification.

Within each regime, an essential variable is the intensity of the forces acting; the viscous stress during laminar flow $\sigma_{viscous}$ is given by,

$$\sigma_{viscous} = \eta G, \qquad (6.44)$$

where G is the velocity gradient.

The intensity in turbulent flow is expressed by the power density ε (the amount of energy dissipated per unit volume per unit time); for turbulent flow,

$$\varepsilon = \eta G^2. \qquad (6.45)$$

The most important regimes are: Laminar/Viscous (LV), Turbulent/Viscous (TV), Turbulent/Inertial (TI). For water as the continuous phase, the regime is always TI. For higher viscosity of the continuous phase ($\eta_C = 0.1$ Pas), the regime is TV. For still higher viscosity or a small apparatus (small l), the regime is LV. For very small apparatus (as is the case with most laboratory homogenizers), the regime is nearly always LV.

For the above regimes, a semi-quantitative theory is available that can give the timescale and magnitude of the local stress σ_{ext}, the droplet diameter d, timescale of droplets deformation τ_{def}, timescale of surfactant adsorption, τ_{ads} and mutual collision of droplets.

Laminar flow can be of a variety of types, purely rotational to purely extensional. For simple shear the flow consists of equal parts of rotation and elongation. The velocity gradient G (in reciprocal seconds) is equal to the shear rate γ. For hyperbolic flow, G is equal to the elongation rate. The strength of a flow is generally expressed by the stress it exerts on any plane in the direction of flow. It is simply equal to Gη (η is simply the shear viscosity).

For elongational flow, the elongational viscosity η_{el} is given by,

$$\eta_{el} = Tr\eta \qquad (6.46)$$

Where Tr is the dimensionless Trouton number which is equal to 2 for Newtonian liquids in two-dimensional uneasily elongation flow. Tr = 3 for axisymmetric uniaxial flow and is equal to 4 and for biaxial flows. Elongational flows exert higher stresses for the same value of G than simple shear. For non-Newtonian liquids, the relationships are more complicated and the values of Tr tends to be much higher.

An important parameter that describes droplet deformation is the Weber number We (which gives the ratio of the external stress over the Laplace pressure),

$$We = \frac{G\eta_C R}{2\gamma}. \qquad (6.47)$$

The deformation of the drop increases with increasing We and above a critical value We_{cr} the drop bursts forming smaller droplets. We_{cr} depends on two parameters: (i) the velocity vector α (α = 0 for simple shear and α = 1 for hyperbolic flow) and (ii) the viscosity ratio λ of the oil η_D and the external continuous phase η_C,

$$\lambda = \frac{\eta_D}{\eta_C}. \qquad (6.48)$$

As mentioned above, the viscosity of the oil plays an important role in the break-up of droplets; the higher the viscosity, the longer it will take to deform a drop. The deformation time τ_{def} is given by the ratio of oil viscosity to the external stress acting on the drop,

$$\tau_{def} = \frac{\eta_D}{\sigma_{ext}}. \qquad (6.49)$$

The above ideas for simple laminar flow were tested using emulsions containing 80 % oil in water stabilized with egg yolk. A colloid mill and static mixers were used to prepare the emulsion. The results are shown in Fig. 6.19 which gives the number of droplets n into which a parent drop is broken down when it is suddenly extended into a long thread, corresponding to We_b which is larger than We_{cr}. The number of drops increases with increasing We_b/We_{cr}. The largest number of drops, i.e. the smaller the droplet size, is obtained when λ = 1, i.e. when the viscosity of the oil phase is closer to that of the continuous phase. In practice, the resulting drop size distribution is of greater importance than the critical drop size for break-up.

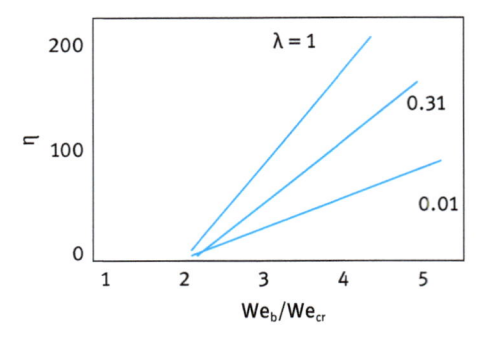

Fig. 6.19: Variation of n with We_b/We_{cr}.

Turbulent flow is characterized by the presence of eddies, which means that the average local flow velocity u generally differs from the time average value ū. The velocity fluctuates in a chaotic way and the average difference between u and u′ equals zero; however, the root mean square average u′ is finite [5–8],

$$u' = \langle(u - \bar{u})^2\rangle^{1/2}. \tag{6.50}$$

The value of u′ generally depends on direction, but for very high Re (> 50 000) and at small length scales the turbulence flow can be isotropic, and u′ does not depend on direction. Turbulent flow shows a spectrum of eddy sizes; the largest eddies have the highest u′, they transfer their kinetic energy to smaller eddies, which have a smaller u′ but larger velocity gradient u′/l.

The break-up of droplets in turbulent flow due to inertial forces may be represented by local pressure fluctuations near energy-bearing eddies,

$$\Delta p(x) = \rho[u'(x)]^2 = \varepsilon^{2/3} x^{2/3} \rho^{1/3}, \tag{6.51}$$

where ε is the power density, i.e. the amount of energy dissipated per unit volume, ρ is the density and x is the distance scale. If Δp is larger than the Laplace pressure (p = 2γ/R) near the eddy, the drop would be broken-up. Break-up would be most effective if $d = l_e$.

Putting $x = d_{max}$, the following expression gives the largest drops that are not broken up in the turbulent field,

$$d_{max} = \varepsilon^{-2/5} \gamma^{3/5} \rho^{-1/5}. \tag{6.52}$$

The validity of equation (6.52) is subject to two conditions: (i) The droplet size obtained cannot be much smaller than l_0. This equation is fulfilled for small η_C. (ii) The flow near the droplet should be turbulent. This depends on the droplet Reynolds number given by,

$$Re_{dr} = du'(d)\rho_C/\eta_C. \tag{6.53}$$

The condition $Re_{dr} > 1$ and combination with equation (6.10) leads to,

$$d > \eta_c^2/\gamma\rho. \tag{6.54}$$

Provided that (i) φ is small, (ii) η_C is not much larger than 1 mPas, (iii) η_D is fairly small, (iv) γ is constant, and (v) the machine is fairly small, equation (6.54) seems to hold well even for non-isotropic turbulence with Reynolds number much smaller than 50 000. The smallest drops are produced at the highest power density. Since the power density varies from place to place (particularly if Re is not very high) the droplet size distribution can be very wide. For break-up of drops in TI regime, the flow near the drop is turbulent. For laminar flow, break-up by viscous forces is possible. If the flow rate near the drop (u) varies greatly with distance d, the local velocity gradient is G. A pressure difference is produced over the drop of $(1/2)\Delta\rho(u_2) = \rho G d$. At the same time a shear stress $\eta_C G$ acts on the drop. The viscous forces will be predominant for $\eta_C G > \rho G d$, leading to the condition,

$$\bar{u} d\rho/\eta_C = Re_{dr} < 1. \quad (6.55)$$

The local velocity gradient is $\eta_C G = \varepsilon^{1/2}\eta_C^{1/2}$. This results in the following expression for d_{max},

$$d_{max} = We_{cr}\gamma\varepsilon^{-1/2}\eta_c^{-1/2} \quad (6.56)$$

The value of We_{cr} is rarely > 1, since the flow has an elongational component. For not very small η_C, d_{max} is smaller for TV than TI.

The viscosity of the oil plays an important role in the break-up of droplets; the higher the viscosity, the longer it will take to deform a drop. The deformation time τ_{def} is given by the ratio of oil viscosity to the external stress acting on the drop,

$$\tau_{def} = \frac{\eta_D}{\eta_C}. \quad (6.57)$$

The viscosity of the continuous phase η_C plays an important role in some regimes: For turbulent inertial regime, η_C has no effect on droplets size. For a turbulent viscous regime, larger η_C leads to smaller droplets. For a laminar viscous regime the effect is even stronger.

The value of η_C and the size of the apparatus determine which regime prevails, via the effect on Re. In a large machine and low η_C, Re is always very large and the resulting average droplet diameter d is proportional to $P_H^{-0.6}$ (where P_H is the homogenization pressure). If η_C is higher and $Re_{dr} < 1$, the regime is TV and $d \propto P_H^{-0.75}$. For a smaller machine, as used in the lab, where the slit width of the valve may be of the order of μm, Re is small and the regime is LV; $d \propto P_H^{-1.0}$. If the slit is made very small (of the order of droplet diameter), the regime can become TV.

In membrane emulsification, the disperse phase is passed through a membrane and droplets leaving the pores are immediately taken up by the continuous phase. The membrane is commonly made of porous glass or of ceramic materials. The general configuration is a membrane in the shape of a hollow cylinder; the disperse phase is pressed through it from outside, and the continuous phase pumped through the cylinder (cross-flow). The flow also causes detachment of the protruding droplets from the membrane.

Several requirements are necessary for the process: (i) For a hydrophobic disperse phase (O/W emulsion) the membrane should be hydrophilic, whereas for a hydrophilic disperse phase (W/O emulsion) the membrane should be hydrophobic, since otherwise the droplets cannot be detached. (ii) The pores must be sufficiently far apart to prevent the droplets coming out from touching each other and coalescing. (iii) The pressure over the membrane should be sufficiently high to achieve drop formation. This pressure should be at least of the order of Laplace pressure of a drop of diameter equal to the pore diameter. For example, for pores of 0.4 μm and $\gamma = 5$ mN m^{-1}, the pressure should be of the order of 10^5 Pa, but larger pressures are needed in practice, this would amount to 3×10^5 Pa, also to obtain a significant flow rate of the disperse phase through the membrane.

The smallest drop size obtained by membrane emulsification is about three times the pore diameter. The main disadvantage is its slow process, which can be of the order of 10^{-3} m^3 m^{-2} s^{-1}. This implies that very long circulation times are needed to produce even small volume fractions.

The most important variables that affect the emulsification process are the nature of the oil and emulsifier, the volume fraction of the disperse phase ϕ and the emulsification process. As discussed above, the method of emulsification and the regime (laminar or turbulent) have a pronounced effect on the process and the final droplet size distribution. The effect of the volume fraction of the disperse phase requires special attention. It affects the rate of collision between droplets during emulsification, and thereby the rate of coalescence. As a first approximation, this would depend on the relation between τ_{ads} and τ_{coal} (where τ_{ads} is the average time it takes for surfactant adsorption and τ_{coal} is the average time it takes until a droplet collides with another one). In the various regimes, the hydrodynamic constraints are the same for τ_{ads}. For example, in regime LV, $\tau_{coal} = \pi/8\phi G$. Thus for all regimes, the ratio of τ_{ads}/τ_{coal} is given by,

$$\kappa \equiv \frac{\tau_{ads}}{\tau_{coal}} \propto \frac{\phi \Gamma}{m_C d}, \qquad (6.58)$$

where the proportionality factor would be at least of order 10. For example, for $\phi = 0.1$, $\Gamma/m_C = 10^{-6}$ m and $d = 10^{-6}$ m (total surfactant concentration of the emulsion should then be about 5 %), κ would be of the order of 1. For $\kappa \gg 1$, considerable coalescence is likely to occur, particularly at high ϕ. The coalescence rate would then markedly increase during emulsification, since both m_C and d become smaller during the process. If emulsification proceeds long enough, the droplet size distribution may then be the result of a steady state of simultaneous break-up and coalescence.

The effect of increasing ϕ can be summarized as follows: (i) τ_{coal} is shorter and coalescence will be faster unless κ remains small. (ii) Emulsion viscosity η_{em} increases, hence Re decreases. This implies a change of flow from turbulent to laminar (LV). (iii) In laminar flow, the effective η_C becomes higher. The presence of many droplets means that the local velocity gradients near a droplet will generally be higher than the overall value of G. Consequently, the local shear stress ηG does increase with increasing ϕ, which is as if η_C increases. (iv) In turbulent flow, increasing ϕ will induce turbulence depression leading to larger d. (v) If the mass ratio of surfactant to continuous phase is constant, an increase in ϕ gives a decrease in surfactant concentration; hence an increase in γ_{eq}, an increase in κ, an increase in d produced by an increase in coalescence rate. If the mass ratio of surfactant to disperse phase is kept constant, the above mentioned changes are reversed, unless κ ≪ 1.

It is clear from the above discussion that general conclusions cannot be drawn, since several of the above mentioned mechanisms may come into play. Using a high pressure homogenizer, Walstra [4] compared the values of d with various ϕ values up to 0.4 at constant initial m_C, regime TI probably changing to TV at higher ϕ. With increasing ϕ (> 0.1), the resulting d increased and the dependence on homogenizer pressure p_H (Fig. 6.20). This points to increased coalescence (effects (i) and (v)).

Fig. 6.20 shows a comparison of the average droplet diameter versus power consumption using different emulsifying machines. It can be seen the smallest droplet diameters were obtained when using the high pressure homogenizers.

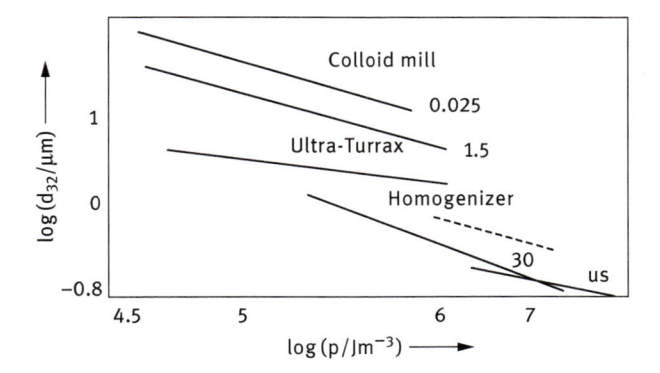

Fig. 6.20: Average droplet diameters obtained in various emulsifying machines as a function of energy consumption p – The number near the curves denote the viscosity ratio λ – the results for the homogenizer are for ϕ = 0.04 (solid line) and ϕ = 0.3 (broken line) – us means ultrasonic generator.

6.6 Rheological properties of cosmetic emulsions

The rheological properties of a cosmetic emulsion that need to be achieved depend on the consumer perspective which is very subjective. However, the efficacy and aesthetic qualities of a cosmetic emulsion are affected by their rheology. For example, with moisturizing creams one requires fast dispersion and deposition of a continuous protective oil film over the skin surface. This requires a shear thinning system (see below).

For characterization of the rheology of a cosmetic emulsion, one needs to combine several techniques, namely steady state, dynamic (oscillatory) and constant stress (creep) measurements [21–23]. A brief description of these techniques is given below.

In steady state measurements one measures the shear stress (τ) – shear rate (γ) relationship using a rotational viscometer. A concentric cylinder or cone and plate geometry may be used depending on the emulsion consistency. Most cosmetic emulsions are non-Newtonian, usually pseudoplastic as illustrated in Fig. 6.21. In this case the viscosity decreases with applied shear rate (shear thinning behaviour, Fig. 6.21), but at very low shear rates the viscosity reaches a high limiting value (usually referred to as the residual or zero shear viscosity).

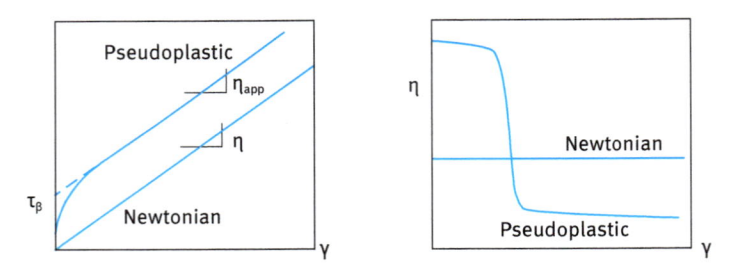

Fig. 6.21: Schematic representation of Newtonian and non-Newtonian (pseudoplastic) flow.

For the above pseudoplastic flow, one may apply a power law fluid model [22], a Bingham model [24] or a Casson model [25]. These models are represented by the following equations respectively,

$$\tau = \eta_{app}\gamma^n \tag{6.59}$$

$$\tau = \tau_\beta + \eta_{app}\gamma \tag{6.60}$$

$$\tau^{1/2} = \tau_c^{1/2} + \eta_c^{1/2}\gamma^{1/2}, \tag{6.61}$$

where n is the power in shear rate that is less than 1 for a shear thinning system (n is sometimes referred to as the consistency index), τ_β is the Bingham (extrapolated) yield value, η is the slope of the linear portion of the τ–γ curve, usually referred to as the plastic or apparent viscosity, τ_c is the Casson's yield value and η_c is the Casson's viscosity.

In dynamic (oscillator) measurements, a sinusoidal strain, with frequency ν in Hz or ω in rad s^{-1} (ω = 2πν) is applied to the cup (of a concentric cylinder) or plate (of a cone and plate) and the stress is measured simultaneously on the bob or the cone which are connected to a torque bar. The angular displacement of the cup or the plate is measured using a transducer. For a viscoelastic system, such as the case with a cosmetic emulsion, the stress oscillates with the same frequency as the strain, but out of phase [21]. This is illustrated in Fig. 6.22 which shows the stress and strain sine waves

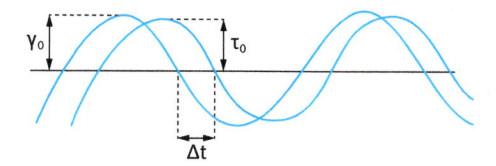

Fig. 6.22: Schematic representation of stress and strain sine waves for a viscoelastic system.

for a viscoelastic system. From the time shift between the sine waves of the stress and strain, Δt, the phase angle shift δ is calculated,

$$\delta = \Delta t \omega. \tag{6.62}$$

The complex modulus, G^*, is calculated from the stress and strain amplitudes (τ_0 and γ_0 respectively), i.e.,

$$G^* = \frac{\tau_0}{\gamma_0}. \tag{6.63}$$

The storage modulus, G', which is a measure of the elastic component is given by the following expression,

$$G' = |G^*| \cos \delta. \tag{6.64}$$

The loss modulus, G'', which is a measure of the viscous component, is given by the following expression,

$$G'' = |G^*| \sin \delta \tag{6.65}$$

and,

$$|G^*| = G' + iG'', \tag{6.66}$$

where i is equal to $(-1)^{-1/2}$.

The dynamic viscosity, η', is given by the following expression,

$$\eta' = \frac{G''}{\omega}. \tag{6.67}$$

In dynamic measurements one carries out two separate experiments. Firstly, the viscoelastic parameters are measured as a function of strain amplitude, at constant frequency, in order to establish the linear viscoelastic region, where G^*, G' and G'' are independent of the strain amplitude. This is illustrated in Fig. 6.23, which shows the variation of G^*, G' and G'' with γ_0. It can be seen that the viscoelastic parameters remain constant up to a critical strain value, γ_{cr}, above which, G^* and G' starts to decrease and G'' starts to increase with a further increase in the strain amplitude. Most cosmetic emulsions produce a linear viscoelastic response up to appreciable strains (> 10 %), indicative of structure build-up in the system ("gel" formation). If the system shows a short linear region (i.e., a low γ_{cr}), it indicates lack of a "coherent" gel structure (in many cases this is indicative of strong flocculation in the system).

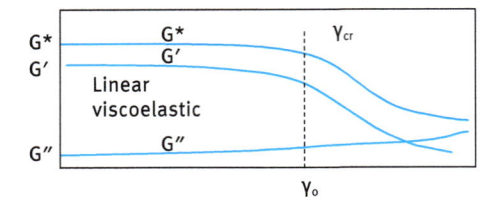

Fig. 6.23: Schematic representation of the variation of G^*, G' and G'' with strain amplitude (at a fixed frequency).

Once the linear viscoelastic region is established, measurements are then made of the viscoelastic parameters, at strain amplitudes within the linear region, as a function of frequency. This is schematically illustrated in Fig. 6.24, which shows the variation of G^*, G' and G'' with v or ω. It can be seen that below a characteristic frequency, v^* or ω^*, $G'' > G'$. In this low frequency regime (long timescale), the system can dissipate energy as viscous flow. Above v^* or ω^*, $G' > G''$, since in this high frequency regime (short timescale) the system is able to store energy elastically. Indeed, at sufficiently high frequency G'' tends to zero and G' approaches G^* closely, showing little dependency on frequency.

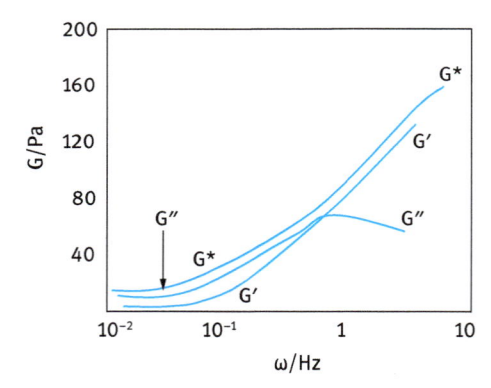

Fig. 6.24: Schematic representation of the variation of G^*, G' and G'' with ω for a viscoelastic system.

The relaxation time of the system can be calculated from the characteristic frequency (the crossover point) at which $G' = G''$, i.e.,

$$t^* = \frac{1}{\omega^*}. \tag{6.68}$$

Many cosmetic emulsions behave as semi-solids with long t^*. They show only elastic response within the practical range of the instrument, i.e. $G' \gg G''$ and they show a small dependence on frequency. Thus, the behaviour of many emulsions creams is similar to many elastic gels. This is not surprising, since in most cosmetic emulsion systems, the volume fraction of the disperse phase of most cosmetic emulsions is fairly high (usually > 0.5) and in many systems a polymeric thickener is added to the contin-

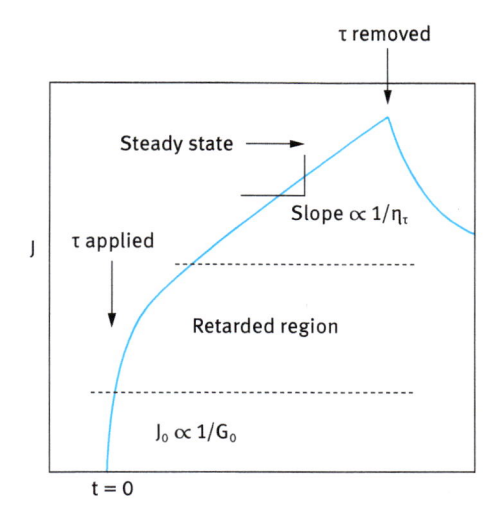

Fig. 6.25: Typical creep curve for a viscoelastic system.

uous phase for stabilization of the emulsion against creaming (or sedimentation) and to produce the right consistency for application.

In creep (constant stress) measurements [21], a stress τ is applied on the system and the deformation γ or the compliance $J = \gamma/\tau$ is followed as a function of time. A typical example of a creep curve is shown in Fig. 6.25. At $t = 0$, i.e. just after the application of the stress, the system shows a rapid elastic response characterized by an instantaneous compliance J_0 which is proportional to the instantaneous modulus G. Clearly at $t = 0$, all the energy is stored elastically in the system. At $t > 0$, the compliance shows a slow increase, since bonds are broken and reformed but at different rates. This retarded response is the mixed viscoelastic region. At sufficiently large timescales, that depend on the system, a steady state may be reached with a constant shear rate. In this region J shows a linear increase with time and the slope of the straight line gives the viscosity, η_τ, at the applied stress. If the stress is removed, after the steady state is reached, J decreases and the deformation reverses sign, but only the elastic part is recovered.

By carrying out creep curves at various stresses (starting from very low values depending on the instrument sensitivity), one can obtain the viscosity of the emulsion at various stresses. A plot of η_τ versus τ shows the typical behaviour shown in Fig. 6.26. Below a critical stress, τ_β, the system shows a Newtonian region with a very high viscosity, usually referred to as the residual (or zero shear) viscosity. Above τ_β, the emulsion shows a shear thinning region and ultimately another Newtonian region with a viscosity that is much lower than $\eta(0)$ is obtained. The residual viscosity gives information on the stability of the emulsion on storage. The higher the value of $\eta(0)$, the lower the creaming or sedimentation of the emulsion. The high stress viscosity gives information on the applicability of the emulsion such as its spreading and film formation. The critical stress τ_β gives a measure of the true yield value of

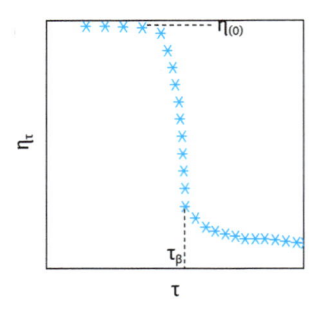

Fig. 6.26: Variation of viscosity with applied stress for a cosmetic emulsion.

the system, which is an important parameter both for application purposes and the long-term physical stability of the cosmetic emulsion.

It is clear from the above discussion that rheological measurements of cosmetic emulsions are very valuable in determining the long-term physical stability of the system as well as its application. This subject has attracted considerable interest in recent years with many cosmetic manufacturers. Apart from its value in the above mentioned assessment, one of the most important considerations is to relate the rheological parameters to the consumer perception of the product. This requires careful measurement of the various rheological parameters for a number of cosmetic products and relating these parameters to the perception of expert panels who assess the consistency of the product, its skin feel, spreading, adhesion, etc. It is claimed that the rheological properties of an emulsion cream determine the final thickness of the oil layer, the moisturizing efficiency and its aesthetic properties such as stickiness, stiffness and oiliness (texture profile). Psychophysical models may be applied to correlate rheology with consumer perception and a new branch of psychorheology may be introduced.

References

[1] Tadros, Th. F. (ed.), "Colloids in Cosmetics and Personal Care", Wiley-VCH, Germany (2008).
[2] Tadros, Th. F., "Cosmetics" in "Encyclopedia of Colloid and Interface Science", Th. F. Tadros (ed.), Springer, Germany (2013).
[3] Tadros, Th. F. and Vincent, B., in "Encyclopedia of Emulsion Technology", P. Becher (ed.), Marcel Dekker, NY (1983).
[4] Walstra, P. and Smoulders, P. E. A., in "Modern Aspects of Emulsion Science", B. P. Binks (ed.), The Royal Society of Chemistry, Cambridge (1998).
[5] Tadros, Th. F., "Applied Surfactants", Wiley-VCH, Germany (2005).
[6] Tadros, Th. F., in "Emulsion Formation Stability and Rheology" in "Emulsion Formation and Stability", Th. F. Tadros (ed.), Wiley-VCH, Germany (2013), Chapter 1.
[7] Tadros, Th. F., "Emulsions", De Gruyter, Germany (2016).
[8] Griffin, W.C., J. Cosmet. Chemists, **1**, 311 (1949); **5**, 249 (1954).
[9] Davies, J. T., Proc. Int. Congr. Surface Activity, Vol.1, p 426 (1959).

[10] Davies, J. T. and Rideal, E. K., "Interfacial Phenomena", Academic Press, NY (1961).
[11] Shinoda, K., J. Colloid Interface Sci., **25**, 396 (1967).
[12] Shinoda, K. and Saito, H., J. Colloid Interface Sci., **30**, 258 (1969).
[13] Beerbower, A. and Hill, M. W., Amer. Cosmet. Perfum., **87**, 85 (1972).
[14] Hildebrand, J. H., "Solubility of Non-Electrolytes", 2nd Ed., Reinhold, NY (1936).
[15] Hansen, C. M., J. Paint Technol., **39**, 505 (1967).
[16] Barton, A. F. M., "Handbook of Solubility Parameters and Other Cohesion Parameters", Boca Raton, Florida, CRC Press, Inc. (1983).
[17] Israelachvili, J. N., Mitchell, D. J. and Ninham, B. W., J. Chem. Soc., Faraday Trans. II, **72**, 1525 (1976).
[18] Lucassen-Reynders, E. H., Colloids and Surfaces, **A91**, 79 (1994).
[19] Lucassen, J., in "Anionic Surfactants", E. H. Lucassen-Reynders (ed.), Marcel Dekker, NY (1981).
[20] van den Tempel, M., Proc. Int. Congr. Surf. Act., **2**, 573 (1960).
[21] Tadros, Th. F., "Rheology of Dispersions", Wiley-VCH, Germany (2010).
[22] Tadros, Th. F., "Rheological Properties of Emulsion Systems" in "Emulsions – A Fundamental and Practical Approach", J. Sjoblom (ed.), NATO ASI Series, 363, Kluwer Academic Publishers, London (1991).
[23] Tadros, Th. F., Colloids and Surfaces, **A91**, 215 (1994).
[24] Bingham, E. C., "Fluidity and Plasticity", McGraw Hill, NY (1922).
[25] Casson, N., "Rheology of Disperse Systems", C. C. Mill (ed.), Pergamon Press, NY (1959), pp. 84–104.

7 Formulation of nanoemulsions in cosmetics

7.1 Introduction

Nanoemulsions are transparent or translucent systems in the size range 20–200 nm [1, 2]. Whether the system is transparent or translucent depends on the droplet size, the volume fraction of the oil and the refractive index difference between the droplets and the medium. Nanoemulsions having diameters < 50 nm appear transparent when the oil volume fraction is < 0.2 and the refractive index difference between the droplets and the medium is not large. With increasing droplet diameter and oil volume fraction the system may appear translucent and at higher oil volume fractions the system may become turbid.

Nanoemulsions are only kinetically stable. They have to be distinguished from microemulsions (that cover the size range 5–50 nm) which are mostly transparent and thermodynamically stable. The long-term physical stability of nanoemulsions (with no apparent flocculation or coalescence) makes them unique and they are sometimes referred to as "approaching thermodynamic stability". The inherently high colloid stability of nanoemulsions can be well understood from a consideration of their steric stabilization (when using nonionic surfactants and/or polymers) and how this is affected by the ratio of the adsorbed layer thickness to droplet radius as will be discussed below.

Unless adequately prepared (to control the droplet size distribution) and stabilized against Ostwald ripening (that occurs when the oil has some finite solubility in the continuous medium), nanoemulsions may show an increase in the droplet size and an initially transparent system may become turbid on storage.

The attraction of nanoemulsions for applications in personal care and cosmetics is due to the following advantages [1]: (i) The very small droplet size causes a large reduction in the gravity force and the Brownian motion may be sufficient for overcoming gravity. This means that no creaming or sedimentation occurs on storage. (ii) The small droplet size also prevents any flocculation of the droplets. Weak flocculation is prevented and this enables the system to remain dispersed with no separation. (iii) The small droplets also prevent their coalescence, since these droplets are non-deformable and hence surface fluctuations are prevented. In addition, the significant surfactant film thickness (relative to droplet radius) prevents any thinning or disruption of the liquid film between the droplets. (iv) Nanoemulsions are suitable for efficient delivery of active ingredients through the skin. – The large surface area of the emulsion system allows rapid penetration of actives. (v) Due to their small size, nanoemulsions can penetrate through the "rough" skin surface and this enhances penetration of actives. (vi) The transparent nature of the system, their fluidity (at reasonable oil concentrations) as well as the absence of any thickeners may give them a pleasant aesthetic character and skin feel. (vii) Unlike microemulsions (which require a high surfactant concentration, usually in the region of 20 % and higher), nanoemulsions can be prepared using reasonable surfactant concentration. For a 20 % O/W nanoemulsion, a surfactant concentration in the region of 5–10 % may be sufficient. (viii) The small size of the droplets allows them to deposit uniformly on substrates. Wetting, spreading and penetration may be also enhanced as a result of the low surface tension of the whole system and the low interfacial tension of the O/W droplets. (ix) Nanoemulsions can be applied for delivery of fragrants which may be incorporated in many personal care products. This could also be applied in perfumes which are desirable to be formulated alcohol free. (x) Nanoemulsions may be applied as a substitute for liposomes and vesicles (which are much less stable) and it is possible in some cases to build lamellar liquid crystalline phases around the nanoemulsion droplets.

The inherently high colloid stability of nanoemulsions when using polymeric surfactants is due to their steric stabilization [3]. The mechanism of steric stabilization was discussed in Chapter 5. As was shown in Fig. 5.8, the energy–distance curve shows a shallow attractive minimum at separation distance comparable to twice the adsorbed layer thickness 2δ. This minimum decreases in magnitude as the ratio between adsorbed layer thickness to droplet size increases. With nanoemulsions the ratio of adsorbed layer thickness to droplet radius (δ/R) is relatively large (0.1–0.2) when compared with macroemulsions.

These systems approach thermodynamic stability against flocculation and/or coalescence. The very small size of the droplets and the dense adsorbed layers ensures lack of deformation of the interface, lack of thinning and disruption of the liquid film between the droplets and hence coalescence is also prevented.

One of the main problems with nanoemulsions is Ostwald ripening which results from the difference in solubility between small and large droplets [4]. The difference in chemical potential of dispersed phase droplets between different sized droplets was given by Lord Kelvin,

$$c(r) = c(\infty) \exp\left(\frac{2\gamma V_m}{rRT}\right), \tag{7.1}$$

where $c(r)$ is the solubility surrounding a particle of radius r, $c(\infty)$ is the bulk phase solubility and V_m is the molar volume of the dispersed phase. The quantity $(2\gamma V_m/RT)$ is termed the characteristic length. It has an order of ≈ 1 nm or less, indicating that the difference in solubility of a 1 µm droplet is of the order of 0.1 % or less.

Theoretically, Ostwald ripening should lead to condensation of all droplets into a single drop (i.e. phase separation). This does not occur in practice since the rate of growth decreases with increasing droplet size.

For two droplets of radii r_1 and r_2 (where $r_1 < r_2$),

$$\frac{RT}{V_m} \ln\left[\frac{c(r_1)}{c(r_2)}\right] = 2\gamma \left(\frac{1}{r_1} - \frac{1}{r_2}\right). \tag{7.2}$$

Equation (7.19) shows that the larger the difference between r_1 and r_2, the higher the rate of Ostwald ripening.

Ostwald ripening can be quantitatively assessed from plots of the cube of the radius versus time t [4, 5],

$$r^3 = \frac{8}{9}\left[\frac{c(\infty)\gamma V_m}{\rho RT}\right]t, \qquad (7.3)$$

where D is the diffusion coefficient of the disperse phase in the continuous phase.

Ostwald ripening can be reduced by incorporation of a second component which is insoluble in the continuous phase (e.g. squalane) [4–6]. In this case significant partitioning between different droplets occurs, with the component having low solubility in the continuous phase expected to be concentrated in the smaller droplets. During Ostwald ripening in a two component disperse phase system, equilibrium is established when the difference in chemical potential between different sized droplets (which results from curvature effects) is balanced by the difference in chemical potential resulting from partitioning of the two components. If the secondary component has zero solubility in the continuous phase, the size distribution will not deviate from the initial one (the growth rate is equal to zero). In the case of limited solubility of the secondary component, the distribution is the same as governed by equation (7.3), i.e. a mixture growth rate is obtained which is still lower than that of the more soluble component.

The above method is of limited application since one requires a highly insoluble oil as the second phase which is miscible with the primary phase.

Another method for reducing Ostwald ripening depends on modification of the interfacial film at the O/W interface [7]. According to equation (7.3) reduction in γ results in reduction of Ostwald ripening. However, this alone is not sufficient since one has to reduce γ by several orders of magnitude. It has been suggested that by using surfactants which are strongly adsorbed at the O/W interface (i.e. polymeric surfactants) and which do not desorb during ripening, the rate could be significantly reduced. An increase in the surface dilational modulus and decrease in γ would be observed for the shrinking drops. The difference in γ between the droplets would balance the difference in capillary pressure (i.e. curvature effects).

To achieve the above effect it is useful to use A–B–A block copolymers that are soluble in the oil phase and insoluble in the continuous phase. A strongly adsorbed polymeric surfactant that has limited solubility in the aqueous phase can also be used (e.g. hydrophobically modified inulin, INUTEC®SP1 – ORAFTI, Belgium) [8, 9] as will be discussed below.

7.2 Preparation of nanoemulsion by the use of high pressure homogenizers

The production of small droplets (submicron) requires application of high energy [1]; the process of emulsification is generally inefficient. Simple calculations show that the mechanical energy required for emulsification exceeds the interfacial energy by several orders of magnitude. For example to produce an emulsion at $\phi = 0.1$ with a volume to surface diameter (Sauter diameter) $d_{32} = 0.6\,\mu m$, using a surfactant that gives an interfacial tension $\gamma = 10\,mN\,m^{-1}$, the net increase in surface free energy is $A\gamma = 6\phi\gamma/d_{32} = 10^4\,J\,m^{-3}$. The mechanical energy required in a homogenizer is $10^7\,J\,m^{-3}$, i.e. an efficiency of 0.1 %. The rest of the energy (99.9 %) is dissipated as heat [1].

Before describing the methods that can be applied to prepare submicron droplets (nanoemulsions), it is essential to consider the thermodynamics of emulsion formation and breakdown, the role of the emulsifier in preventing coalescence during emulsification and the procedures that can be applied for selection of the emulsifier. This was described in detail in Chapter 6.

The mechanism of emulsification was described in detail in Chapter 6 and only a summary is given here. To prepare an emulsion oil, water, surfactant and energy are needed. This can be analysed from a consideration of the energy required to expand the interface, $\Delta A\gamma$ (where ΔA is the increase in interfacial area when the bulk oil with area A_1 produces a large number of droplets with area A_2; $A_2 \gg A_1$, γ is the interfacial tension). Since γ is positive, the energy to expand the interface is large and positive; this energy term cannot be compensated by the small entropy of dispersion $T\Delta S$ (which is also positive) and the total free energy of formation of an emulsion, ΔG, is positive. Thus, emulsion formation is non-spontaneous and energy is required to produce the droplets.

The formation of large droplets (few μm) as is the case for macroemulsions is fairly easy and hence high speed stirrers such as the Ultra-Turrax or Silverson Mixer are sufficient to produce the emulsion [1]. In contrast, the formation of small drops (submicron as is the case with nanoemulsions) is difficult and this requires a large amount of surfactant and/or energy. The high energy required for formation of nanoemulsions can be understood from a consideration of the Laplace pressure Δp (the difference in pressure between inside and outside the droplet),

$$\Delta p = \frac{2\gamma}{r}. \tag{7.4}$$

To break up a drop into smaller ones, it must be strongly deformed and this deformation increases Δp. Surfactants play major roles in the formation of emulsions: By lowering the interfacial tension, Δp is reduced and hence the stress needed to break up a drop is reduced. Surfactants also prevent coalescence of newly formed drops

To describe emulsion formation one has to consider two main factors: hydrodynamics and interfacial science. In hydrodynamics one has to consider the type of flow: laminar flow and turbulent flow. This depends on the Reynolds number as will be discussed later.

To assess emulsion formation, one usually measures the droplet size distribution using for example laser diffraction techniques. A useful average diameter d is,

$$d_{nm} = \left(\frac{S_m}{S_n}\right)^{1/(n-m)}. \tag{7.5}$$

In most cases d_{32} (the volume/surface average or Sauter mean) is used. The width of the size distribution can be given as the variation coefficient c_m which is the standard deviation of the distribution weighted with d_m divided by the corresponding average d. Generally C_2 will be used which corresponds to d_{32}.

An alternative way to describe the emulsion quality is to use the specific surface area A (surface area of all emulsion droplets per unit volume of emulsion),

$$A = \pi s^2 = \frac{6\phi}{d_{32}}. \tag{7.6}$$

Several procedures may be applied for emulsion preparation [1], these range from simple pipe flow (low agitation energy, L), static mixers and general stirrers (low to medium energy, L–M), high speed mixers such as the Ultra-Turrax (M), colloid mills and high pressure homogenizers (high energy, H), ultrasound generators (M–H). The method of preparation can be continuous (C) or batchwise (B): pipe flow and static mixers – C; stirrers and Ultra-Turrax – B,C; colloid mill and high pressure homogenizers – C; ultrasound – B,C.

In all methods, there is liquid flow; unbounded and strongly confined flow. In the unbounded flow any droplets is surrounded by a large amount of flowing liquid (the confining walls of the apparatus are far away from most of the droplets). The forces can be frictional (mostly viscous) or inertial. Viscous forces cause shear stresses to act on the interface between the droplets and the continuous phase (primarily in the direction of the interface). The shear stresses can be generated by laminar flow (LV) or turbulent flow (TV); this depends on the Reynolds number Re,

$$\text{Re} = \frac{v l \rho}{\eta}, \tag{7.7}$$

where v is the linear liquid velocity, ρ is the liquid density and η is its viscosity. l is a characteristic length that is given by the diameter of flow through a cylindrical tube and by twice the slit width in a narrow slit.

For laminar flow $\text{Re} \lesssim 1000$, whereas for turbulent flow $\text{Re} \gtrsim 2000$. Thus whether the regime is linear or turbulent depends on the scale of the apparatus, the flow rate and the liquid viscosity [8–11].

If the turbulent eddies are much larger than the droplets, they exert shear stresses on the droplets. If the turbulent eddies are much smaller than the droplets, inertial forces will cause disruption (TI).

In bounded flow other relations hold. If the smallest dimension of the part of the apparatus in which the droplets are disrupted (say a slit) is comparable to droplet size, other relations hold (the flow is always laminar). A different regime prevails if the droplets are directly injected through a narrow capillary into the continuous phase (injection regime), i.e. membrane emulsification.

Within each regime, an essential variable is the intensity of the forces acting; the viscous stress during laminar flow $\sigma_{viscous}$ is given by,

$$\sigma_{viscous} = \eta G, \tag{7.8}$$

where G is the velocity gradient.

The intensity in turbulent flow is expressed by the power density ε (the amount of energy dissipated per unit volume per unit time); for turbulent flow [11],

$$\varepsilon = \eta G^2. \tag{7.9}$$

The most important regimes are: Laminar/Viscous (LV) – Turbulent/Viscous (TV) – Turbulent/Inertial (TI). For water as the continuous phase, the regime is always TI. For higher viscosity of the continuous phase ($\eta_C = 0.1$ Pas), the regime is TV. For still higher viscosity or a small apparatus (small l), the regime is LV. For very small apparatus (as is the case with most laboratory homogenizers), the regime is nearly always LV.

For the above regimes, a semi-quantitative theory is available that can give the timescale and magnitude of the local stress σ_{ext}, the droplet diameter d, timescale of droplet deformation τ_{def}, timescale of surfactant adsorption, τ_{ads} and mutual collision of droplets.

An important parameter that describes droplet deformation is the Weber number We (which gives the ratio of the external stress over the Laplace pressure),

$$We = \frac{G\eta_C R}{2\gamma}. \tag{7.10}$$

The viscosity of the oil plays an important role in the break-up of droplets; the higher the viscosity, the longer it will take to deform a drop. The deformation time τ_{def} is given by the ratio of oil viscosity to the external stress acting on the drop,

$$\tau_{def} = \frac{\eta_D}{\sigma_{ext}}. \tag{7.11}$$

The viscosity of the continuous phase η_C plays an important role in some regimes: For a turbulent inertial regime, η_C has no effect on droplets size. For a turbulent viscous regime, larger η_C leads to smaller droplets. For a laminar viscous regime, the effect is even stronger. Surfactants lower the interfacial tension γ and this causes a reduction in droplet size. The latter decrease with decreasing γ. For laminar flow the droplet diameter is proportional to γ; for turbulent inertial regime, the droplet diameter is proportional to $\gamma^{3/5}$.

Another important role of the surfactant is its effect on the interfacial dilational modulus ε [12–15],

$$\varepsilon = \frac{d\gamma}{d\ln A}. \tag{7.12}$$

7.2 Preparation of nanoemulsion by the use of high pressure homogenizers

During emulsification an increase in the interfacial area A takes place and this causes a reduction in Γ. The equilibrium is restored by adsorption of surfactant from the bulk, but this takes time (shorter times occur at higher surfactant activity). Thus ε is small at small a and also at large a. Because of the lack or slowness of equilibrium with polymeric surfactants, ε will not be the same for expansion and compression of the interface.

In practice, surfactant mixtures are used and these have pronounced effects on γ and ε. Some specific surfactant mixtures give lower γ values than either of the two individual components. The presence of more than one surfactant molecule at the interface tends to increase ε at high surfactant concentrations. The various components vary in surface activity. Those with the lowest γ tend to predominate at the interface, but if present at low concentrations, it may take a long time before reaching the lowest value. Polymer-surfactant mixtures may show some synergetic surface activity.

Apart from their effect on reducing γ, surfactants play major roles in deformation and break-up of droplets; this is summarized as follows. Surfactants allow the existence of interfacial tension gradients which are crucial for formation of stable droplets. In the absence of surfactants (clean interface), the interface cannot withstand a tangential stress; the liquid motion will be continuous.

Interfacial tension gradients are very important in stabilizing the thin liquid film between the droplets which is very important during the beginning of emulsification (films of the continuous phase may be drawn through the disperse phase and collision is very large). The magnitude of the γ-gradients and of the Marangoni effect depends on the surface dilational modulus ε given by equation (7.12).

For conditions that prevail during emulsification, ε increases with increasing surfactant concentration, m_C, and it is given by the relationship,

$$\varepsilon = \frac{d\pi}{d\ln \Gamma}, \qquad (7.13)$$

where π is the surface pressure ($\pi = \gamma_0 - \gamma$). Fig. 7.1 shows the variation of π with ln Γ; ε is given by the slope of the line [16].

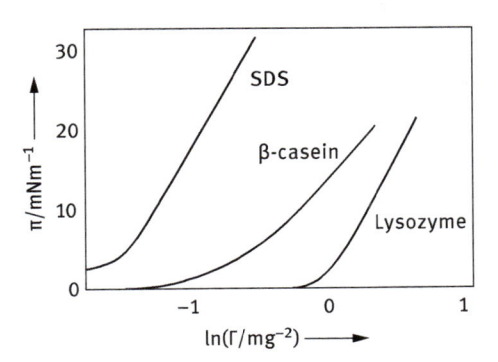

Fig. 7.1: π versus ln Γ for various emulsifiers.

The SDS shows a much higher ε value when compared with β-casein and lysozome. This is because the value of Γ is higher for SDS. The two proteins show a difference in their ε values which may be attributed to the conformational change that occurs upon adsorption.

Another important role of the emulsifier is to prevent coalescence during emulsification. This is certainly not due to the strong repulsion between the droplets, since the pressure at which two drops are pressed together is much greater than the repulsive stresses. The counteracting stress must be due to the formation of γ-gradients. When two drops are pushed together, liquid will flow out from the thin layer between them, and the flow will induce a γ-gradient. This produces a counteracting stress given by,

$$\tau_{\Delta\gamma} \approx \frac{2|\Delta\gamma|}{(1/2)d}. \tag{7.14}$$

The factor 2 follows from the fact that two interfaces are involved. Taking a value of $\Delta\gamma = 10$ mN m^{-1}, the stress amounts to 40 kPa (which is of the same order of magnitude as the external stress).

Closely related to the above mechanism, is the Gibbs–Marangoni effect [12–15], schematically represented in Fig. 7.2. The depletion of surfactant in the thin film between approaching drops results in γ-gradient without liquid flow being involved. This results in an inward flow of liquid that tends to drive the drops apart.

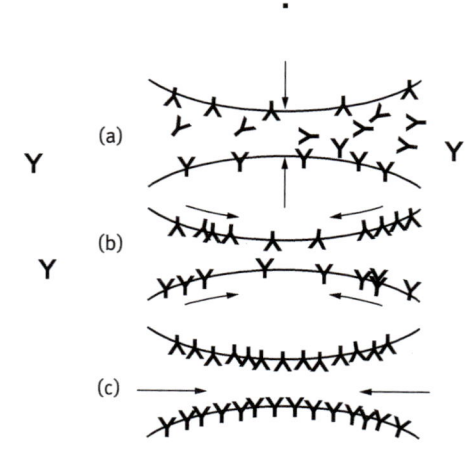

Fig. 7.2: Schematic representation of the Gibbs–Marangoni effect for two approaching drops.

The Gibbs–Marangoni effect also explains the Bancroft rule which states that the phase in which the surfactant is most soluble forms the continuous phase. If the surfactant is in the droplets, a γ-gradient cannot develop and the drops would be prone to coalescence. Thus, surfactants with HLB > 7 tend to form O/W emulsions and HLB < 7 tend to form W/O emulsions.

7.2 Preparation of nanoemulsion by the use of high pressure homogenizers

The Gibbs–Marangoni effect also explains the difference between surfactants and polymers for emulsification – polymers give larger drops when compared with surfactants. Polymers give a smaller value of ε at small concentrations when compared to surfactants (Fig. 7.1).

Various other factors should also be considered for emulsification: The disperse phase volume fraction ϕ. An increase in ϕ leads to an increase in droplet collision and hence coalescence during emulsification. With increasing ϕ, the viscosity of the emulsion increases and could change the flow from being turbulent to being laminar (LV regime).

The presence of many particles results in a local increase in velocity gradients. This means that G increases. In turbulent flow, increasing ϕ will induce turbulence depression. This will result in larger droplets. Turbulence depression by added polymers tends to remove the small eddies, resulting in the formation of larger droplets.

If the mass ratio of surfactant to continuous phase is kept constant, increasing ϕ results in decreasing surfactant concentration and hence an increase in γ_{eq}, resulting in larger droplets. If the mass ratio of surfactant to disperse phase is kept constant, the above changes are reversed.

General conclusions cannot be drawn since several of the above mentioned mechanism may come into play. Experiments using a high pressure homogenizer at various ϕ values at constant initial m_C (regime TI changing to TV at higher ϕ) showed that with increasing ϕ (> 0.1) the resulting droplet diameter increased and the dependence on energy consumption became weaker. Fig. 7.3 shows a comparison of the average droplet diameter versus power consumption using different emulsifying machines. It can be seen the smallest droplet diameters were obtained when using the high pressure homogenizers.

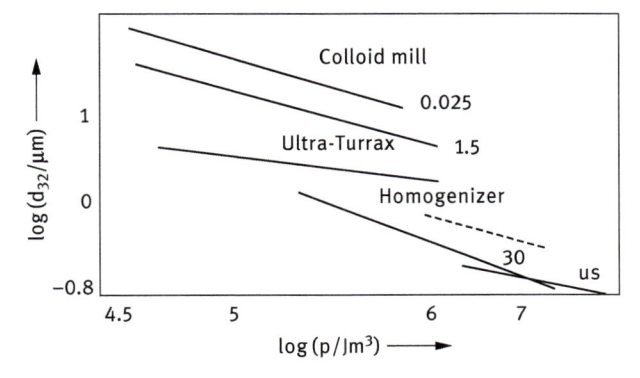

Fig. 7.3: Average droplet diameters obtained in various emulsifying machines as a function of energy consumption p – The number near the curves denote the viscosity ratio λ – the results for the homogenizer are for ϕ = 0.04 (solid line) and ϕ = 0.3 (broken line) – us means ultrasonic generator.

The selection of different surfactants in the preparation of either O/W or W/O was described in detail in Chapter 6. Four methods can be used for selection of emulsifiers, namely the hydrophilic–lipophilic balance (HLB), the phase inversion temperature (PIT), the cohesive energy ratio (CER) and the critical packing parameter concepts. These methods were described in detail in Chapter 6.

As mentioned above, emulsification combines the creation of fine droplets and their stabilization against coalescence. The emulsion droplets are created by premixing the lipophilic and hydrophilic phases. The coarse droplets are then finely dispersed in the µm range or even smaller by deforming and disrupting them at high specific energy. These droplets have to be stabilized against coalescence by using an efficient emulsifier. The latter must adsorb quickly at the oil/water interface to prevent droplet coalescence during emulsification. In most cases, a synergistic mixture of emulsifiers is used.

In most cases the nanoemulsions is produced in two stages, firstly by using a rotor-stator mixer (such as an Ultra-Turrax or Silverson) that can produce droplets in the µm range, followed by high pressure homogenization (reaching 3000 bar) to produce droplets in the nanometre size (as low as 50 nm).

The rotor-stator mixer consist of a rotating and fixed machine part [1]. Different geometries are available with various sizes and gaps between the rotor and stator. The simplest rotor stator machine is a vessel with a stirrer, which is used to produce the emulsion batchwise or quasicontinuously. The power density is relatively low and broadly distributed. Therefore, small mean droplet diameter (< 1 µm) can rarely be produced. In addition, a long residence time and emulsification for several minutes are required, often resulting in broad droplet size distribution. Some of these problems can be overcome by reducing the disruption zone that enhances the power density, e.g. using colloid mills or toothed-disc dispersing machines.

To produce submicron droplets, high pressure homogenization is commonly used [1]. These homogenizers are operated continuously and throughputs up to several thousand liters per hour can be achieved. The homogenizer consists of essentially a high pressure pump, and a homogenization nozzle. The pump creates the pressure which is then transferred within the nozzle to kinetic energy that is responsible for droplet disintegration. The design of the homogenization nozzle influences the flow pattern of the emulsion in the nozzle and hence droplet disruption. A good example of efficient homogenization nozzles are opposing jets that operate in the Microfluidizer. Other examples are the jet disperser (designed by Bayer) and the simple orifice valve. Droplet disruption in high pressure homogenizers is predominantly due to inertial forces in turbulent flow, shear forces in laminar elongational flow, as well as cavitation.

Droplets can also be disrupted by means of ultrasonic waves (frequency > 18 kHz) which cause cavitation that induces micro-jets and zones of high microturbulence [1]. A batchwise operation at small scale has been applied in the laboratory, especially for low viscosity systems. Continuous application requires the use of a flow chamber of special design into which the ultrasound waves are introduced. Due to the limited power of sound inducers, there are technical limits for high throughput.

Another method that can be applied for droplet disintegration is the use of microchannel systems (membrane emulsification). This can be realized by pressing the disperse phase through microporous membrane pores [1]. Droplets are formed at the membrane surface and detached from it by wall shear stress of the continuous phase. Besides tubular membranes made from ceramics like aluminium oxide, special porous glasses and polymers like polypropylene, polytetrafluoroethylene (PTFE), nylon and silicon have been used. The membrane's surface wetting behaviour is of major influence; if the membrane is wetted by the continuous phase only, emulsions of very narrow droplet size distribution are produced with mean droplet sizes in the range of three times the mean droplet diameter of the pore. The pressure to be applied should ideally be a little above the capillary pressure. Membrane emulsification reduces the shear forces acting in droplet formation.

The droplets are disrupted if they are deformed over a period of time t_{def} that is longer than a critical deformation time $t_{defcrit}$ and if the deformation described by the Weber number We, equation (7.10), exceeds a critical value We_{cr}. The droplet-deforming tensions are supplied by the continuous phase.

In turbulent flow, the droplets are disrupted mostly by inertial forces that are generated by energy dissipating small eddies. Due to internal viscous forces the droplets try to regain their initial form and size. Two dimensionless numbers, the turbulent Weber number We_{turb} and the Ohnsorge number Oh, characterize the tensions working on droplets in deformation and breakup [17].

$$We_{turb} = \frac{C^2 P_v^{2/3} \rho_c^{1/3} x^{5/3}}{\gamma} \tag{7.15}$$

$$Oh = \frac{\eta_d}{(\gamma \rho_d x)^{1/2}} \cdot \tag{7.16}$$

C is a constant, P_v is the volumetric power density, ρ_c the continuous phase viscosity, ρ_d the droplet density and x is the droplet diameter.

Droplet disruption in laminar shear flow is restricted to a narrow range of viscosity ratio between the disperse phase and continuous phase (η_d/η_c) for single droplet disruption, or between the disperse phase and emulsion (η_d/η_e) for emulsions. For laminar shear flow,

$$x_{3,2} \propto E_v^{-1} f(\eta_d/\eta_e), \tag{7.17}$$

and for laminar elongational flow,

$$x_{3,2} \propto E_v^{-1}, \tag{7.18}$$

where E_v is the volumetric energy density or specific disruption energy.

Laminar elongational is successfully applied in innovative high pressure homogenization valves, where it adds to the effect of turbulent droplet disruption by predeforming the droplets. Thus, the droplet disruption efficiency of high pressure homogenization can be significantly increased, especially for droplets with high viscosities.

The intensity of the process or the effectiveness in making small droplets is often governed by the net power density ($\varepsilon(t)$).

$$p = \varepsilon(t)\,dt, \tag{7.19}$$

where t is the time during which emulsification occurs.

Break-up of droplets will only occur at high ε values, which means that the energy dissipated at low ε levels is wasted. Batch processes are generally less efficient than continuous processes. This shows why with a stirrer in a large vessel, most of the energy applied at low intensity is dissipated as heat. In a homogenizer, p is simply equal to the homogenizer pressure [4, 5].

Several procedures may be applied to enhance the efficiency of emulsification when producing nanoemulsions: One should optimize the efficiency of agitation by increasing ε and decreasing dissipation time. The emulsion is preferably prepared at high volume faction ϕ of the disperse phase and diluted afterwards. However, very high ϕ values may result in coalescence during emulsification. Add more surfactant, whereby creating a smaller γ_{eff} and possibly diminishing recoalescence. Use a surfactant mixture that shows more reduction in γ of the individual components. If possible dissolve the surfactant in the disperse phase rather than the continuous phase; this often leads to smaller droplets. It may be useful to emulsify in steps of increasing intensity, particularly with emulsions having highly viscous disperse phase.

7.3 Low-energy methods for preparation of nanoemulsions

The low-energy methods for preparation on nanoemulsions are of particular interest, since they can make production more economical and offer the possibility to produce narrow droplet distribution nanoemulsions. In these methods, the chemical energy of the components is the key factor for the emulsification. The most well-known low-energy emulsification methods are direct or self-emulsification [18–20] and phase inversion methods [21–23]. Generally, emulsification by low-energy methods allows obtaining smaller and more uniform droplets.

In the so-called direct or self-emulsification methods, emulsification is achieved by a dilution process at a constant temperature, without any phase transitions (no change in the spontaneous curvature of the surfactant) taking place in the system during emulsification [18–20]. In this case, oil-in-water nanoemulsions (O/W) are obtained by the addition of water over a direct microemulsion phase, whereas water-in-oil nanoemulsions (W/O) are obtained by the addition of oil over an indirect microemulsion phase. This method is described in detail below. This self-emulsification method uses the chemical energy of dissolution in the continuous phase of the solvent present in the initial system (which is going to constitute the disperse phase). When the intended continuous phase and the intended disperse phase are mixed, the solvent present in the later phase is dissolved into the continuous phase, dragging and dispersing the micelles of the initial system, thus giving rise to the nanoemulsion droplets.

Phase inversion methods make use of the chemical energy released during the emulsification process as a consequence of a change in the spontaneous curvature of surfactant molecules, from negative to positive (obtaining oil-in-water, O/W, nanoemulsions) or from positive to negative (obtaining water-in-oil, W/O, nanoemulsions). This change of the surfactant curvature can be achieved by a change in composition keeping the temperature constant (Phase Inversion Composition method, PIC) [21, 22], or by a rapid change in temperature with no variation in composition (Phase Inversion Temperature method, PIT) [23]. The PIT method can only be applied to systems with surfactants sensitive to changes in temperature, i.e. the POE-type surfactants, in which changes in temperature induce a change in the hydration of the poly(oxyethylene) chains, and thus, a change in their curvature [23, 24]. In the PIC method, the change in curvature is induced by the progressive addition of the intended continuous phase, which may be pure water or oil [21, 22] over the mixture of the intended disperse phase (oil or water and surfactant/s).

Studies on surfactant phase behaviour are important when the low-energy emulsification methods are used, since the phases involved during emulsification are crucial in order to obtain nanoemulsions with small droplet size and low polydispersity. In contrast, if shear methods are used, only phases present at the final composition are important.

7.3.1 Phase Inversion Composition (PIC) principle

A study of the phase behaviour of water/oil/surfactant systems demonstrated that emulsification can be achieved by three different low-energy emulsification methods, as schematically shown in Fig. 7.4. (A) stepwise addition of oil to a water surfactant mixture. (B) stepwise addition of water to a solution of the surfactant in oil. (C) Mixing all the components in the final composition, pre-equilibrating the samples prior to emulsification. In these studies, the system water/Brij 30 (polyoxyethylene lauryl ether with an average of 4 mol of ethylene oxide)/decane was chosen as a model to obtain O/W emulsions. The results showed that nanoemulsions with droplet sizes of the order of 50 nm were formed only when water was added to mixtures of surfactant and oil (method B) whereby inversion from W/O emulsion to O/W nanoemulsion occurred.

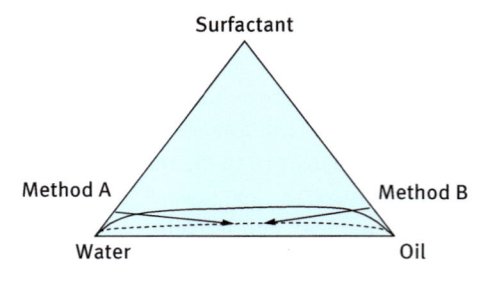

Fig. 7.4: Schematic representation of the experimental path in two emulsification methods: Method A, addition of decane to water/surfactant mixture; method B, addition of water to decane/Brij 30 solutions.

7.3.2 Phase Inversion Temperature (PIT) principle

Phase inversion in emulsions can be one of two types: Transitional inversion induced by changing factors which affect the HLB of the system, e.g. temperature and/or electrolyte concentration, or catastrophic inversion which is induced by increasing the volume fraction of the disperse phase.

Transitional inversion can also be induced by changing the HLB number of the surfactant at constant temperature using surfactant mixtures. This is illustrated in Fig. 7.5 which shows the average droplet diameter and rate constant for attaining constant droplet size as a function of the HLB number. It can be seen that the diameter decreases and the rate constant increases as inversion is approached.

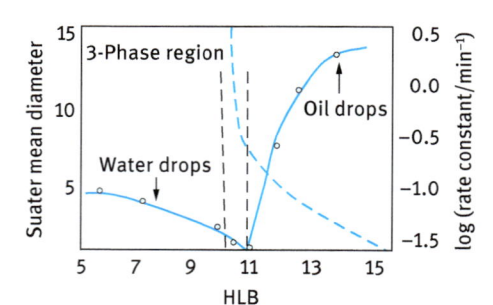

Fig. 7.5: Emulsion droplet diameters (circles) and rate constant for attaining steady size (squares) as function of HLB – cyclohexane/nonylphenol ethoxylate.

For application of the phase inversion principle one uses the transitional inversion method which has been demonstrated by Shinoda and co-workers [23, 24] when using nonionic surfactants of the ethoxylate type. These surfactants are highly dependent on temperature, becoming lipophilic with increasing temperature due to the dehydration of the polyethyleneoxide chain. When an O/W emulsion is prepared using a nonionic surfactant of the ethoxylate type and is heated, then at a critical temperature (the PIT), the emulsion inverts to a W/O emulsion. At the PIT the droplet size reaches a minimum and the interfacial tension also reaches a minimum. However, the small droplets are unstable and they coalesce very rapidly. By rapid cooling of the emulsion that is prepared at a temperature near the PIT, very stable and small emulsion droplets could be produced.

A clear demonstration of the phase inversion that occurs on heating an emulsion is illustrated from a study of the phase behaviour of emulsions as a function of temperature. This is illustrated in Fig. 7.6 which shows schematically what happens when the temperature is increased [6, 25]. At low temperature, over the Winsor I region, O/W macroemulsions can be formed and are quite stable. On increasing the temperature, the O/W emulsion stability decreases and the macroemulsion finally resolves when the system reaches the Winsor III phase region (both O/W and W/O emulsions are unstable). At higher temperature, over the Winsor II region, W/O emulsions become stable.

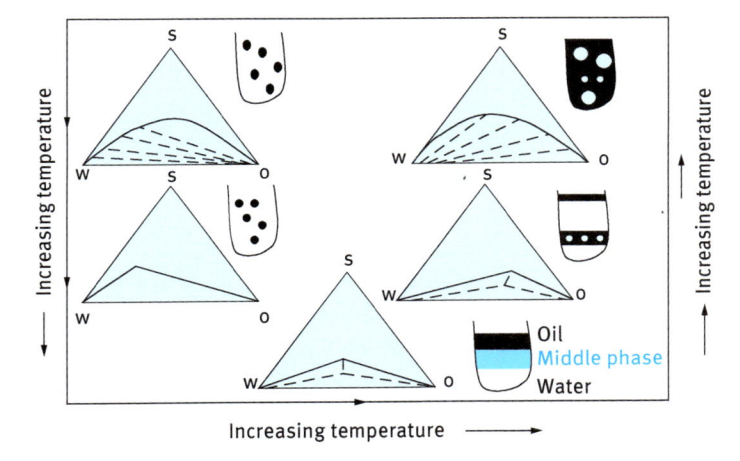

Fig. 7.6: The PIT concept.

Near the HLB temperature, the interfacial tension reaches a minimum. This is illustrated in Fig. 7.7. Thus by preparing the emulsion at a temperature 2–4 °C below the PIT (near the minimum in y) followed by rapid cooling of the system, nanoemulsions may be produced. The minimum in y can be explained in terms of the change in curvature H of the interfacial region, as the system changes from O/W to W/O. For an O/W system and normal micelles, the monolayer curves towards the oil and H is given a positive value. For a W/O emulsion and inverse micelles, the monolayer curves towards the water and H is assigned a negative value. At the inversion point (HLB temperature) H becomes zero and y reaches a minimum.

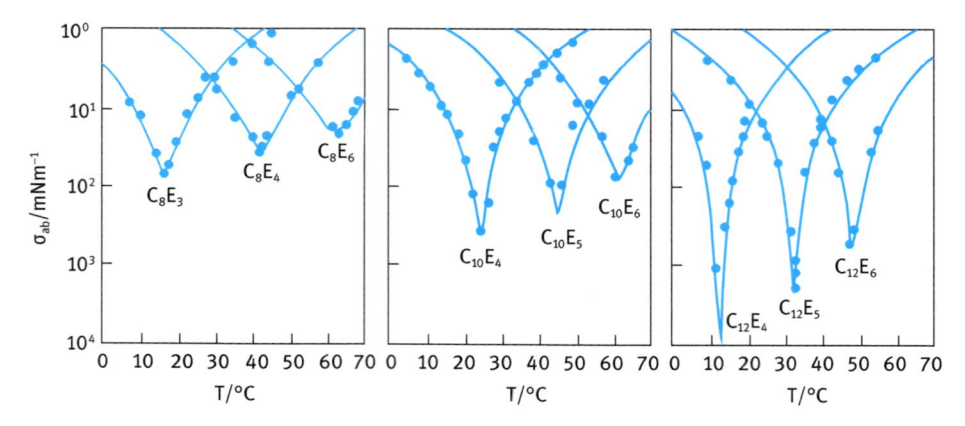

Fig. 7.7: Interfacial tensions of n-octane against water in the presence of various $C_n E_m$ surfactants above the cmc as a function of temperature.

7.3.3 Preparation of nanoemulsions by dilution of microemulsions

A common way to prepare nanoemulsions by self-emulsification is to dilute an O/W microemulsion with water. When diluting a microemulsion with water, part of the surfactant and/or cosurfactant diffuses to the aqueous phase. The droplets are no longer thermodynamically stable, since the surfactant concentration is not high enough to maintain the ultra-low interfacial tension ($< 10^{-4}$ mN m^{-1}) for thermodynamic stability. The system becomes unstable and the droplets show a tendency to grow by coalescence and/or Ostwald ripening forming a nanoemulsion. This is illustrated in Fig. 7.8

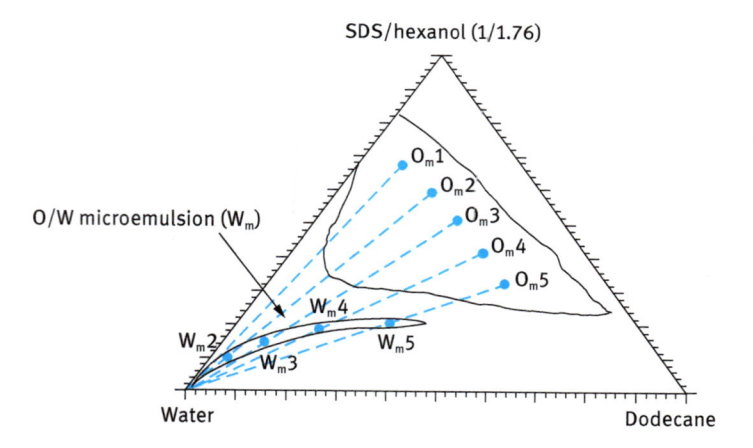

Fig. 7.8: Pseudoternary phase diagram of water/SDS/hexanol/dodecane with SDS:hexanol ratio of 1:1.76. Solid and dashed lines indicate the emulsification paths followed starting from both O/W (W_m) and W/O (O_m) microemulsion domains.

which shows the phase diagram of the system water/SDS-hexanol (ratio of 1 : 1.76)/dodecane.

Nanoemulsions can be prepared starting from microemulsions located in the inverse microemulsion domain, O_m, and in the direct microemulsion domain, W_m, at different oil:surfactant ratios ranging from 12 : 88 to 40 : 60, and coincident for both types of microemulsions. The water concentration is fixed at 20 % for microemulsions in the O_m domain labelled as O_m1, O_m2, O_m3, O_m4, O_m5. The microemulsions in the W_m region are accordingly W_m2, W_m3, W_m4, W_m5 and their water content decreased from W_m2 to W_m5.

Several emulsification methods can be applied: (i) addition of microemulsion into water in one step; (ii) addition of microemulsion into water stepwise; (iii) addition of water into microemulsion in one step; (iv) addition of water into microemulsion stepwise. The final water content is kept constant at 98 wt%.

Starting emulsification from W_m microemulsions, low-polydispersed nanoemulsions with droplet sizes within the range 20–40 nm are obtained regardless of the emulsification method used. When starting from O_m microemulsions the nanoemulsion formation and properties depend on the emulsification method. From O_m1 microemulsion, a turbid emulsion with rapid creaming is obtained whatever method is used. In this case the direct microemulsion region W_m is not crossed. Starting from O_m2 to O_m5 and using emulsification method d in which water is gradually added to the microemulsion, the nanoemulsion droplet sizes coincide with those obtained starting from microemulsions in the W_m domain for the corresponding O : S ratio. Methods (i), (ii) and (iii) produce coarse emulsions.

7.4 Practical examples of nanoemulsions

Several experiments were carried out to investigate the methods of preparation of nanoemulsions and their stability [25]. The first method applied the PIT principle for preparation of nanoemulsions. Experiments were carried out using hexadecane and isohexadecane (Arlamol HD) as the oil phase and Brij 30 ($C_{12}EO_4$) as the nonionic emulsifier. The phase diagrams of the ternary system water–$C_{12}EO_4$–hexadecane and water–$C_{12}EO_4$–isohexadecane are shown in Figs. 7.9 and 7.10. The main features of the pseudoternary system are as follows: (i) O_m isotropic liquid transparent phase, which extends along the hexadecane–$C_{12}EO_4$ or isohexadecane–$C_{12}EO_4$ axis, corresponding to inverse micelles or W/O microemulsions; (ii) L_α lamellar liquid crystalline phase extending from the W–$C_{12}EO_4$ axis toward the oil vertex; (iii) the rest of the phase diagram consists of two- or three-phase regions: (W_m + O) two-liquid-phase region, which appears along the water-oil axis; (W_m + L_α + O) three-phase region, which consists of a bluish liquid phase (O/W microemulsion), a lamellar liquid crystalline phase (L_α) and a transparent oil phase; (L_α + O_m) two-phase region consisting of an oil and a liquid crystalline region; MLC a multiphase region containing a lamellar liquid crystalline

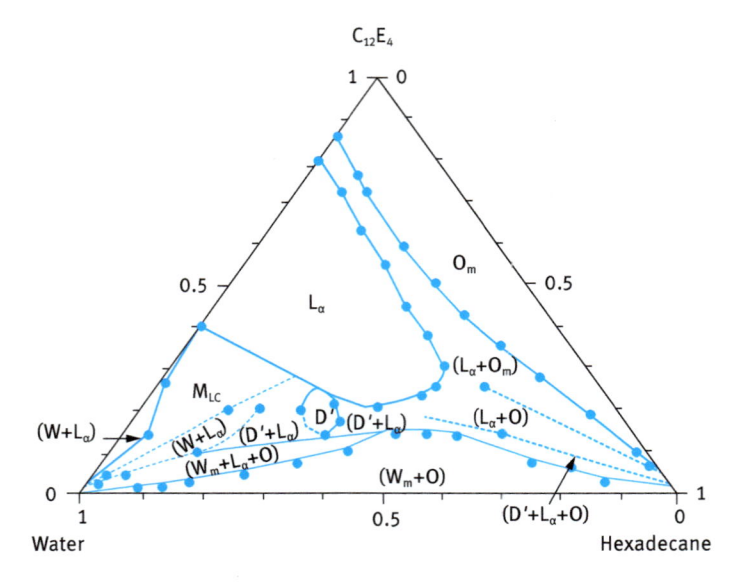

Fig. 7.9: Pseudoternary phase diagram at 25 °C of the system water–$C_{12}EO_4$–hexadecane.

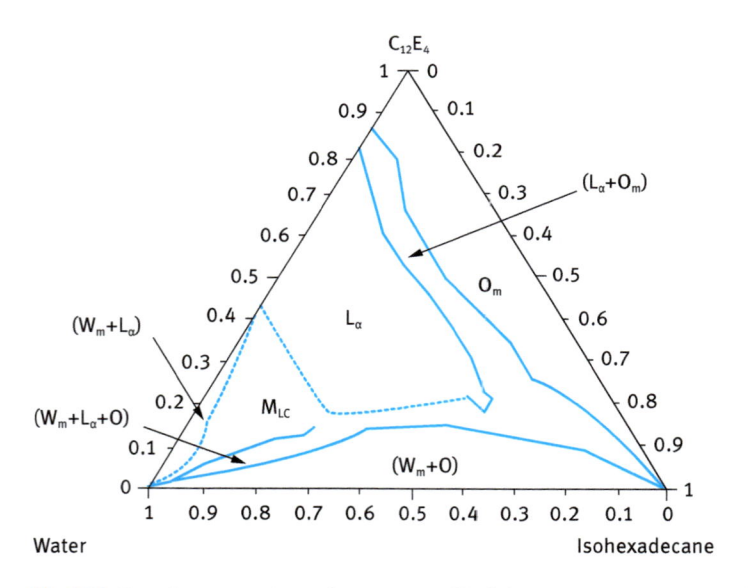

Fig. 7.10: Pseudoternary phase diagram at 25 °C of the system water–$C_{12}EO_4$–isohexadecane.

phase (L_α). The HLB temperature was determined using conductivity measurements, whereby 10^{-2} mol dm^{-3} NaCl was added to the aqueous phase (to increase the sensitivity of the conductibility measurements). The concentration of NaCl was low and hence it had little effect on phase behaviour.

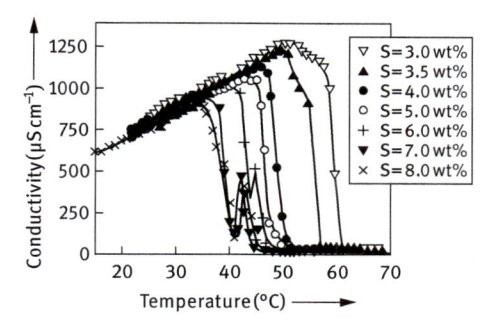

Fig. 7.11: Conductivity versus temperature for a 20 : 80 hexadecane: water emulsion at various $C_{12}EO_4$ concentrations.

Fig. 7.11 shows the variation of conductivity versus temperature for 20% O/W emulsions at different surfactant concentrations. It can be seen that there is a sharp decrease in conductivity at the PIT or HLB temperature of the system.

The HLB temperature decreases with increasing surfactant concentration – this could be due to the excess nonionic surfactant remaining in the continuous phase. However, at a concentration of surfactant higher than 5%, the conductivity plots showed a second maximum (Fig. 7.11). This was attributed to the presence of L_α phase and bicontinuous L3 or D' phases [25].

Nanoemulsions were prepared by rapid cooling of the system to 25 °C. The droplet diameter was determined using photon correlation spectroscopy (PCS). The results are summarized in Tab. 7.1, which shows the exact composition of the emulsions, HLB temperature, z-average radius and polydispersity index.

O/W nanoemulsions with droplet radii in the range 26–66 nm could be obtained at surfactant concentrations between 4 and 8%. The nanoemulsion droplet size and polydispersity index decreases with increasing surfactant concentration. The decrease in droplet size with increasing surfactant concentration is due to the increase in surfactant interfacial area and the decrease in interfacial tension, γ.

Tab. 7.1: Composition, HLB temperature (T_{HLB}), droplet radius r and polydispersity index (pol.) for the system water–$C_{12}EO_4$–hexadecane at 25 °C.

Surfactant / wt%	Water / wt%	Oil/Water	T_{HLB} / °C	r / nm	pol.
2.0	78.0	20.4/79.6	—	320	1.00
3.0	77.0	20.6/79.4	57.0	82	0.41
3.5	76.5	20.7/79.3	54.0	69	0.30
4.0	76.0	20.8/79.2	49.0	66	0.17
5.0	75.0	21.2/78.9	46.8	48	0.09
6.0	74.0	21.3/78.7	45.6	34	0.12
7.0	73.0	21.5/78.5	40.9	30	0.07
8.0	72.0	21.7/78.3	40.8	26	0.08

As mentioned above, γ reaches a minimum at the HLB temperature. Therefore, the minimum in interfacial tension occurs at lower temperature as the surfactant concentration increases. This temperature becomes closer to the cooling temperature as the surfactant concentration increases and this results in smaller droplet sizes.

All nanoemulsions showed an increase in droplet size with time, as a result of Ostwald ripening. Fig. 7.12 shows plots of r^3 versus time for all the nanoemulsions studied. The slope of the lines gives the rate of Ostwald ripening ω ($m^3\,s^{-1}$) and this showed an increase from 2×10^{-27} to $39.7 \times 10^{-27}\,m^3\,s^{-1}$ as the surfactant concentration is increased from 4 to 8 wt%. This increase could be due to a number of factors: (i) Decrease in droplet size increases the Brownian diffusion and this enhances the rate. (ii) Presence of micelles, which increases with increasing surfactant concentration. This has the effect of increasing the solubilization of the oil into the core of the micelles. This results in an increase of the flux J of diffusion of oil molecules from different size droplets. Although the diffusion of micelles is slower than the diffusion of oil molecules, the concentration gradient (δC/δX) can be increased by orders of magnitude as a result of solubilization. The overall effect will be an increase in J and this may enhance Ostwald ripening. (iii) Partition of surfactant molecules between the oil and aqueous phases. With higher surfactant concentrations, the molecules with shorter EO chains (lower HLB number) may preferentially accumulate at the O/W interface and this may result in reduction of the Gibbs elasticity, which in turn results in an increase in the Ostwald ripening rate.

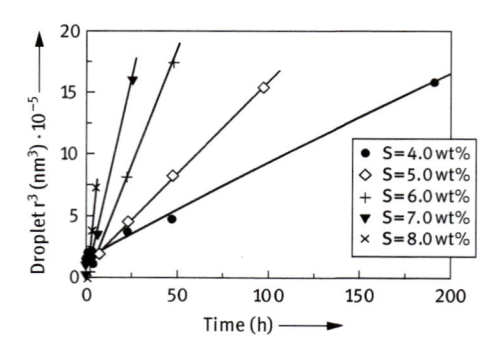

Fig. 7.12: r^3 versus time at 25 °C for nanoemulsions prepared using the system water–$C_{12}EO_4$–hexadecane.

The results with isohexadecane are summarized in Tab. 7.2. As with the hexadecane system, the droplet size and polydispersity index decreased with increasing surfactant concentration. Nanoemulsions with droplet radii of 25–80 nm were obtained at 3–8 % surfactant concentration. It should be noted, however, that nanoemulsions could be produced at lower surfactant concentration when using isohexadecane, when compared with the results obtained with hexadecane. This could be attributed to the higher solubility of the isohexadecane (a branched hydrocarbon), the lower HLB temperature and the lower interfacial tension.

Tab. 7.2: Composition, HLB temperature (T_{HLB}), droplet radius r and polydispersity index (pol.) at 25 °C for emulsions in the system water–$C_{12}EO_4$–isohexadecane.

Surfactant / wt%	Water / wt%	Oil/Water	T_{HLB} / °C	r / nm	pol.
2.0	78.0	20.4/79.6	—	97	0.50
3.0	77.0	20.6/79.4	51.3	80	0.13
4.0	76.0	20.8/79.2	43.0	65	0.06
5.0	75.0	21.2/78.9	38.8	43	0.07
6.0	74.0	21.3/78.7	36.7	33	0.05
7.0	73.0	21.3/78.7	33.4	29	0.06
8.0	72.0	21.7/78.3	32.7	27	0.12

The stability of the nanoemulsions prepared using isohexadecane was assessed by following the droplet size as a function of time. Plots of r3 versus time for four surfactant concentrations (3, 4, 5 and 6 wt%) are shown in Fig. 7.13. The results show an increase in Ostwald ripening rate as the surfactant concentration is increased from 3 to 6 % (the rate increased from 4.1×10^{-27} to 50.7×10^{-27} m^3 s^{-1}). The nanoemulsions prepared using 7 wt% surfactant were so unstable that they showed significant creaming after 8 hours. However, when the surfactant concentration was increased to 8 wt%, a very stable nanoemulsion could be produced with no apparent increase in droplet size over several months. This unexpected stability was attributed to the phase behaviour at such surfactant concentrations. The sample containing 8 wt% surfactant showed birefringence to shear when observed under polarized light. It seems that the ratio between the phases ($W_m + L_\alpha + O$) may play a key factor in nanoemulsion stability. Attempts were made to prepare nanoemulsions at higher O/W ratios (hexadecane being the oil phase), while keeping the surfactant concentration constant at 4 wt%. When the oil content was increased to 40 and 50 %, the droplet radius increased to 188 and 297 nm respectively. In addition, the polydispersity index also increased to 0.95. These systems become so unstable that they showed creaming within few hours. This is not surprising, since the surfactant concentration is not sufficient to produce nanoemulsion droplets with high surface area. Similar results were obtained with isohexadecane. However, nanoemulsions could be produced using 30/70 O/W ratio (droplet size being 81 nm), but with high polydispersity index (0.28). The nanoemulsions showed significant Ostwald ripening.

The effect of changing the alkyl chain length and branching was investigated using decane, dodecane, tetradecane, hexadecane and isohexadecane. Plots of r3 versus time are shown in Fig. 7.14 for 20/80 O/W ratio and surfactant concentration of 4 wt%. As expected, by reducing the oil solubility from decane to hexadecane, the rate of Ostwald ripening decreases. The branched oil isohexadecane also shows a higher Ostwald ripening rate when compared with hexadecane. A summary of the results is shown in Tab. 7.3 which also shows the solubility of the oil $C(\infty)$.

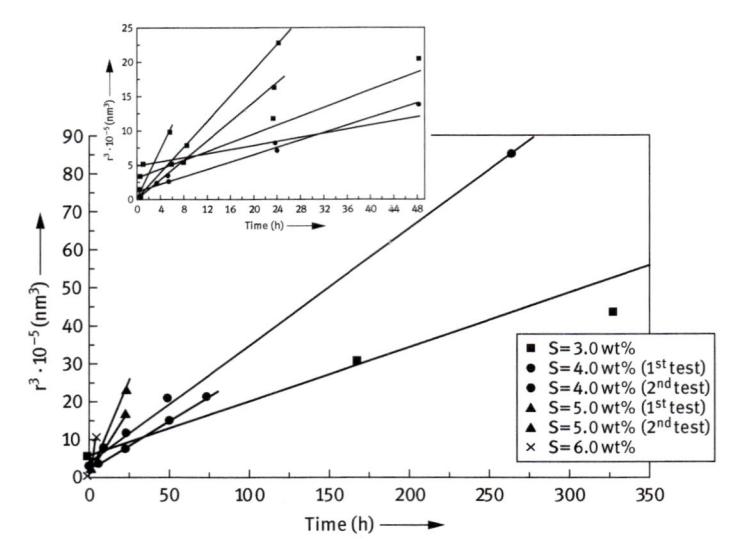

Fig. 7.13: r^3 versus time at 25 °C for the system water–$C_{12}EO_4$–isohexadecane at various surfactant concentration; O/W ratio 20/80.

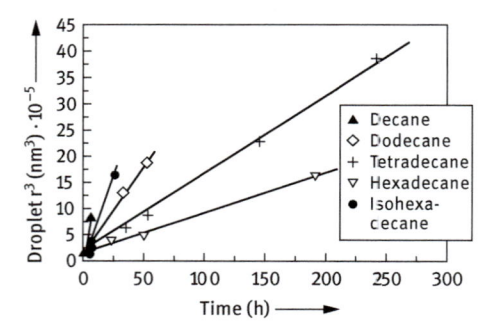

Fig. 7.14: r^3 versus time at 25 °C for nanoemulsions (O/W ratio 20/80) with hydrocarbons of various alkyl chain lengths. System water–$C_{12}EO_4$–hydrocarbon (4 wt% surfactant).

Tab. 7.3: HLB temperature (T_{HLB}), droplet radius r, Ostwald ripening rate ω and oil solubility for nanoemulsions prepared using hydrocarbons with different alkyl chain length.

Oil	T_{HLB} / °C	r / nm	ω / 10^{27} m^3 s^{-1}	$C(\infty)$ / ml ml^{-1}
Decane	38.5	59	20.9	710.0
Dodecane	45.5	62	9.3	52.0
Tetradecane	49.5	64	4.0	3.7
Hexadecane	49.8	66	2.3	0.3
Isohexadecane	43.0	60	8.0	—

As expected from the Ostwald ripening theory (see Chapter 6), the rate of Ostwald ripening decreases as the oil solubility decreases. Isohexadecane has a rate of Ostwald ripening similar to that of dodecane.

As discussed before, one would expect that the Ostwald ripening of any given oil should decrease on incorporation of a second oil with much lower solubility. To test this hypothesis, nanoemulsions were made using hexadecane or isohexadecane to which various proportions of a less soluble oil, namely squalane, was added. The results using hexadecane did significantly decrease in stability on addition of 10 % squalane. This was thought to be due to coalescence rather than an increase in the Ostwald ripening rate. In some cases addition of a hydrocarbon with a long alkyl chain can induce instability as a result of change in the adsorption and conformation of the surfactant at the O/W interface.

In contrast to the results obtained with hexadecane, addition of squalane to the O/W nanoemulsion system based on isohexadecane showed a systematic decrease in the Ostwald ripening rate as the squalane content was increased. The results are shown in Fig. 7.15 which shows plots of r3 versus time for nanoemulsions containing varying amounts of squalane. Addition of squalane up to 20 % based on the oil phase showed a systematic reduction in the rate (from 8.0×10^{-27} to 4.1×10^{-27} m^3 s^{-1}). It should be noted that when squalane alone was used as the oil phase, the system was very unstable and it showed creaming within 1 h. This shows that the surfactant used is not suitable for emulsification of squalane.

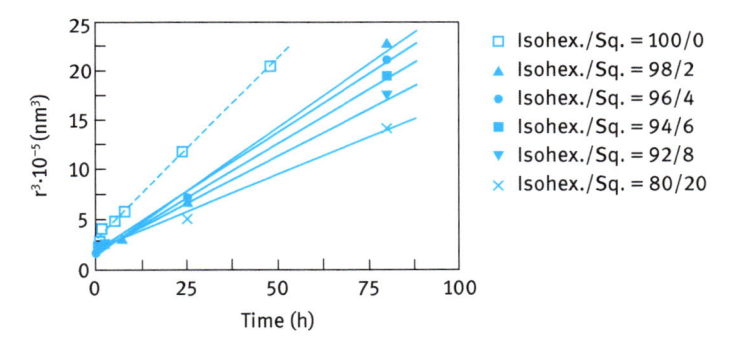

Fig. 7.15: r^3 versus time at 25 °C for the system water–$C_{12}EO_4$–isohexadecane–squalane (20/80 O/W ratio and 4 wt% surfactant).

The effect of HLB number on nanoemulsion formation and stability was investigated by using mixtures of $C_{12}EO_4$ (HLB = 9.7) and $C_{12}EO_6$ (HLB = 11.7). Two surfactant concentrations (4 and 8 wt%) were used and the O/W ratio was kept at 20/80. Fig. 7.16 shows the variation of droplet radius with HLB number. This figure shows that the droplet radius remain virtually constant in the HLB range 9.7–11.0, after which there is a gradual increase in droplet radius with increasing HLB number of the surfactant

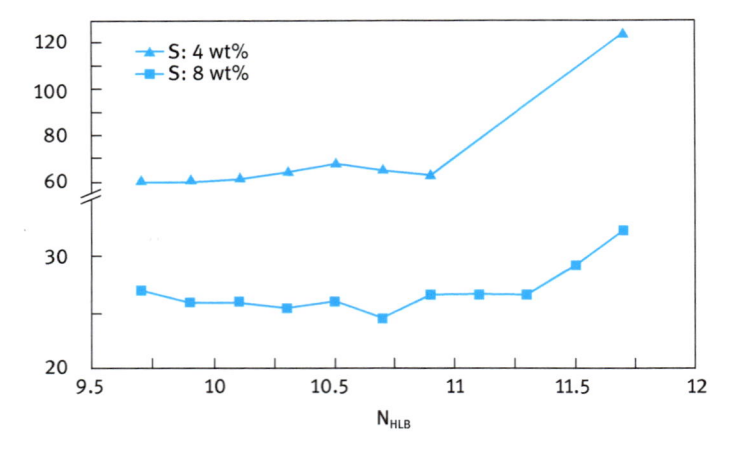

Fig. 7.16: r versus HLB number at two different surfactant concentrations (O/W ratio 20/80).

mixture. All nanoemulsions showed an increase in droplet radius with time, except for the sample prepared at 8 wt% surfactant with an HLB number of 9.7 (100 % $C_{12}EO_4$). Fig. 7.17 shows the variation of Ostwald ripening rate constant ω with HLB number of surfactant. The rate seems to decrease with increasing surfactant HLB number and when the latter is > 10.5, the rate reaches a low value (< 4×10^{-27} m^3 s^{-1}).

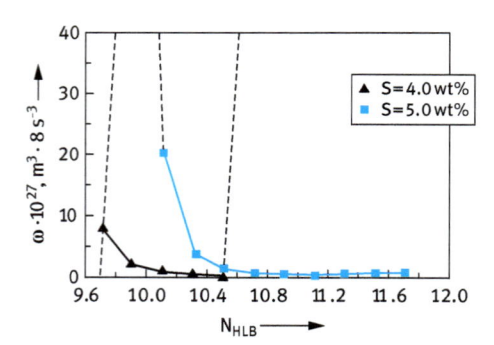

Fig. 7.17: ω versus HLB number in the system water–$C_{12}EO_4$–$C_{12}EO_6$–isohexadecane at two surfactant concentrations.

As discussed above, with the incorporation of an oil soluble polymeric surfactant that adsorbs strongly at the O/W interface, one would expect a reduction in the Ostwald ripening rate. To test this hypothesis, an A–B–A block copolymer of polyhydroxystearic acid (PHS, the A chains) and polyethylene oxide (PEO, the B chain) PHS–PEO–PHS (Arlacel P135) was incorporated in the oil phase at low concentrations (the ratio of surfactant to Arlacel was varied between 99 : 1 to 92 : 8). For the hexadecane system, the Ostwald ripening rate showed a decrease with the addition of Arlacel P135 surfactant at ratios lower than 94 : 6. Similar results were obtained using isohexadecane. However, at higher polymeric surfactant concentrations, the nanoemulsion became unstable.

As mentioned above, the nanoemulsions prepared using the PIT method are relatively polydisperse and they generally give higher Ostwald ripening rates when compared to nanoemulsions prepared using high pressure homogenization techniques. To test this hypothesis, several nanoemulsions were prepared using a Microfluidizer (that can apply pressures in the range 5000–15 000 psi or 350–1000 bar). Using an oil:surfactant ratio of 4 : 8 and O/W ratios of 20 : 80 and 50 : 50, emulsions were prepared first using the Ultra-Turrax followed by high pressure homogenization (ranging from 1500 to 15 000 psi). The best results were obtained using a pressure of 15 000 psi (one cycle of homogenization). The droplet radius was plotted versus the oil : surfactant ratio, R(O/S) as shown in Fig. 7.18.

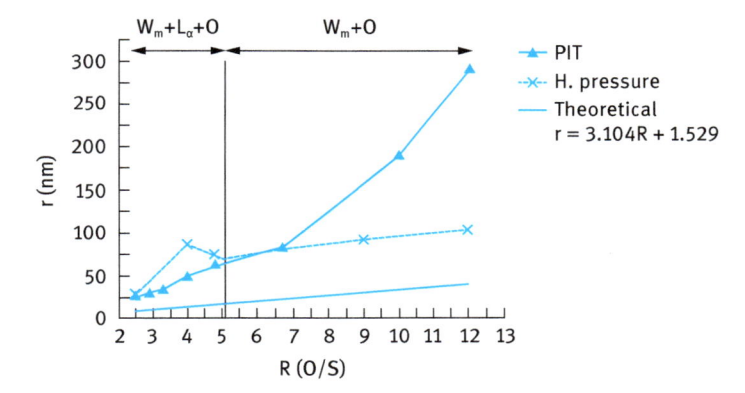

Fig. 7.18: r versus R(O/S) at 25 °C for the system water–$C_{12}EO_4$–hexadecane. W_m = micellar solution or O/W microemulsion, L_α = lamellar liquid crystalline phase; O = oil phase.

For comparison, the theoretical radii values calculated by assuming that all surfactant molecules are at the interface was calculated using the Nakajima equation [26, 27],

$$r = \left(\frac{3M_b}{AN\rho_a}\right)R + \left(\frac{3\alpha M_b}{AN\rho_b}\right) + d, \tag{7.20}$$

where M_b is the molecular weight of the surfactant, A is the area occupied by a single molecule, N is Avogadro's number, ρ_a is the oil density, ρ_b is the density of the surfactant alkyl chain, α is the alkyl chain weight fraction and d is the thickness of the hydrated layer of PEO.

In all cases, there is an increase in nanoemulsion radius with increasing R(O/S). However, when using the high pressure homogenizer, the droplet size can be maintained to values below 100 nm at high R(O/S) values. With the PIT method, there is a rapid increase in r with increasing R(O/S) when the latter exceeds 7.

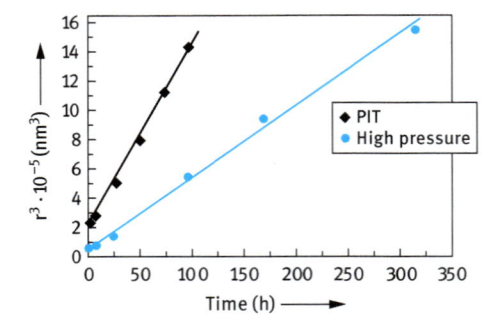

Fig. 7.19: r^3 versus time for nanoemulsion systems prepared using the PIT and Microfluidizer. 20 : 80 O/W ratio and 4 wt% surfactant.

As expected, the nanoemulsions prepared using high pressure homogenization showed a lower Ostwald ripening rate when compared to the systems prepared using the PIT method. This is illustrated in Fig. 7.19 which shows plots of r^3 versus time for the two systems.

7.5 Nanoemulsions based on polymeric surfactants

The use of polymeric surfactants for preparation of nanoemulsions is expected to significantly reduce Ostwald ripening due to the high interfacial elasticity produced by the adsorbed polymeric surfactant molecules [28]. To test this hypothesis, several nanoemulsions were formulated using a graft copolymer of hydrophobically modified inulin. The inulin backbone consists of polyfructose with a degree of polymerization > 23. This hydrophilic backbone is hydrophobically modified by attachment of several C_{12} alkyl chains [28]. The polymeric surfactant (with a trade name of INUTEC®SP1) adsorbs with several alkyl chains that can be soluble in the oil phase or strongly attached to the oil surface, leaving the strongly hydrated hydrophilic polyfructose loops and tails "dangling" in the aqueous phase. These hydrated loops and tails (with a hydrodynamic thickness > 5 nm) provide effective steric stabilization.

Oil/Water (O/W) nanoemulsions were prepared by two-step emulsification processes. In the first step, an O/W emulsion was prepared using a high speed stirrer, namely an Ultra-Turrax [28]. The resulting coarse emulsion was subjected to high pressure homogenization using a Microfluidizer (Microfluidics, USA). In all case, the pressure used was 700 bar and homogenization was carried out for 1 minute. The z-average droplet diameter was determined using PCS measurements as discussed before.

Fig. 7.20 shows plots of r^3 versus t for nanoemulsions of the hydrocarbon oils that were stored at 50 °C. It can be seen that both paraffinum liquidum with low and high viscosity gives almost a zero-slope indicating absence of Ostwald ripening in this case. This is not surprising since both oils have very low solubility and the hydrophobically modified inulin, INUTEC®SP1, strongly adsorbs at the interface giving high elasticity that reduces both Ostwald ripening and coalescence. However with the more soluble

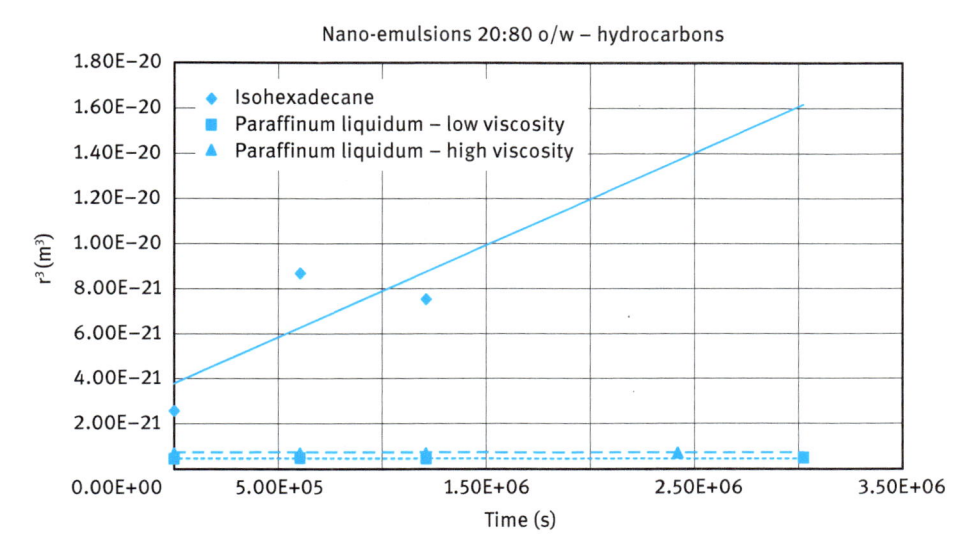

Fig. 7.20: r^3 versus t for nanoemulsions based on hydrocarbon oils.

hydrocarbon oils, namely isohexadecane, there is an increase in r^3 with time, giving a rate of Ostwald ripening of 4.1×10^{-27} m^3 s^{-1}. The rate for this oil is almost three orders of a magnitude lower than that obtained with a nonionic surfactant, namely laureth-4 (C$_{12}$-alkyl chain with 4 mol ethylene oxide) when stored at 50 °C. This clearly shows the effectiveness of INUTEC®SP1 in reducing Ostwald ripening. This reduction can be attributed to the enhancement of the Gibbs dilational elasticity [27] which results from the multipoint attachment of the polymeric surfactant with several alkyl groups to the oil droplets. This results in a reduction of the molecular diffusion of the oil from the smaller to the larger droplets.

Fig. 7.21 shows the results for isopropylalkylate O/W nanoemulsions. As with the hydrocarbon oils, there is a significant reduction in the Ostwald ripening rate with increasing alkyl chain length of the oil. The rate constants are 1.8×10^{-27}, 1.7×10^{-27} and 4.8×10^{-28} m^3 s^{-1} respectively.

Fig. 7.22 shows the r^3–t plots for nanoemulsions based on natural oils. In all cases, the Ostwald ripening rate is very low. However, a comparison between squalene and squalane shows that rate is relatively higher for squalene (unsaturated oil) when compared with squalane (with lower solubility). The Ostwald ripening rate for these natural oils is given in Tab. 7.4.

Fig. 7.23 shows the results based on silicone oils. Both dimethicone and phenyl trimethicone give an Ostwald ripening rate close to zero, whereas cyclopentasiloxane gives a rate of 5.6×10^{-28} m^3 s^{-1}.

Fig. 7.24 shows the results for nanoemulsions based on esters and the Ostwald ripening rates are given in Tab. 7.5. C$_{12-15}$ alkylbenzoate seems to give the highest rate.

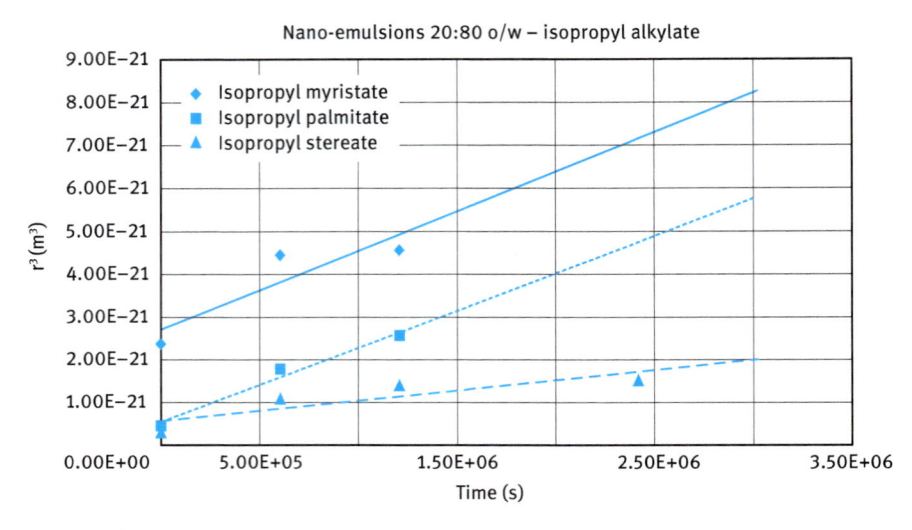

174 — 7 Formulation of nanoemulsions in cosmetics

Fig. 7.21: r^3 versus t for nanoemulsions based on isopropylalkylate.

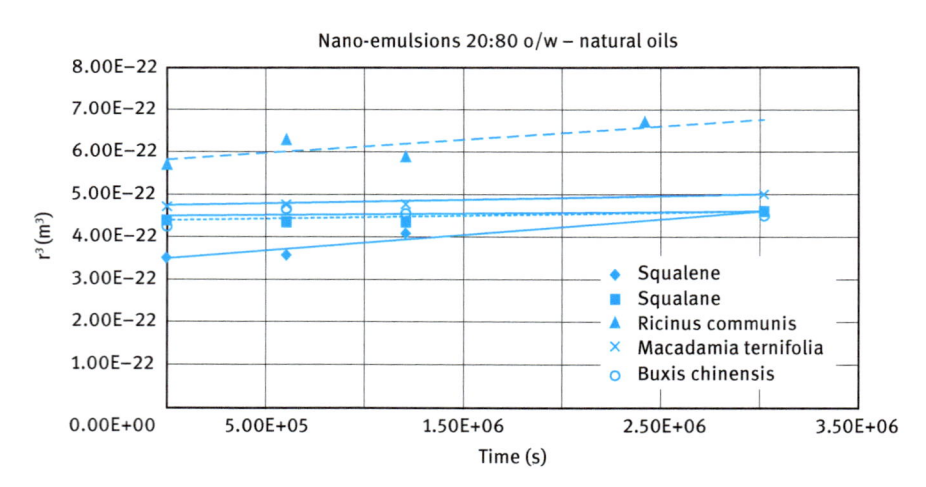

Fig. 7.22: r^3 versus t for nanoemulsions based on natural oils.

Tab. 7.4: Ostwald ripening rates for nanoemulsions based on natural oils.

Oil	Ostwald ripening rate / $m^3 s^{-1}$)
Squalene	2.9×10^{-28}
Squalane	5.2×10^{-30}
Ricinus Communis	3.0×10^{-29}
Macadamia Ternifolia	4.4×10^{-30}
Buxis Chinensis	≈ 0

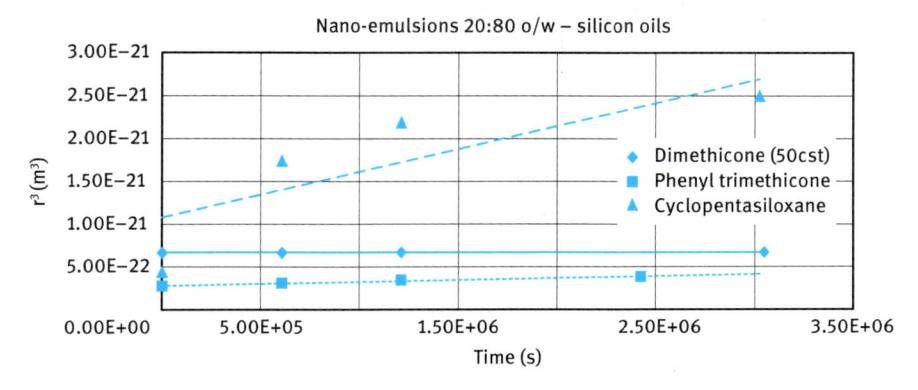

7.5 Nanoemulsions based on polymeric surfactants — 175

Fig. 7.23: r^3 versus t for nanoemulsions based on silicone oils.

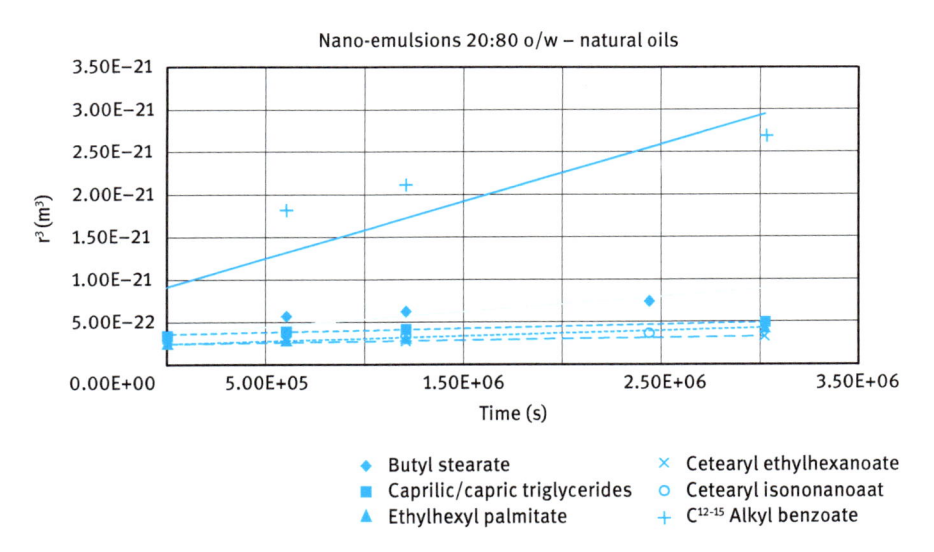

Fig. 7.24: r^3 versus t for nanoemulsions based on esters.

Tab. 7.5: Ostwald ripening rates for nanoemulsions based on esters.

Oil	Ostwald ripening rate / m^3 s^{-1})
Butyl stearate	1.8×10^{-28}
Caprylic Capric triglyceride	4.9×10^{-29}
Cetearyl ethylhexanoate	1.9×10^{-29}
Ethylhexyl palmitate	5.1×10^{-29}
Cetearyl isononanoate	1.8×10^{-29}
C_{12-15} alkyl benzoate	6.6×10^{-28}

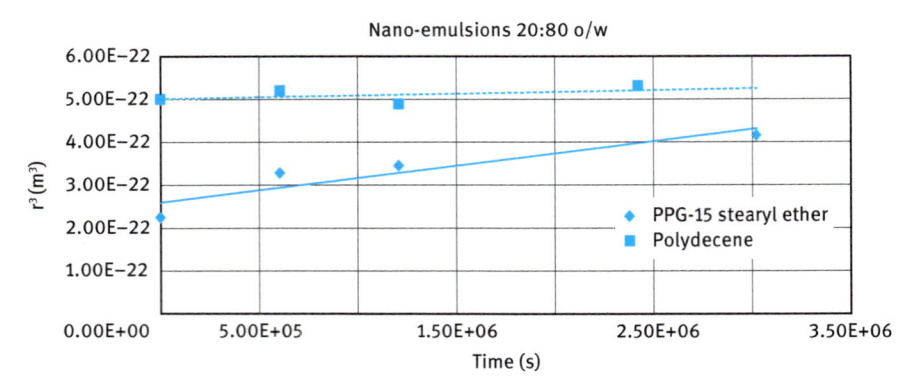

Fig. 7.25: r^3 versus t for nanoemulsions based on PPG-15 stearyl ether and polydecene.

Fig. 7.25 gives a comparison for two nanoemulsions based on polydecene, a highly insoluble nonpolar oil and PPG-15 stearyl ether which is relatively more polar. Polydecene gives a low Ostwald ripening rate of 6.4×10^{-30} m^3 s^{-1} which is one order of magnitude lower than that of PPG-15 stearyl ether (5.5×10^{-29} m^3 s^{-1}).

The influence of addition of glycerol (which is sometimes added to personal care formulations as a humectant) which can be used to prepare transparent nanoemulsions (by matching the refractive index of the oil and the aqueous phase) on the Ostwald ripening rate is shown in Fig. 7.26. With the more insoluble silicone oil, addition

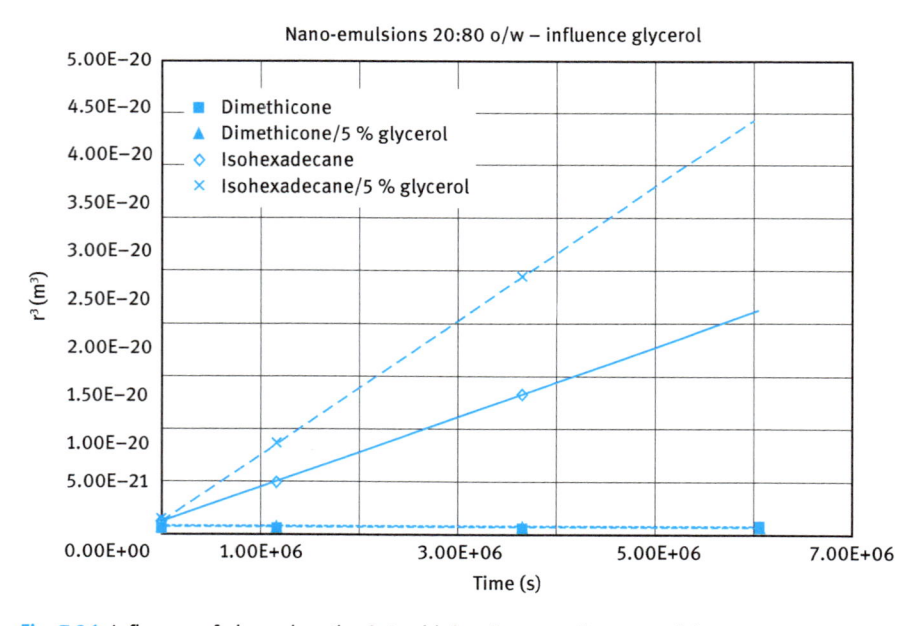

Fig. 7.26: Influence of glycerol on the Ostwald ripening rate of nanoemulsions.

of 5% glycerol does not show an increase in the Ostwald ripening rate, whereas for the more soluble isohexadecane oil, glycerol increases the rate.

It can be seen that hydrophobically modified inulin, HMI (INUTEC®SP1), reduces the Ostwald ripening rate of nanoemulsions when compared with nonionic surfactants such as laureth-4. This is due to the strong adsorption of INUTEC®SP1 at the oil-water interface (by multipoint attachment) and enhancement of the Gibbs dilational elasticity, both reducing the diffusion of oil molecules from the smaller to the larger droplets [7]. The present study also showed a big influence of the nature of the oil-phase with the more soluble and more polar oils giving the highest Ostwald ripening rate. However in all cases, when using INUTEC®SP1, the rates are reasonably low allowing one to use this polymeric surfactant in formulation of nanoemulsions for personal care applications.

References

[1] Tadros, Th. F., "Nanodispersions", De Gruyter, Germany (2016).
[2] Solans, C., Izquierdo, P., Nolla, J., Azemar, N., García-Celma, M. J., Nanoemulsions. Current Opinion in Colloid & Interface Science, 2005, **10**, (3/4), 102–110.
[3] Tadros, T., Izquierdo, P., Esquena, J. and Solans, C., Formation and stability of nano-emulsions, Adv. Colloid Interface Sci., **108/109**, 303–318 (2004).
[4] Thomson, W., (Lord Kelvin), Phil. Mag., **42**, 448 (1871).
[5] Kabalnov, A. S. and Schukin, E. D., Adv. Colloid Interface Sci., **38**, 69 (1992).
[6] Kabalnov, A. S., Langmuir, **10**, 680 (1994).
[7] Walstra, P., Chem. Eng. Sci., **48**, 333 (1993).
[8] Stone, H. A., Ann. Rev. Fluid Mech., **226**, 95 (1994).
[9] Wierenga, J. A., ven Dieren, F., Janssen, J. J. M. and Agterof, W. G. M., Trans. Inst. Chem. Eng., **74-A**, 554 (1996).
[10] Levich, V. G., "Physicochemical Hydrodynamics", Prentice-Hall, Englewood Cliffs (1962).
[11] Davis, J. T., "Turbulent Phenomena", Academic Press, London (1972).
[12] Lucassen-Reynders, E. H., in "Encyclopedia of Emulsion Technology", P. Becher (ed.), Marcel Dekker, NY (1996).
[13] Lucassen-Reynders, E. H., Colloids and Surfaces, **A91**, 79 (1994).
[14] Lucassen, J., in "Anionic Surfactants", E. H. Lucassen-Reynders (ed.), Marcel Dekker, NY (1981).
[15] van den Tempel, M., Proc. Int. Congr. Surf. Act., **2**, 573 (1960).
[16] Walstra, P. and Smoulders, P. E. A., in "Modern Aspects of Emulsion Science", B. P. Binks (ed.), The Royal Society of Chemistry, Cambridge (1998).
[17] Tadros, Th. F., "Emulsions", De Gruyter, Germany (2016).
[18] Ganachaud, F. and Katz J. L., Nanoparticles and nanocapsules created using the ouzo effect: Spontaneous emulsification as an alternative to ultrasonic and high-shear devices, Chem. Phys. Chem., **6**, 209–216 (2005).
[19] Bouchemal, K., Briançon, S., Perrier E. and Fessi, H., Nanoemulsion formulation using spontaneous emulsification: Solvent, oil and surfactant optimization, Int. J. Pharm., **280**, 241–251 (2004).
[20] Vitale, S. A. and Katz, J. L., Liquid droplet dispersions formed by homogeneous liquid-liquid nucleation: "The ouzo effect", Langmuir, **19**, 4105–4110 (2003).

[21] Forgiarini, A., Esquena, J., Gonzalez, C. and Solans, C., Formation of nano-emulsions by low-energy emulsification methods at constant temperature, Langmuir, **17**, (7), 2076–2083 (2001).
[22] Izquierdo, P., Esquena, J., Tadros, Th. F., Dederen, C., Garcia, M. J., Azemar, N. and Solans, C., Formation and stability of nanoemulsions prepared using the phase inversion temperature method, Langmuir, **18**, (1), 26–30 (2002).
[23] Shinoda K. and Saito, H., J. Colloid Interface Sci., **26**, 70 (1968).
[24] Shinoda, K. and Saito, H., The stability of O/W type emulsions as functions of temperature and the HLB of emulsifiers: The emulsification by PIT-method, J. Colloid Interface Sci., **30**, 258–263 (1969).
[25] Izquierdo, P., Thesis "Studies on Nano-Emulsion Formation and Stability", University of Barcelona, Spain (2002).
[26] Nakajima, H., Tomomossa, S. and Okabe, M., First Emulsion Conference, Paris (1993).
[27] Nakajima, H., in "Industrial Applications of Microemulsions", C. Solans and H. Konieda (eds.), Marcel Dekker (1997).
[28] Tadros, Th. F. (ed.), "Colloids in Cosmetics and Personal Care", Wiley-VCH, Germany (2008).

8 Formulation of multiple emulsions in cosmetics

8.1 Introduction

Multiple emulsions are complex systems of "Emulsions of Emulsions" [1–3]. Two main types can be distinguished: (i) Water-in-Oil-in-Water (W/O/W) multiple emulsions whereby the dispersed oil droplets contain emulsified water droplets. (ii) Oil-in-Water-in-Oil (O/W/O) multiple emulsions whereby the dispersed water droplets contain emulsified oil droplets. The most commonly used multiple emulsions are the W/O/W emulsions. The W/O/W multiple emulsion may be considered as water/water emulsion whereby the internal water droplets are separated by an "oily layer" (membrane). The internal droplets could also consist of a polar solvent such glycol or glycerol which may contain a dissolved or dispersed active ingredient (AI). The O/W/O multiple emulsion can be considered as an oil/oil emulsion separated by an "aqueous layer" (membrane). Application of multiple emulsions in pharmacy for control and sustained release of drugs has been investigated over several decades using animal studies. The only successful application of multiple emulsions in industry was in the field of personal care and cosmetics. Products based on W/O/W systems have been introduced by several cosmetic companies.

Due to the oily liquid or aqueous membrane formed, multiple emulsions ensure complete protection of the entrapped active ingredient used in many cosmetic systems (e.g. anti-wrinkle agents) and controlled release of this active ingredient from the internal to the external phase. In addition, multiple emulsions offer several advantages such as protection of fragile ingredients, separation of incompatible ingredients, prolonged hydration of the skin and in some cases formation of a firm gelled structure. In addition, a pleasant skin feels like that of an O/W emulsion combined with the well-known moisturizing properties of W/O emulsions are obtained with W/O/W multiple emulsions. Multiple emulsions can be usefully applied for controlled release by controlling the rate of the breakdown process of the multiple emulsions on application. Initially, one prepares a stable multiple emulsion (with a shelf life of two years) which on application breaks down in a controlled manner thus releasing the active ingredient in a controlled manner (slow or sustained release).

For applications in personal care and cosmetics, a wider range of surfactants can be used provided these molecules satisfy some essential criteria such as lack of skin irritation, lack of toxicity on application and safety to the environment (biodegradability of the molecule is essential in this case).

8.2 Types of multiple emulsions

Florence and Whitehall [1] distinguished between three types of multiple emulsions (W/O/W) that were prepared using isopropyl microstate as the oil phase, 5% Span 80 to prepare the primary W/O emulsion and various surfactants to prepare the secondary emulsion: (A) Brij 30 (polyoxyethylene 4 Lauryl ether) 2%. (B) Triton X-165 (polyoxyethylene 16.5 nonyl phenyl ether (2%). (C) 3:1 Span 80 : Tween 80 mixtures. A schematic picture of the three structures is shown in Fig. 8.1. Type A contains one large internal droplet similar to that described by Matsumoto et al. [2]. This type was produced when polyoxyethylene (4) lauryl ether (Brij 30) was used as emulsifier at 2%. Type B contains several small internal droplets. These were prepared using 2% polyoxyethylene (16.5) nonyl phenyl ether (Triton X-165). Type C drops entrapped a very large number of small internal droplets. These were prepared using a 3:1 Span 80 : Tween 80 mixture. It should be mentioned that type A multiple emulsions are not encountered much in practice. Type C is difficult to prepare since a large number of small water internal droplets (which are produced in the primary emulsification process) results in a large increase in viscosity. Thus, the most common multiple emulsions used in practice are those presented by type B, whereby the large size multiple emulsion droplets (10–100 μm) contain water droplets ≈ 1 μm.

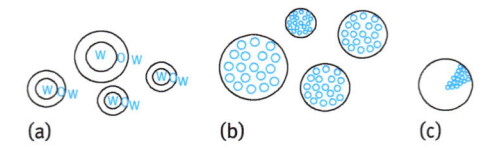

Fig. 8.1: Schematic representation of three structures of W/O/W multiple emulsions: (a) one large internal (Brij 30); (b) several small internal (Triton X-165); (c) large number of very small droplets (3:1 Span 80 : Tween 80).

8.3 Breakdown processes of multiple emulsions

A schematic representation of some breakdown pathways that may occur in W/O/W multiple emulsions is shown in Fig. 8.2. One of the main instabilities of multiple emulsions is the osmotic flow of water from the internal to the external phase or vice versa [1, 2]. This leads to shrinkage or swelling of the internal water droplets respectively. This process assumes the oil layer to act as a semi-permeable membrane (permeable to water but not to solute). The volume flow of water, J_W, may be equated with the change of droplet volume with time dv/dt,

$$J_W = \frac{dv}{dt} = -L_p A \, RT(g_2 c_2 - g_1 c_1) \tag{8.1}$$

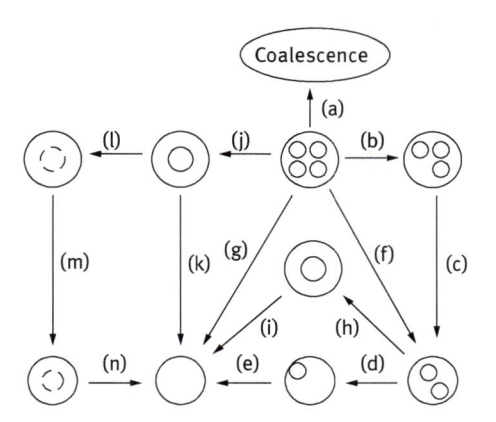

Fig. 8.2: Schematic representation of the possible breakdown pathways in W/O/W multiple emulsions: (a) coalescence; (b)–(f) expulsion of one or more internal aqueous droplets; (g) less frequent expulsion; (h), (i) coalescence of water droplets before expulsion; (j), (k) diffusion of water through the oil phase; (l)–(n) shrinking of internal droplets.

L_p is the hydrodynamic coefficient of the oil "membrane", A is the cross-sectional area, R is the gas constant and T is the absolute temperature. g is the osmotic coefficient of electrolyte solution with concentration c.

The flux of water ϕ_W is,

$$\phi_W = \frac{J_W}{V_m}, \tag{8.2}$$

where V_m is the partial molar volume of water.

An osmotic permeability coefficient P_o can be defined,

$$P_o = \frac{L_p RT}{V_m}. \tag{8.3}$$

Combining equations (8.1)–(8.3),

$$\phi_W = -P_o A(g_2 c_2 - g_1 c_1). \tag{8.4}$$

The diffusion coefficient of water D_W can be obtained from P_o and the thickness of the diffusion layer Δx,

$$-P_o = \frac{D_W}{\Delta x}. \tag{8.5}$$

For isopropyl myristate W/O/W emulsions, $\Delta x \approx 8.2\ \mu m$ and $D_W \approx 5.15 \times 10^{-8}\ cm^2\ s^{-1}$, the value expected for diffusion of water in reverse micelles.

8.4 Preparation of multiple emulsions

Two main criteria are essential for the preparation of stable multiple emulsions: (i) Two emulsifiers: with low and high HLB numbers. Emulsifier 1 should prevent coalescence of the internal water droplets, preferably producing a viscoelastic film which also reduces water transport. The secondary emulsifier should also produce an effective steric barrier at the O/W interface to prevent any coalescence of the multiple emulsion droplet. (ii) Optimum osmotic balance: This is essential to reduce water transport. This is achieved by addition of electrolytes or nonelectrolytes. The osmotic pressure in the external phase should be slightly lower than that of the internal phase to compensate for curvature effects.

Multiple emulsions are usually prepared in a two-stage process. For example a W/O/W multiple emulsion is formulated by first preparing a W/O emulsion using a surfactant with a low HLB number (5–6) using a high speed mixer (e.g. an Ultra-Turrax or Silverson). The resulting W/O emulsion is further emulsified in aqueous solution containing a surfactant with a high HLB number (9–12) using a low speed stirrer (e.g. a paddle stirrer). A schematic representation of preparation of multiple emulsions is given in Fig. 8.3.

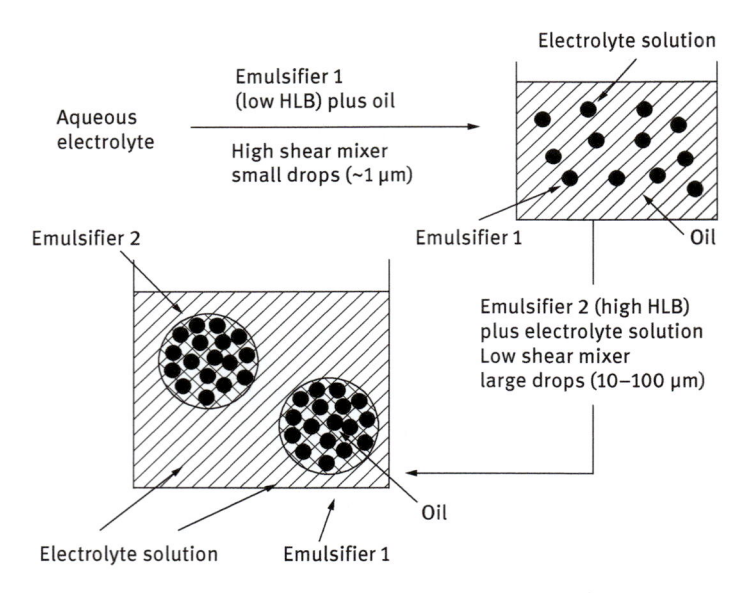

Fig. 8.3: Scheme for preparation of W/O/W multiple emulsion.

The yield of the multiple emulsion can be determined using dialysis for W/O/W multiple emulsions. A water soluble marker is used and its concentration in the outside phase is determined.

$$\% \text{ Multiple Emulsion} = \frac{C_i}{C_i + C_e} \times 100, \tag{8.6}$$

where C_i is the amount of marker in the internal phase and C_e is the amount of marker in the external phase. It has been suggested that if a yield of more than 90 % is required, the lipophilic (low HLB) surfactant used to prepare the primary emulsion must be ≈ 10 times higher in concentration than the hydrophilic (high HLB) surfactant.

The oils that can be used for the preparation of multiple emulsions must be cosmetically acceptable (no toxicity). Most convenient oils are vegetable oils such as soybean or safflower oil. Paraffinic oils with no toxic effect may be used. Also some polar oils such as isopropyl myristate can be applied; silicone oils can also be used. The low HLB emulsifiers (for the primary W/O emulsion) are mostly the sorbitan esters (Spans), but these may be mixed with other polymeric emulsifiers such as silicone emulsifiers. The high HLB surfactant can be chosen from the Tween series, although the block copolymers PEO–PPO–PEO (Poloxamers or Pluronics) may give much better stability. The polymeric surfactant INUTEC®SP1 can also give much higher stability. For controlling the osmotic pressure of the internal and external phases, electrolytes such as NaCl or nonelectrolytes such as sorbitol may be used.

In most cases, a "gelling agent" is required both for the oil and the outside external phase. For the oil phase, fatty alcohols may be used. For the aqueous continuous phase one can use the same "thickeners" that are used in emulsions, e.g. hydroxyethyl cellulose, Xanthan gum, alginates, carrageenans, etc. Sometimes liquid crystalline phases are applied to stabilize the multiple emulsion droplets. These can be generated using a nonionic surfactant and long chain alcohol. "Gel" coating around the multiple emulsion droplets may also be formed to enhance stability.

As an illustration a typical formulation of a W/O/W multiple emulsion is described below, using two different thickeners, namely Keltrol (Xanthan Gum from Kelco) and Carbopol 980 (a crosslinked polyacrylate gel produced by BF Goodrich). These thickeners were added to reduce creaming of the multiple emulsion. A two-step process was used in both cases.

The primary W/O emulsion was prepared using an A–B–A block copolymer (where A is polyhydroxystearic acid, PHS, and B is polyethylene oxide, PEO), i.e. PHS–PEO–PHS. Four grams of PHS–PEO–PHS were dissolved in 30 g of a hydrocarbon oil. For quick dissolution, the mixture was heated to 75 °C. The aqueous phase consisted of 65.3 g water, 0.7 g $MgSO_4 \cdot 7H_2O$ and a preservative. This aqueous solution was also heated to 75o. The aqueous phase was added to the oil phase slowly while stirring intensively using a high speed mixer. The W/O emulsion was homogenized for 1 minute and allowed to cool to 40–45 °C followed by further homogenization for another minute and stirring was continued until the temperature reached ambient.

The primary W/O emulsion was emulsified in an aqueous solution containing the polymeric surfactant PEO–PPO–PEO, namely Pluronic PEF127. Two grams of the polymeric surfactant were dissolved in 16.2 g water containing a preservative by stirring at 5 °C. Then 0.4 g $MgSO_4 \cdot 7H_2O$ were added to the aqueous polymeric surfactant solution. Sixty grams of the primary W/O emulsion were slowly added to the aqueous PEF127 solution while stirring slowly at 700 rpm (using a paddle stirrer). An aqueous Keltrol solution was prepared by slowly adding 0.7 g Keltrol powder to 20.7 g water, while stirring. The resulting thickener solution was further stirred for 30–40 minutes until a homogeneous gel was produced. The thickener solution was slowly added to the multiple emulsion while stirring at low speed (400 rpm) and the whole system was homogenized for 1 minute followed by gentle stirring at 300 rpm until the thickener completely dispersed in the multiple emulsion (about 30 minutes stirring was sufficient). The final system was investigated using optical microscopy to ensure that the multiple emulsion was produced. The formulation was left standing for several months and the droplets of the multiple emulsion were investigated using optical microscopy (see below). The rheology of the multiple emulsion was also measured (see below) at various intervals to ensure that the consistency of the product remained the same on long storage.

The above multiple emulsion was made again under the same conditions except using Carbopol 980 as a thickener (gel). In this case, no $MgSO_4$ was added, since the Carbopol gel was affected by electrolytes. The aqueous PEF127 polymeric surfactant solution was made by dissolving 2 g of the polymer in 23 g water. Fifteen grams of 2 % master gel of Carbopol were added to the PEF127 solution while stirring until the Carbopol was completely dispersed. Sixty grams of the primary W/O emulsion were slowly added to the aqueous solution of PEF127/Carbopol solution, while stirring thoroughly at 700 rpm. Triethanolamine was added slowly, while gently stirring until the pH of the system reached 6.0–6.5.

Another example of a W/O/W multiple emulsion was prepared using two polymeric surfactants. A W/O emulsion was prepared using an A–B–A block copolymer of PHS–PEO–PHS. This emulsion was prepared using a high speed mixer giving droplet sizes in the region of 1 μm. The W/O emulsion was then emulsified in an aqueous solution of hydrophobically modified inulin (INUTEC®SP1) using low speed stirring to produce multiple emulsion droplets in the range 10–100 μm. The osmotic balance was achieved using $0.1\,mol\,dm^{-3}$ $MgCl_2$ in the internal water droplets and outside continuous phase. The multiple emulsion was stored at room temperature and 50 °C and photomicrographs were taken at various intervals of time. The multiple emulsion was very stable for several months. A photomicrograph of the W/O/W multiple emulsion is shown in Fig. 8.4. An O/W/O multiple emulsion was made by first preparing a nanoemulsion using INUTEC®SP1. The nanoemulsion was then emulsified into an oil solution of PHS–PEO–PHS using a low speed stirrer. The O/W/O multiple emulsion was stored at room temperature and 50 °C and photomicrographs taken at various intervals of time. The O/W/O multiple emulsion was stable for several months both at room temperature and 50 °C. A photomicrograph of the O/W/O multiple emulsion is shown in Fig. 8.5. A schematic representation of the W/O/W multiple emulsion drop is shown in Fig. 8.6.

8.4 Preparation of multiple emulsions — 185

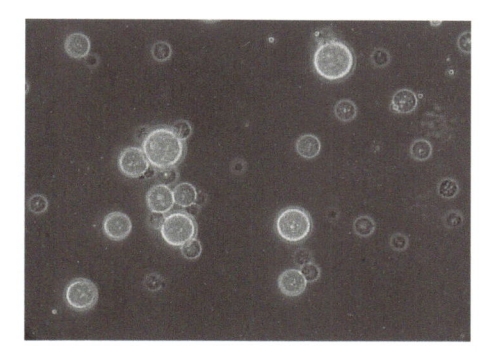

Fig. 8.4: Photomicrograph of the W/O/W multiple emulsion.

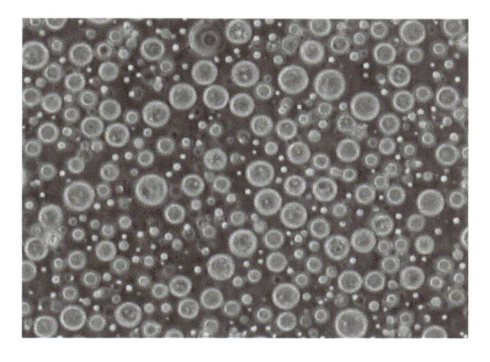

Fig. 8.5: Photomicrograph of O/W/O multiple emulsion.

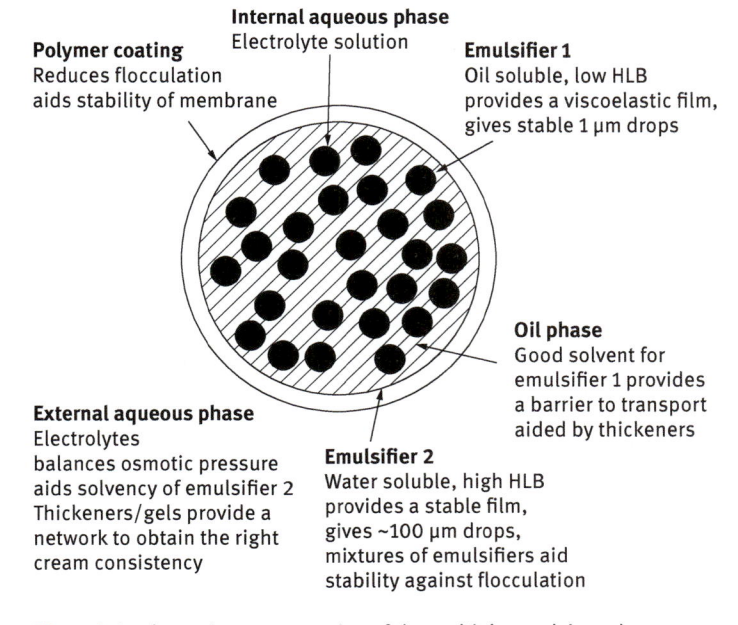

Internal aqueous phase
Electrolyte solution

Polymer coating
Reduces flocculation
aids stability of membrane

Emulsifier 1
Oil soluble, low HLB
provides a viscoelastic film,
gives stable 1 μm drops

Oil phase
Good solvent for
emulsifier 1 provides
a barrier to transport
aided by thickeners

External aqueous phase
Electrolytes
balances osmotic pressure
aids solvency of emulsifier 2
Thickeners/gels provide a
network to obtain the right
cream consistency

Emulsifier 2
Water soluble, high HLB
provides a stable film,
gives ~100 μm drops,
mixtures of emulsifiers aid
stability against flocculation

Fig. 8.6: A schematic representation of the multiple emulsions drop.

8.5 Characterization of multiple emulsions

Several methods can be applied for characterization of multiple emulsions.

8.5.1 Droplet size analysis

The droplet size distribution of the primary emulsion (internal droplets of the multiple emulsion) is usually in the region 0.5–2 µm, with an average of ≈ 0.5–1.0 µm. The droplet size distribution of this primary emulsion can be determined using photon correlation spectroscopy (PCS). This depends on measuring the intensity fluctuation of scattered light by the droplets as they undergo Brownian motion. Alternatively, light diffraction techniques, e.g. using the Master Sizer (Malvern, UK) can be used. The multiple emulsion droplets cover a wide range of sizes, usually 5–100 µm, with an average in the region of 5–20 µm. Optical microscopy (differential interference contrast) can be used to assess the droplets of the multiple emulsion. Optical micrographs may be taken at various storage times to assess the stability.

Freeze fracture and electron microscopy can give a quantitative assessment of the structure of the multiple emulsion droplets. Techniques can be applied to measure the droplet size of the multiple emulsion. Since the particle size is > 5 µm (i.e. the diameter is much greater than the wavelength of light), they show light diffraction (Fraunhofer's diffraction) and the Master Sizer could also be used.

8.5.2 Dialysis

As mentioned above, this could be used to measure the yield of the multiple emulsion; it can also be applied to follow any solute transfer from the inner droplets to the outer continuous phase.

8.5.3 Rheological techniques

Three rheological techniques may be applied.

8.5.3.1 Steady state shear stress (τ) – shear rate (γ) measurements

A pseudoplastic flow is obtained as is illustrated in Fig. 8.7. This flow curve can be analysed using, for example, the Herschley–Bulkley equation [4, 5],

$$\tau = \tau_\beta + k\gamma^n \tag{8.7}$$

where τ_β is the "yield value", k is the consistency index and n is the shear thinning index. This equation can be used to obtain the viscosity η as a function of shear rate.

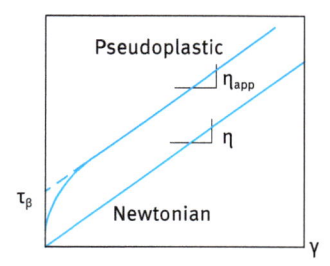

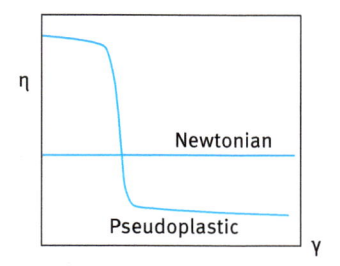

Fig. 8.7: Flow curves for Newtonian and pseudoplastic systems.

By following the change in viscosity with time, one can obtain information on multiple emulsion stability. For example, if there is water flow from the external phase to the internal water droplets ("swelling"), the viscosity will increase with time. If after some time, the multiple emulsion droplets begin to disintegrate forming O/W emulsion, the viscosity will drop.

8.5.3.2 Constant stress (creep) measurements

In this case, a constant stress is applied and the strain γ (or compliance $J = \gamma/\tau$) is followed as a function of time as shown in Fig. 8.8. If the applied stress is below the yield stress, the strain will initially show a small increase and then it remains virtually constant. Once the stress exceeds the yield value, the strain shows a rapid increase with time and eventually it reaches a steady state (with constant slope). From the slopes of the creep tests one can obtain the viscosity at any applied stress as is illustrated in Fig. 8.9 which shows a plateau high value below the yield stress (residual or zero shear

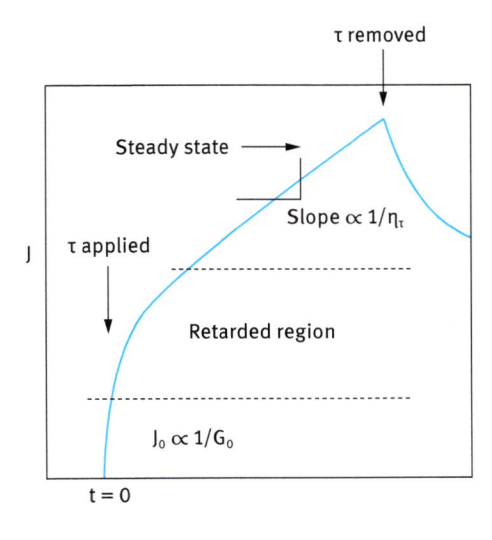

Fig. 8.8: Typical creep curve for a viscoelastic system.

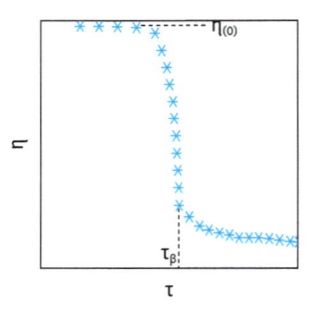

Fig. 8.9: Variation of viscosity with applied stress.

viscosity) followed by a rapid decrease when the yield stress is exceeded. By following the creep curves as a function of storage time one can assess the stability of the multiple emulsion. Apart from swelling or shrinking of the droplets which cause reduction in zero shear viscosity and yield value, any separation will also show a change in the rheological parameters.

8.5.3.3 Dynamic or oscillatory measurements

In dynamic (oscillator) measurements, a sinusoidal strain, with frequency ν in Hz or ω in rad s^{-1} (ω = 2πν) is applied to the cup (of a concentric cylinder) or plate (of a cone and plate) and the stress is measured simultaneously on the bob or the cone which are connected to a torque bar [5]. The angular displacement of the cup or the plate is measured using a transducer. For a viscoelastic system, such as the case with a multiple emulsion, the stress oscillates with the same frequency as the strain, but out of phase [5]. This is illustrated in Fig. 8.10 which shows the stress and strain sine waves for a viscoelastic system. From the time shift between the sine waves of the stress and strain, Δt, the phase angle shift δ is calculated,

$$\delta = \Delta t \omega. \tag{8.8}$$

The complex modulus, G^*, is calculated from the stress and strain amplitudes (τ_0 and γ_0 respectively), i.e.,

$$G^* = \frac{\tau_0}{\gamma_0}. \tag{8.9}$$

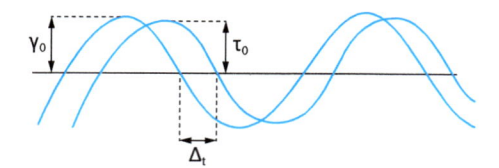

Fig. 8.10: Schematic representation of stress and strain sine waves for a viscoelastic system.

The storage modulus, G', which is a measure of the elastic component is given by the following expression,

$$G' = |G^*|\cos\delta. \tag{8.10}$$

The loss modulus, G'', which is a measure of the viscous component, is given by the following expression,

$$G'' = |G^*|\sin\delta \tag{8.11}$$

and,

$$|G^*| = G' + iG'', \tag{8.12}$$

where i is equal to $(-1)^{1/2}$.

The dynamic viscosity, η', is given by the following expression,

$$\eta' = \frac{G''}{\omega}. \tag{8.13}$$

In dynamic measurements one carries out two separate experiments. Firstly, the viscoelastic parameters are measured as a function of strain amplitude, at constant frequency, in order to establish the linear viscoelastic region, where G^*, G' and G'' are independent of the strain amplitude. This is illustrated in Fig. 8.11, which shows the variation of G^*, G' and G'' with γ_0. It can be seen that the viscoelastic parameters remain constant up to a critical strain value, γ_{cr}, above which, G^* and G' start to decrease and G'' starts to increase with a further increase in the strain amplitude. Most multiple emulsions produce a linear viscoelastic response up to appreciable strains (> 10 %), indicative of structure build-up in the system ("gel" formation). If the system shows a short linear region (i.e., a low γ_{cr}), it indicates lack of a "coherent" gel structure (in many cases this is indicative of strong flocculation in the system).

Once the linear viscoelastic region is established, measurements are then made of the viscoelastic parameters, at strain amplitudes within the linear region, as a function of frequency. This is schematically illustrated in Fig. 8.12, which shows the variation of G^*, G' and G'' with ν or ω. It can be seen that below a characteristic frequency, ν^* or ω^*, $G'' > G'$. In this low frequency regime (long timescale), the system can dissipate energy as viscous flow. Above ν^* or ω^*, $G' > G''$, since in this high frequency regime (short timescale) the system is able to store energy elastically. Indeed, at sufficiently high frequency G'' tends to zero and G' approaches G^* closely, showing little dependency on frequency. The relaxation time of the system can be calculated from the characteristic frequency (the crossover point) at which $G' = G''$, i.e.,

$$t^* = \frac{1}{\omega^*}. \tag{8.14}$$

It is clear from the above discussion that rheological measurements of multiple emulsions are very valuable in determining the long-term physical stability of the system as well as its application. This subject has attracted considerable interest in recent years

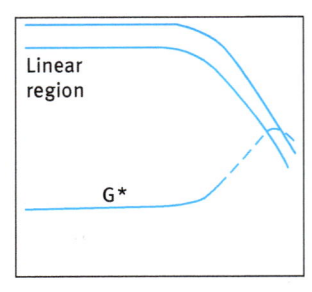

Fig. 8.11: Schematic representation of the variation of G^*, G' and G'' with strain amplitude (at a fixed frequency).

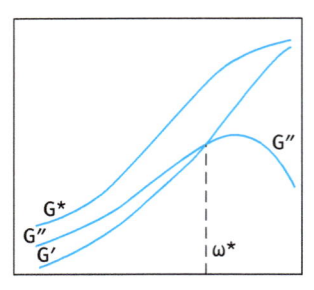

Fig. 8.12: Schematic representation of the variation of G^*, G' and G'' with ω for a viscoelastic system.

with many cosmetic manufacturers. Apart from its value in the above mentioned assessment, one of the most important considerations is to relate the rheological parameters to the consumer perception of the product. This requires careful measurement of the various rheological parameters for a number of multiple emulsions and relating these parameters to the perception of expert panels who assess the consistency of the product, its skin feel, spreading, adhesion, etc. It is claimed that the rheological properties of an emulsion cream formulated as a multiple emulsion determine the final thickness of the oil layer, the moisturizing efficiency and its aesthetic properties such as stickiness, stiffness and oiliness (texture profile). Psychophysical models may be applied to correlate rheology with consumer perception.

8.6 Summary of the factors affecting stability of multiple emulsions and criteria for their stabilization

As discussed above, the stability of the multiple emulsion is influenced by the nature of the two emulsifiers used for preparation of the multiple emulsion. Most papers published in the literature on multiple emulsions are based on conventional nonionic surfactants. Unfortunately, most of these surfactant systems produce multiple emulsions with limited shelf life, particularly if the system is subjected to large temperature variations. As mentioned above, we have formulated multiple emulsions using polymeric surfactants for both the primary and multiple emulsion preparation. These polymeric surfactants proved to be superior over the conventional nonionic surfactants in maintaining the physical stability of the multiple emulsion and they now could be successfully applied for formulation of cosmetic multiple emulsions. The key is to use polymeric surfactants that are approved by the CTA for cosmetics

The stability of the resulting multiple emulsion depends on a number of factors: (i) the nature of the emulsifiers used for preparation of the primary and multiple emulsion; (ii) the osmotic balance between the aqueous droplets in the multiple emulsion drops and that in the external aqueous phase; (iii) the volume fractions of the disperse water droplets in the multiple emulsion drops and the final volume fraction of the multiple emulsions; (iv) the temperature range to which the multiple emulsion is subjected; (v) the process used to prepare the system; (vi) the rheology of the whole system which can be modified by the addition of thickeners in the external aqueous phase.

As discussed above, the main criteria for preparation of a stable multiple emulsion are: (i) Two emulsifiers, one with low (emulsifier I) and one with high (emulsifier II) HLB number. (ii) Emulsifier I should provide a very effective barrier against coalescence of the water droplets in the multiple emulsion drop. Emulsifier II should also provide an effective barrier against flocculation and/or coalescence of the multiple emulsion drops. (iii) The amount of emulsifiers used in preparation of the primary and multiple emulsions is critical. Excess emulsifier I in the oil phase may result in further emulsification of the aqueous phase into the multiple emulsion with the ultimate production of a W/O emulsion. Excess emulsifier II in the aqueous phase may result in solubilization of the low HLB number surfactant with the ultimate formation of an O/W emulsion. (iv) Optimum osmotic balance of the internal and external aqueous phases. If the osmotic pressure of the internal aqueous droplets is higher than the external aqueous phase, water will flow into the internal droplets resulting in "swelling" of the multiple emulsion drops with the ultimate production of a W/O emulsion. In contrast, if the osmotic pressure in the outside external phase is higher, water will diffuse in the opposite direction and the multiple emulsion will revert to an O/W emulsion.

Various formulation variables must be considered: (i) primary W/O emulsifier; various low HLB number surfactants are available of which the following may be mentioned: decaglycerol decaoleate; mixed triglycerol trioleate and sorbitan trioleate; ABA block copolymers of PEO and PHS; (ii) primary volume fraction of the W/O or O/W emulsion; usually volume fractions between 0.4 and 0.6 are produced, depending on the requirements; (iii) nature of the oil phase; various paraffinic oils (e.g. heptamethyl nonane), silicone oil, soybean and other vegetable oils may be used; (iv) secondary O/W emulsifier; high HLB number surfactants or polymers may be used, e.g. Tween 20, polyethylene oxide-polypropylene oxide block copolymers (Pluronics) may be used; (v) secondary volume fraction; this may be varied between 0.4 and 0.8 depending on the consistency required; (vi) electrolyte nature and concentration; e.g. NaCl, $CaCl_2$, $MgCl_2$ or $MgSO_4$; (vii) thickeners and other additives; in some cases a gel coating for the multiple emulsion drops may be beneficial, e.g. polymethacrylic acid or carboxymethyl cellulose. Gels in the outside continuous phase for a W/O/W multiple emulsion may be produced using xanthan gum (Keltrol or Rhodopol), Carbopol or alginates; (viii) process; for the preparation of the primary emulsion, high speed mixers such as Ultra-Turrax or Silverson may be used. For the secondary emulsion preparation, a low shear mixing regime is required, in which case paddle stirrers are probably the most convenient. The mixing times, speed and order of addition need to be optimized.

References

[1] Florence, A. T. and Whitehill, D., J. Colloid Interface Sci., **79**, 243 (1981).
[2] Matsumoto, S., Kita, Y. and Yonezawa, D., J. Colloid Interface Sci., **57**, 353 (1976).
[3] Tadros, Th. F., Int. J. Cosmet. Sci., **14**, 93 (1992).
[4] Whorlow, R. W., "Rheological Techniques", Ellis Hoorwood, Chichester (1980).
[5] Tadros, Th. F., "Rheology of Dispersions", Wiley-VCH, Germany (2010).

9 Liposomes and vesicles in cosmetic formulations

9.1 Introduction

Liposomes are spherical phospholipid liquid crystalline phases (smectic mesophases) that are simply produced by dispersion of phospholipid (such as lecithin) in water by simple shaking [1, 2]. This results in the formation of multilayer structures consisting of several bilayers of lipids (several µm). When sonicated, these multilayer structures produce unilamellar structures (with size range of 25–50 nm) that are referred to as vesicles. A schematic picture of liposomes and vesicles is given in Fig. 9.1.

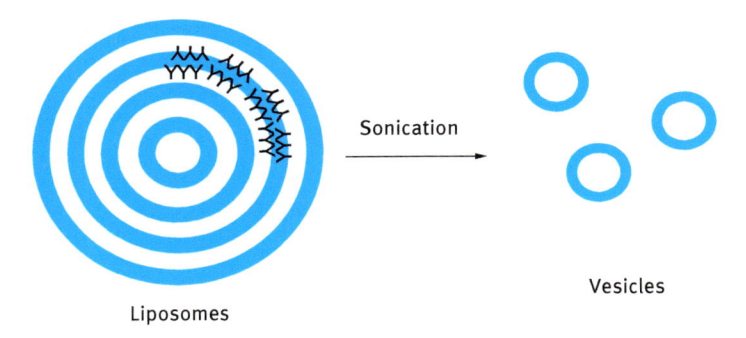

Fig. 9.1: Schematic representation of liposomes and vesicles.

Glycerol-containing phospholipids are used for the preparation of liposomes and vesicles. The structure of some lipids is shown in Fig. 9.2. The most widely used lipid for cosmetic formulations is phosphatidylcholine that can be obtained from eggs or soybean. In most preparations, a mixture of lipids is used to obtain the optimum structure. These liposome bilayers can be considered as mimicking models of biological membranes. They can solubilize both lipophilic active ingredients in the lipid bilayer phase, as well as hydrophilic molecules in the aqueous layers between the lipid bilayers as well as the inner aqueous phase. For example, addition of liposomes to cosmetic formulations can be applied for enhancement of the penetration of anti-wrinkle agents [3]. They will also form lamellar liquid crystalline phases and they do not disrupt the stratum corneum. No facilitated transdermal transport is possible, thus eliminating skin irritation. Phospholipid liposomes can be used as in vitro indicators for studying skin irritation by surfactants.

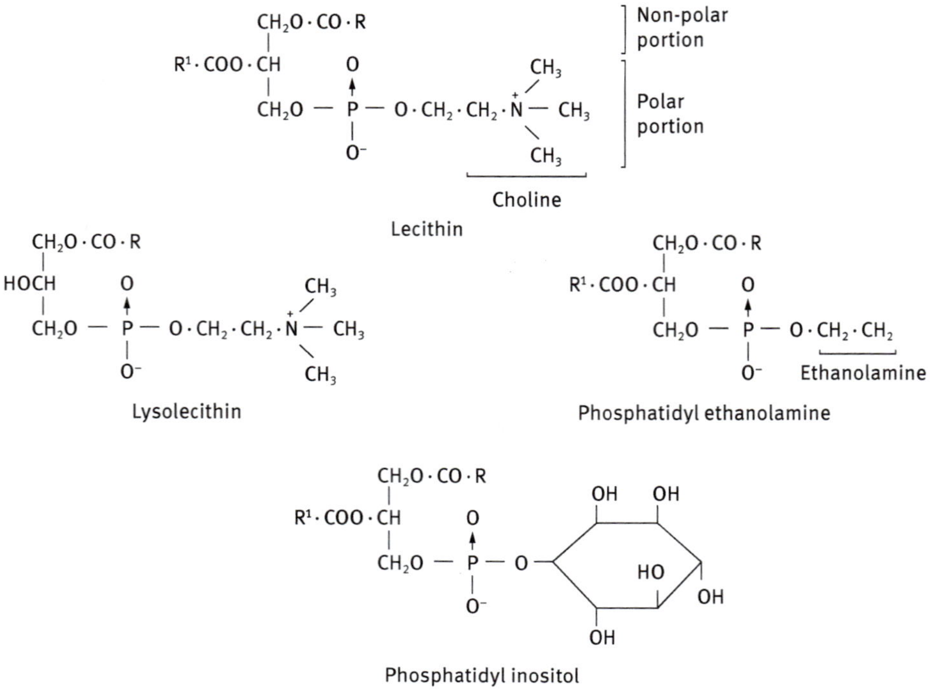

Fig. 9.2: Structure of lipids.

9.2 Nomenclature of liposomes and their classification

The nomenclature for liposomes is far from being clear; it is now generally accepted that "All types of lipid bilayers surrounding an aqueous phase are in the general category of liposomes" [3, 4]. The term "liposome" is usually reserved for vesicles composed, even partly, by phospholipids. The more generic term "vesicle" is to be used to describe any structure consisting of one or more bilayers of various other surfactants. In general the names "liposome" and "phospholipid vesicle" are used interchangeably. Liposomes are classified in terms of the number of bilayers, as Multilamellar Vesicles (MLVs > 400 nm), Large Unilamellar Vesicles (LUVs > 100 nm) and Small Unilamellar Vesicles (SUVs < 100 nm). Other types reported are the Giant Vesicles (GV), which are unilamellar vesicles of diameter between 1–5 μm and Large Oligolamellar Vesicles (LOV) where a few vesicles are entrapped in the LUV or GV.

9.3 Driving force for formation of vesicles

The driving force for formation of vesicles has been described in detail by Israelachvili et al. [5–7]. From equilibrium thermodynamics, small aggregates, or even monomers, are entropically favoured over larger ones. This entropic force explains the aggregation of single-chain amphiphiles into small spherical micelles instead to bilayers or cylinders, as the aggregation number of the latter aggregates is much higher. Israelachvili et al. [5–7] attempted to describe the thermodynamic drive for vesicle formation by biological lipids. From equilibrium thermodynamics of self-assembly, the chemical potential of all molecules in a system of aggregated structures such as micelles or bilayers will be the same,

$$\mu_N^0 + \frac{kT}{N} \ln\left(\frac{X_N}{N}\right) = \text{constant}; \quad N = 1, 2, 3, \ldots, \tag{9.1}$$

where μ_N^0 is the free energy per molecule in the aggregate, X_N is the mole fraction of molecules incorporated into the aggregate, with an aggregation number N, k is the Boltzmann constant and T is the absolute temperature.

For monomers in solution with $N = 1$,

$$\mu_N^0 + \frac{kT}{N} \ln\left(\frac{X_N}{N}\right) = \mu_1^0 + kT \ln X_1. \tag{9.2}$$

Equation (9.1) can be written as,

$$X_N = N \left(\frac{X_M}{M}\right)^{N/M} \exp\left(\frac{N(\mu_M^0 - \mu_N^0)}{kT}\right), \tag{9.3}$$

where M is any arbitrary state of reference of aggregation number M.

The following assumptions are made to obtain the free energy per molecule: (i) the hydrocarbon interior of the aggregate is considered to be in a fluid-like state; (ii) geometric consideration and packing constrains in term of aggregate formation are excluded; (iii) strong long-range forces (van der Waals and electrostatic) are neglected. By considering the "opposing forces" approach of Tanford [8], the contributions to the chemical potential, μ_N^0, can be estimated. A balance exists between the attractive forces mainly of hydrophobic (and interfacial tension) nature and the repulsive forces due to steric repulsion (between the hydrated head group and alkyl chains), electrostatic and other forces [9]. The free energy per molecule is thus,

$$\mu_N^0 = \gamma a + \frac{C}{a}. \tag{9.4}$$

The attractive contribution (the hydrophobic free energy contribution) to μ_N^0 is γa where γ is the interfacial free energy per unit area and a is the molecular area measured at the hydrocarbon/water interface. C/a is the repulsive contribution where C is a constant term used to incorporate the charge per head group, e, and includes terms such as the dielectric constant at the head group region, ε, and curvature corrections.

This fine balance yields the optimum surface area, a_o, for the polar head groups of the amphiphile molecules at the water interface, at which the total interaction free energy per molecule is a minimum,

$$\mu_N^0(\min) = \gamma a + \frac{C}{a} = 0 \tag{9.5}$$

$$\frac{\partial \mu_N^0}{\partial a} = \gamma - \frac{C}{a^2} = 0 \tag{9.6}$$

$$a = a_o = \left(\frac{C}{\gamma}\right)^{1/2}. \tag{9.7}$$

Using the above equations, the general form relating the free energy per molecule μ_N^0 with a_o can be expressed as,

$$\mu_N^0 = \gamma\left(a + \frac{a_o^2}{a}\right) = 2a_o\gamma + \frac{\gamma}{a}(a - a_o)^2. \tag{9.8}$$

Equation (9.8) shows that: (i) μ_N^0 has a parabolic (elastic) variation about the minimum energy; (ii) amphiphilic molecules, including phospholipids, can pack in a variety of structures in which their surface areas will remain equal or close to a_o. Both single-chain and double-chain amphiphiles have very much the same optimum surface area per head group ($a_o \approx 0.5$–0.7 nm^2), i.e. a_o is not dependent on the nature of the hydrophobe. Thus, by considering the balance between entropic and energetic contributions to the double-chain phospholipid molecule one arrives at the conclusion that the aggregation number must be as low as possible and a_o for each polar group is of the order of 0.5–0.7 nm^2 (almost the same as that for a single-chain amphiphile). For phospholipid molecules containing two hydrocarbon chains of 16–18 carbon atoms per chain, the volume of the hydrocarbon part of the molecule is double the volume of a single-chain molecule, while the optimum surface area for its head group is of the same order as that of a single-chain surfactant ($a_o \approx 0.5$–0.7 nm^2). Thus, the only way for this double-chain surfactant is to form aggregates of the bilayer sheet or the close bilayer vesicle type. This will be further explained using the critical packing parameter concept (CPP) described by Israelachvili et al. [5–7]. The CPP is a geometric expression given by the ratio of the cross-sectional area of the hydrocarbon tail(s) a to that of the head group a_o. a is equal to the volume of the hydrocarbon chain(s), v, divided by the critical chain length, l_c, of the hydrocarbon tail. Thus the CPP is given by [10],

$$CPP = \frac{v}{a_o l_c}. \tag{9.9}$$

Regardless of shape, any aggregated structure should satisfy the following criterion: no point within the structure can be farther from the hydrocarbon-water surface than l_c which is roughly equal to, but less than the fully extended length l of the alkyl chain.

Tab. 9.1: CPP concept and various shapes of aggregates.

Lipid	Critical packing parameter $v/a_0 l_c$	Critical packing shape	Structures formed
Single-chained lipids (surfactants) with large head-group areas: – SDS in low salt	< 1/3	Cone	Spherical micelles
Single-chained lipids with small head-group areas: – SDS and CTAB in high salt – nonionic lipids	1/3–1/2	Truncated cone	Cylindrical micelles
Double-chained lipids with large head-group areas, fluid chains: – Phosphatidyl choline (lecithin) – phosphatidyl serine – phosphatidyl glycerol – phosphatidyl inositol – phosphatidic acid – sphingomyelin, DGDG[a] – dihexadecyl phosphate – dialkyl dimethyl ammonium – salts	1/2–1	Truncated one	Flexible bilayers, vesicles
Double-chained lipids with small head-group areas, anionic lipids in high salt, saturated frozen chains: – phosphatidyl ethanaiamine – phosphatidyl serine + Ca^{2+}	≈ 1	Cylinder	Planar bilayers
Double-chained lipids with small head-group areas, nonionic lipids, poly(cis) unsaturated chains, high T: – unsat. phosphatidyl ethanolamine – cardiolipin + Ca^{2+} – phosphatidic acid + Ca^{2+} – cholesterol, MGDG[b]	> 1	Inverted truncated cone or wedge	Inverted micelles

a DGDG: digalactosyl diglyceride, diglucosyldiglyceride
b MGDG: monogalactosyl diglyceride, monoglucosyl diglyceride

For a spherical micelle, the radius $r = l_c$ and from simple geometry CPP = $v/a_0 l_c \leq 1/3$. Once $v/a_0 l_c > 1/3$, spherical micelles cannot be formed and when $1/2 \geq$ CPP $> 1/3$ cylindrical micelles are produced. When the CPP $> 1/2$ but < 1, vesicles are produced. These vesicles will grow until CPP ≈ 1 when planer bilayers will start forming. A schematic representation of the CPP concept is given in Tab. 9.1.

According to Israelachvili et al. [5–7], the bilayer sheet lipid structure is energetically unfavourable to the spherical vesicle, because of the lower aggregation number of the spherical structure. Without introduction of packing constrains (described above), the vesicles should shrink to such a small size that they would actually form micelles. For double-chain amphiphiles three considerations must be considered: (i) an optimum a_0 (almost the same as that for single-chain surfactants) must be achieved by considering the various opposing forces; (ii) structures with minimum aggregation number N must be formed; (iii) aggregates into bilayers must be the favourite structure. A schematic picture of the formation of bilayer vesicle and tubule structures was introduced by Israelachvili and Mitchell [10] is shown in Fig. 9.3.

Israelachvili et al. [5–7] believe that steps A and B are energetically favourable. They considered step C to be governed by packing constraints and thermodynamics in terms of the least aggregation number. They concluded that the spherical vesicle is an equilibrium state of the aggregate in water and it is certainly more favoured over extended bilayers.

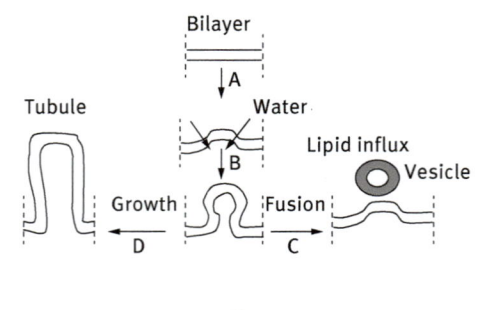

Fig. 9.3: Bilayer vesicle and tubule formation [10].

The main drawback of application of liposomes in cosmetic formulations is their metastability. On storage, the liposomes tend to aggregate and fuse to form larger polydisperse systems and finally the system reverses into a phospholipid lamellar phase in water. This process takes place relatively slowly because of the slow exchange between the lipids in the vesicle and the monomers in the surrounding medium. Therefore, it is essential to investigate both the chemical and physical stability of the liposomes. Examination of the process of aggregation can be obtained by measuring their size as a function of time. Maintenance of the vesicle structure can be assessed using freeze fracture and electron microscopy.

Several methods have been applied to increase the rigidity and physicochemical stability of the liposome bilayer of which the following methods are the most commonly used: hydrogenation of the double bonds within the liposomes, polymerization of the bilayer using synthesized polymerizable amphiphiles and inclusion of cholesterol to rigidify the bilayer [3].

Other methods to increase the stability of the liposomes include modification of the liposome surface, for example by physical adsorption of polymeric surfactants onto the liposome surface (e.g. proteins and block copolymers). Another approach is to covalently bond the macromolecules to the lipids and subsequent formation of vesicles. A third method is to incorporate the hydrophobic segments of the polymeric surfactant within the lipid bilayer. This latter approach has been successfully applied by Kostarelos et al. [4] who used A–B–A block copolymers of polyethylene oxide (A) and polypropylene oxide (PPO), namely Poloxamers (Pluronics). Two different techniques for adding the copolymer were attempted [4]. In the first method (A), the block copolymer was added after formation of the vesicles. In the second method, the phospholipid and copolymer are first mixed together and this is followed by hydration and formation of SUV vesicles. These two methods are briefly described below.

The formation of small unilamellar vesicles (SUVs) was carried out by sonication of 2 w/w% of the hydrated lipid (for about 4 h). This produced SUV vesicles with a mean vesicle diameter of 45 nm (polydispersity index of 1.7–2.4). This is followed by the addition of the block copolymer solution and dilution of ×100 times to obtain a lipid concentration of 0.02 % (method A). In the second method (I) SUV vesicles were prepared in the presence of the copolymer at the required molar ratio.

In method (A), the hydrodynamic diameter increases with increasing block copolymer concentration, particularly those with high PEO content, reaching a plateau at a certain concentration of the block copolymer. The largest increase in hydrodynamic diameter (from ≈ 43 nm to ≈ 48 nm) was obtained using Pluronic F127 (that contains a molar mass of 8330 PPO and molar mass of 3570 PEO). In method I the mean vesicle diameter showed a sharp increase with increasing w/w% copolymer reaching a maximum at a certain block copolymer concentration, after which a further increase in polymer concentration showed a sharp reduction in average diameter. For example with Pluronic F127, the average diameter increased from ≈ 43 nm to ≈ 78 nm at 0.02 w/w% block copolymer and then it decreased sharply with a further increase in polymer concentration, reaching ≈ 45 nm at 0.06 w/w% block copolymer. This reduction in average diameter at high polymer concentration is due to the presence of excess micelles of the block copolymer.

A schematic representation of the structure of the vesicles obtained on addition of the block copolymer using methods (A) and (I) is shown in Fig. 9.4.

With method (A), the triblock copolymer is adsorbed on the vesicle surface by both PPO and PEO blocks. These "flat" polymer layers are prone to desorption due to the weak binding onto the phospholipid surface. In contrast, with the vesicles prepared using method (I) the polymer molecules are more strongly attached to the lipid bilayer

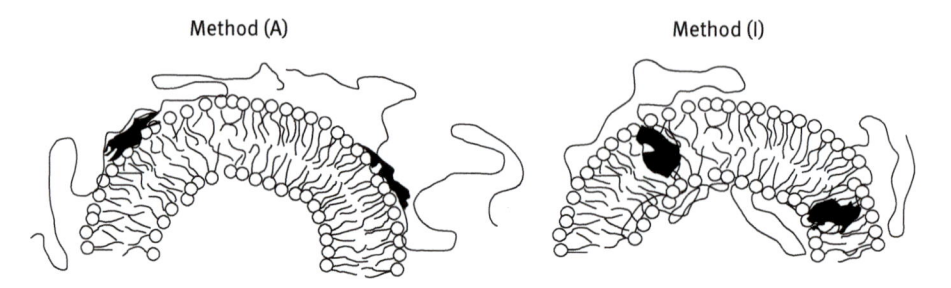

Fig. 9.4: Schematic representation of vesicle structure in the presence of triblock copolymer prepared using methods (A) and (I) [4].

with PPO segments "buried" in the bilayer environment surrounded by the lipid fatty acids. The PEO chains remain at the vesicle surfaces free to dangle in solution and attain the preferred conformation. The resulting sterically stabilized vesicles [(I) system] have several advantages over the (A) system with the copolymer simply coating their outer surface. The anchoring of the triblock copolymer using method (I) results in irreversible adsorption and lack of desorption. This is confirmed by dilution of both systems. With (A), dilution of the vesicles results in reduction of the diameter to its original bare liposome system, indicating polymer desorption. In contrast, dilution of the vesicles prepared by method (I) showed no significant reduction in diameter size indicating strong anchoring of the polymer to the vesicle. A further advantage of constructing the vesicles with bilayer-associated copolymer molecules is the possibility of increased rigidity of the lipid-polymer bilayer [3, 4].

References

[1] Tadros, Th. F., "Colloid Aspects of Cosmetic Formulations", in: Th. F. Tadros, (ed.), "Colloids in Cosmetics and Personal Care", Wiley-VCH, Germany (2008).
[2] Tadros, Th. F., "Nanodispersions", De Gruyter, Germany (2016).
[3] Kostarelos, K., PhD Thesis, Imperial College, London (1995).
[4] Kostarelos, K., Tadros, Th. F. and Luckham, P. F., Langmuir, **15**, 369 (1999).
[5] Israelachvili, J. N., Mitchell, D. J. and Ninham, B. W., J. Chem. Soc., Faraday Trans. II, **72**, 1525 (1976).
[6] Israelachvili, J. N., Marcelja, S. and Horn, R. G., Q. Rev. Biophys, **13**(2), 121 (1980).
[7] Israelachvili, J. N., "Intermolecular and Surface Forces", Academic Press, London (1992).
[8] Tanford, C., "The Hydrophobic Effect", Wiley, NY (1980).
[9] Tanford, C., Biomembranes, Proc. Int. Sch. Phys. Enrico. Ferm., **90**, 547 (1985).
[10] Israelachvili, J. N. and Mitchell, D. J., Biochim. Biophys. Acta., **389**, 13 (1975).

10 Formulation of shampoos

10.1 Introduction

The purpose of a shampoo is to clean the hair from sebum, dead epidermal cells, residues from hair dressing, hair sprays, dust, etc. [1–4]. It must also remove the greasy substances of hair oils, pomades and hair sprays. Soiled hair lacks lustre, becomes oily and unmanageable and develops an unpleasant odour. The shampoo must clean the hair and leave it in a lustrous, manageable condition. This requires the application of surfactants and hair conditioners, the so-called "two-in-one" shampoos. Shampoos can be formulated as clear, pearly or opaque liquids, gels or creams.

Shampoos should possess good, stable foaming action which depends on the surfactants used and the additives that are incorporated. Good shampoos should provide satisfactory cleaning power and easy rinsing without producing soap scum in any water hardness. They also should give a soft touch to the hair, with good manageability after shampooing. The shampoo should also have a low order of irritation to skin and eyes. In addition, for consumer appeal, the product should have an attractive colour and fragrance, and make a rich and mild foam.

Two types of shampoos are marketed, namely powder shampoo and liquid type. The latter can be a clear liquid with low, medium or high (gel form) viscosity. Alternatively the shampoo can be an opaque liquid shampoo consisting of either a pearly liquid or milk lotion shampoo. From the point of view of functions and application, the shampoo can be plain, medicated (anti-dandruff or deodorant), conditioning and low irritation (baby shampoo).

In this chapter I will discuss the following points that are relevant for formulating a conditioning shampoo: (i) The surfactants used in shampoo formulations. (ii) The desirable properties of a shampoo. (iii) The components that are used in the formulation. (iv) The role of the ingredients: mixed surfactant systems and their synergistic action to lower skin irritation, cleansing function, foam boosters, thickening agents as rheology modifiers and silicone oil emulsions in shampoos. The subject of hair conditioners will be dealt with in the next chapter with particular reference to the structure and properties of human hair.

10.2 Surfactants for use in shampoo formulations [2, 3]

10.2.1 Anionic surfactants

Alkyl carboxylates or soaps, with C_{12}–C_{14} chain and counterions of potassium, di- or tri-ethanolamine are sometimes used in shampoos in combination with alkyl sulphates or polyoxyethylene ether sulphate. These carboxylates are applied as foam boosters or foam thickeners. They are seldom used alone due to the disadvantages produced by them. For example, the potassium salt of alkyl carboxylates because of its alkalinity which can make the hair swell. The di- and tri-ethanolamine salts can become discoloured by heat or light. In addition, the soap-based shampoos can leave an insoluble metal salt on the hair after shampooing in hard water and cause unpleasant stickiness. The most commonly used anionic surfactants in shampoos are the alkyl sulphates and their ethoxylates (polyoxyethylene alkyl ether sulphates, AES). The alkyl sulphates are produced by sulphation of higher alcohols (C_{12}–C_{14} chain) using chlorosulphonic acid or sulphuric anhydrate. Since the sodium or potassium salts of alkyl sulphates are not easily soluble in water, their uses are limited to powder or paste shampoo base, although in warm climates they can also be used as liquid shampoo base either alone or in combination with AES. For liquid shampoo, triethanolamine salt and ammonium salt are commonly used.

Alkyl sulphates exhibit good creamy foaming even with oily hair and a good soft feel after shampooing. The performance of alkyl sulphates depends on the length and the distribution of the alkyl chain as well as the nature of the counterion. The alkyl sulphate based on coconut alcohol is the most popular type. It is difficult to thicken shampoo based on alkyl sulphate by simply adding NaCl and this requires the addition of polymer thickener such as Xanthan gum.

The most commonly used anionic surfactants are the ether sulphates, referred to as polyoxyethylene alkyl ether sulphate or AES. They are obtained by sulphating ethoxylated alcohols based mainly on lauryl (dodecyl) alcohol obtained from coconut or synthetic material. Unlike alkyl sulphate based shampoos, liquid shampoos based on AES can be easily thickened by addition of inorganic salts such as NaCl. The maximum viscosity can be obtained by adding a certain amount of NaCl, regardless of the content of AES.

The solubility in water and the foaming ability of NaAES vary with the average moles of ethylene oxide (EO) and the linearity of the alcohol. The higher the number of EO units, the better the solubility and the lower the foaming ability. The more linear the alkyl group of the alcohol, the higher the foaming ability.

10.2.2 Amphoteric surfactants

The most commonly used amphoteric surfactants used in combination with anionics (AES) are the fatty alkyl betaines, e.g. lauryl amidopropyl dimethyl betaine $C_{12}H_{25}CON(CH_3)_2COOH$ (dimethyl lauryl betaine). The addition of the amphoteric surfactant to the anionic surfactant lowers the critical micelle concentration (cmc) of the latter (see below) and this significantly reduces skin irritation. In addition, the amphoteric surfactant acts as a foam booster and thickener. It produces lighter and more voluminous foam.

A suitable basic ingredient for baby shampoos with low irritation is imidazoline and its derivatives. It is also used as a conditioning booster when combined with a cationic polymer such as Polymer JR (see Chapter 11)

10.2.3 Nonionic surfactants

The most commonly used nonionic surfactants in shampoos are the fatty acid alkanolamides which improve foaming ability and solubility in water as well as increase viscosity when combined with anionic surfactants. Another commonly used nonionic surfactant in shampoos is the fatty acid amine oxide which is used as foam stabilizer, thickener and for the improvement of tactile feeling of hair for shampoos based on alkyl sulphates or alkyl ether sulphates (AES). When the pH is in the acid range, it tends to behave like a cationic surfactant and compatibility with anionic surfactants becomes poor. Occasionally, high HLB (hydrophilic-lipophilic balance) surfactants such as Tween 80 (Sorbitan mono-oleate with 20 mol EO) are used as solubilizing agents.

10.3 Properties of a shampoo

Several desirable properties of a shampoo can be listed [1]: (i) Ease of application; the shampoo should have the desirable rheology profile, enough viscosity and elasticity (reasonably high yield value) to stay in the hand before application to the hair. During application, the shampoo must spread easily and disperse quickly over the head and hair, i.e. a shear thinning system is required. This rheological profile can be achieved when using a concentrated surfactant solution that contains liquid crystalline structures (rod-shaped micelles), but in most cases a thickener (high molecular weight material) is included to arrive at the desirable high viscosity at low shear rates. (ii) Dense and luxurious lather: This requires the presence of a foam booster. The surfactant used for cleaning develops an abundant lacy foam in soft water, but the foam quality drops in the presence of oily soils such as sebum. A foam stabilizer is required and this could be a mixture of more than one surfactant. (iii) Ease of rinsing; the shampoo should not leave a residual tackiness or stickiness and it should not precipitate in hard water. (iv) Ease of wet combing; after rinsing the hair should comb through easily without entanglement. Hair conditioners that are cationically modified polymers neutralize the charge on the hair surface (which is negatively charged) and this helps in combing the hair (see Chapter 11). With long hair, a cationic cream rinse after shampooing is more effective. (v) Manageability; when combed dry the hair should be left in a manageable condition (not "fly away" or frizzy). Again charge neutralization of the hair surface by the conditioner helps in this respect. (vi) Lustrous; the hair should be left in a lustrous condition. (vii) Body; the hair should have "body" when dry, i.e. it should not be limp or over-conditioned. (viii) Fragrance; this should not have any objectionable odour. (ix) Low level of irritation; this is the most important factor in any shampoo and for this purpose amphoteric surfactants are preferred over anionics which are more irritable to the skin. As will be discussed later, the use of amphoteric surfactants in combination with anionics reduces the skin irritation of the latter. (x) Preservatives; these should be effective against microbial and fungal contamination. (xi) Good stability; the product should remain stable for at least two or three years at ambient temperatures (both low and high for various regions) as well as when stored in daylight. Both physical and chemical stability should be maintained (no separation, no change in the rheology of the system and no chemical degradation on storage).

10.4 Components of a shampoo

10.4.1 Cleansing agents

Several surfactant systems are used in formulations of shampoos as discussed in Section 10.2. These are mostly anionic surfactants which are usually mixed with amphoteric molecules. As mentioned in the introduction, the main criteria required are good cleansing from sebum, scales and other residues, as well as developing an acceptable lather. For the latter purpose foam boosters or lather enricheners are added. The surfactant concentration in a typical shampoo is in the region of 10–20 %. This concentration is far in excess of that required to clean the hair; the sebum and other oily materials that inhibit foam formation require the use of such a high concentration. As mentioned in Section 10.2, the most widely used anionic surfactants are the alkyl sulphates R–O–SO$_3$–M$^+$ with R being a mixture of C$_{12}$ and C$_{14}$ and M$^+$ being sodium, ammonium, triethanolamine, diethanolamine or monoethanolamine. These anionic surfactants hydrolyse and produce the corresponding alcohol and this may result in the separation of the shampoo. The rate of hydrolysis depends on the pH of the system and this should remain in the range 5–9 to reduce the rate of hydrolysis. The sodium salt has a high Krafft temperature (> 20 °C) and separation (cloudiness) may occur when the temperature is reduced below 15 °C. The ammonium and triethanolamine surfactant has a much lower Krafft temperature and this ensures good stability at low temperatures. Monoethanolamine lauryl sulphate produces very viscous shampoo and this could be considered for formulating a clear gel product. The low temperature stability can also be improved by using ether sulphates R–O–(CH$_2$–CH$_2$–O)$_n$SO$_4$ (with n = 1–5) which also reduce irritancy. Sulphosuccinates, e.g. disodium monococamido sulphosuccinate, disodium monolauramido sulphosuccinate, disodium monooleamido sulphosuccinate (and its PEG modified molecule) are commonly used in shampoos in combination with anionic surfactants. The sulphosuccinates alone do not lather well, but in combination with the anionics they result in excellent shampoos with good foam and reduced eye and skin irritation. Several other surfactants are used in combination with the anionics such as sarcosinates, glutamates, etc. As mentioned in Section 10.2, the most important class of surfactants that are used in combination with anionics are the amphoterics, e.g. amphoteric glycinates/propionates, betaines, amino/imino propionates, etc. These amphoteric surfactants impart mildness and hair conditioning properties to shampoos. Due to their low degree of eye irritation they are used to develop baby shampoos. The pH of the system must be carefully adjusted to 6.9–7.5 (near the isoelectric point of the surfactant), since at low pH the surfactant acquires a positive charge and this leads to an increase in irritation. Several classes of amphoterics have been developed and these will be discussed in the section on role of ingredients. Nonionic surfactants are not used alone in shampoos due to their poor foaming properties. However, they are used in mixtures with anionics to modify the primary cleansing agent, as viscosity builders, solubilizing agents, emulsifiers, lime soap dispersants, etc. They are also incorporated to reduce eye and skin irritation. The most commonly used nonionics are the polysorbates (Tweens) but in some cases Pluronics (or Poloxamers) (A–B–A block copolymers of polyethylene oxide (A) and polypropylene oxide (B)) are also used.

10.4.2 Foam boosters

Most of the surfactants used as cleansing agents develop an abundant lacy foam in soft water. However, in the presence of oily soils such as sebum the abundance and quality of the lather drops drastically. Accordingly, one or more ingredients are added to the shampoo to improve the quality, volume and characteristics of the lather. Examples are fatty acid alkanolamides and amine oxides. As will be discussed later these molecules stabilize the foams by strengthening the surfactant film at the air/water interface (by enhancing the Gibbs elasticity).

10.4.3 Thickening agents

As mentioned before, the viscosity of the shampoo must be carefully adjusted to give a shear thinning system. The most commonly used materials to enhance the viscosity of a shampoo are simple salts such as sodium or ammonium chloride. As will be discussed later, these salts enhance the viscosity simply by producing rod-shaped micelles which have much higher viscosity than the spherical units. Some nonionic surfactants such as PEG distearate or PEG dioleate can also enhance the viscosity of many anionic surfactant solutions. Several other polymeric thickeners can also be used to enhance the viscosity, e.g. hydroxyethylcellulose, xanthan gum, Carbomers (crosslinked polyacrylate), etc. The mechanism of their action will be discussed later.

10.4.4 Preservatives

As shampoos are directly applied to human hair and scalp, they must be completely hygienic. Preservatives are necessary to prevent the growth of germs which can be caused by contamination during preparation or use. The most commonly used preservatives are benzoic acid (0.1–0.2%), sodium benzoate (0.5–1.0%), salicylic acid (0.1–0.2%), sodium salicylate (0.5–1%) and methyl para-hydroxy benzoate (0.2–0.5%). The effects of preservatives depend upon concentration, pH and ingredients of the shampoo. Generally, shampoos of higher concentration are more resistant to germ contamination.

10.4.5 Miscellaneous additives

Many other components are also included in shampoos: Opacifying agents, e.g. ethylene glycol stearate, glyceryl monostearate, cetyl and stearyl alcohol, etc. These materials produce rich, lustrous, pearlescent texture. Clarifying agents; in many cases the perfume added may result in a slight haze and a solubilizer is added to clarify the shampoo. Buffers; these need to be added to control the pH to a value around 7 to avoid production of cationic charges.

10.5 Role of the components

10.5.1 Behaviour of mixed surfactant systems [2, 3]

As mentioned above, most shampoo formulations contain a mixed surfactant system, mostly anionic and amphoteric. For a surfactant mixture with no net interaction, mixed micelles are produced and the critical micelle concentration (cmc) of the mixture is an average of the two cmcs of the single components,

$$\text{cmc} = x_1 \text{cmc}_1 + x_2 \text{cmc}_2 . \tag{10.1}$$

With most surfactant systems, there is a net interaction between the two molecules and the cmc of the mixture is not given by simple additivity. The interaction between surfactant molecules is described by an interaction parameter β which is positive when there is net repulsion and negative when there is net attraction between the molecules. In these cases the cmc of the mixture is given by the following expression,

$$\text{cmc} = x_1^m f_1^m \text{cmc}_1 + x_2^m f_2^m \text{cmc}_2 , \tag{10.2}$$

where f_1^m and f_2^m are the activity coefficients which are related to the interaction parameter β,

$$\ln f_1^m = (x_1^m)^2 \beta \qquad (10.3)$$
$$\ln f_2^m = (x_2^m)^2 \beta . \qquad (10.4)$$

With mixtures of anionic and amphoteric surfactants (near the isoelectric point) there will be net attraction between the molecules and β is negative. This means that addition of the amphoteric surfactant to the anionic surfactant results in lowering of the cmc and the mixture gives a better foam stabilization. In addition, the irritation of the mixture decreases when compared with that of the anionic surfactant alone. As mentioned, above the amphoteric surfactant that contains a nitrogen group is more substantive for the hair (better deposition).

10.5.2 Cleansing function

The main function of the surfactants in the shampoo is to clean the hair from sebum, scales, residues, dust and any oily deposits. The principal action is to remove any soil by the same mechanism as for detergency [1]. For removal of solid particles one has to replace the soil/surface interface (characterized by a tension γ_{SD}) with a solid/water interface (characterized by a tension γ_{SW}) and dirt/water interface (characterized by a tension γ_{DW}). The work of adhesion between a particle of dirt and a solid surface, W_{SD}, is given by,

$$W_{SD} = \gamma_{DW} + \gamma_{SW} - \gamma_{SD} . \qquad (10.5)$$

Fig. 10.1 gives a schematic representation of dirt removal. The task of the surfactant in the shampoo is to lower γ_{DW} and γ_{SW} which decreases W_{SD} and facilitates the removal of dirt by mechanical agitation. Nonionic surfactants are generally less effective in removal of dirt than anionic surfactants. In practice, a mixture of anionic and nonionic surfactants is used. If the dirt is a liquid (oil or fat) its removal depends on the balance of contact angles. The oil or fat forms a low contact angle with the substrate (as illustrated in Fig. 10.2). To increase the contact angle between the oil and the substrate (with its subsequent removal), one has to increase the substrate/water interfacial tension, γ_{SW}. The addition of surfactant increases the contact angle at the dirt/substrate/water interface so that the dirt "rolls up" and off the substrate. Surfactants that adsorb

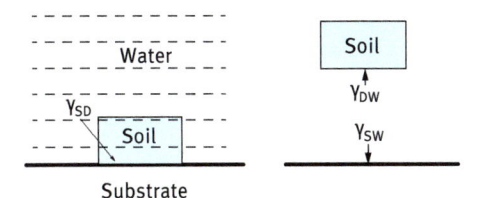

Fig. 10.1: Scheme of dirt removal.

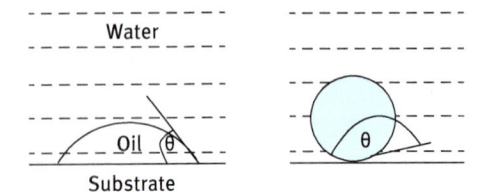

Fig. 10.2: Scheme of oil removal.

both at the substrate/water and the dirt/water interfaces are the most effective. If the surfactant adsorbs only at the dirt/water interface and lowers the interfacial tension between the oil and substrate (γ_{SD}) dirt removal is more difficult. Nonionic surfactants are the most effective in liquid dirt removal since they reduce the oil/water interfacial tension without reducing the oil/substrate tension.

10.5.3 Foam boosters

As mentioned before, with many shampoo formulations the abundance and quality of the lather may drop drastically in the presence of oily soils such as sebum and this requires the addition of a foam booster. Addition of a conditioner such as Polymer JR 400 (cationically modified hydroxyethyl cellulose) will cause a significant reduction of the surface tension of an anionic surfactant such as SDS below its cmc. This occurs even in the precipitation zone and it illustrates the high surface activity of the polymer-surfactant complex.

The polymer-surfactant complex has high surface viscosity and elasticity (i.e. surface viscoelasticity), both will enhance foam stability (see below). The amphoteric surfactants such as betaines and the phospholipid surfactants when used in conjunction with alkyl sulphates or alkyl ether sulphates can also enhance foam stability. All these molecules strengthen the film of surfactant at the air/water interface, thus modifying the lather from a loose lacy structure to a rich, dense, small bubble size, luxurious foam. Several foam boosters have been suggested and these include fatty acid alkanolamide and amine oxides. Fatty alcohol and fatty acids can also act as foam boosters when used at levels of 0.25–0.5 %. Several approaches have been considered to explain foam stability: (i) Surface viscosity and elasticity theory: the adsorbed surfactant film is assumed to control the mechanical-dynamical properties of the surface layers by virtue of its surface viscosity and elasticity. This may be true for thick films (> 100 nm) whereby intermolecular forces are less dominant. Some correlations have been found between surface viscosity and elasticity and foam stability, e.g. when adding lauryl alcohol to sodium lauryl sulphate. This explains why mixed surfactant films are more effective in stabilizing foam as discussed above. (ii) Gibbs–Marangoni effect theory: the Gibbs coefficient of elasticity, ε, was introduced as a variable resistance to surface deformation during thinning [2, 3],

$$\varepsilon = 2(d\gamma/d\ln A) = -2(d\gamma/d\ln h), \tag{10.6}$$

where γ is the surface tension, A is the area of the interface and d ln h is the relative change in lamella thickness. ε is the "film elasticity of compression modulus" and it is a measure of the ability of the film to adjust its surface tension in a constant stress. The higher the value of ε the more stable the film; ε depends on surface concentration and film thickness and this explains the advantage of using mixed surfactant films. The diffusion of surfactant from the bulk solution, i.e. the Marangoni effect, also plays a major role in stabilizing the film. The Marangoni effect opposes any rapid displacement of the surface and this leads to a more stable foam. (iii) Surface forces theory (disjoining pressure) [2, 3]: this theory operates under static (equilibrium) conditions particularly for thin liquid films (< 100 nm) in relatively dilute surfactant concentrations (e.g. during rinsing). The disjoining pressure π is made of three contributions, namely electrostatic repulsion π_{el}, steric repulsion π_{st} (both are positive) and van der Waals attraction π_{vdw} (which is negative),

$$\pi = \pi_{el} + \pi_{st} + \pi_{vdw}. \tag{10.7}$$

For a stable film to form $\pi_{el} + \pi_{st} \gg \pi_{vdw}$. This explains the stability of foams in which both electrostatic and steric repulsion exist. (iv) Stabilization by micelles and liquid crystalline phases: this occurs at high surfactant concentrations and in the presence of surfactant systems that can produce lamellar liquid crystalline phases. The latter, which are formed from several surfactant bilayers, "wrap" around the air bubbles and this can produce a very stable foam. This concept is very important in formulation of shampoos which contain high surfactant concentrations and several components that can produce the lamellar phases.

10.5.4 Thickeners and rheology modifiers

As mentioned above, the shampoo should be viscous enough to stay in the hand before application, but during application the viscosity must decrease enough for good spreading and dispersion over the hair and the head. This requires a shear thinning system (reduction of viscosity on application of shear). Several methods can be applied to increase the viscosity of the shampoo at low shear rates and its reduction on application of shear and these are summarized below.

10.5.4.1 Addition of electrolytes

Many surfactant systems increase their viscosity on addition of electrolytes at an optimum concentration, e.g. sodium chloride, ammonium chloride, sodium sulphate, monoethanolamine chloride, ammonium or sodium phosphate, etc. Of these, sodium chloride and ammonium chloride are the most commonly used. The mechanism by which these electrolytes increase the viscosity of the shampoo can be related to the micellar structure of the surfactant system. Before addition of electrolytes, the micelles are most likely spherical in nature, but when electrolytes are added at an optimum level, the micelles may change to cylindrical (rod-shaped) structures and the viscosity increases. This can be understood when considering the packing parameter of the surfactant system P. The packing parameter P is given by the ratio of the cross-sectional area of the alkyl chain (v/l$_c$, where v is the volume of the hydrocarbon chain and l$_c$ its extended length) to the cross-sectional area of the head group a [5–7],

$$P = v/l_c a \qquad (10.8)$$

For a spherical micelle P ≤ (1/3), whereas for a cylindrical (rod-shaped) micelle P ≤ (1/2). Addition of electrolyte reduces a (by screening the charge) and the spherical micelles change to rod-shaped micelles. This leads to an increase in viscosity. A schematic representation of the rod-shaped (thread-like) micelles and their overlap is given in Fig. 10.3. The viscosity increases gradually by increasing electrolyte concentration, reaches a maximum at an optimum electrolyte concentration and then decreases on a further increase in electrolyte concentration (due to salting-out of the surfactant). The concentration of electrolyte required to reach maximum viscosity depends on the nature of the electrolyte and temperature. These surfactant systems produce viscoelastic solutions that occur at a critical surfactant concentration at which the rod-shaped micelles begin to overlap (similar to the case of polymer solutions). However, these viscoelastic solutions may not have sufficient viscosity to stay on the hand before application. This may be due to their insufficient relaxation times (note that the relaxation time is given by the ratio of viscosity to the modulus). For this reason, many shampoos contain high molecular weight polymers such as hydroxyethyl cellulose (HEC) or xanthan gum and these thickeners are discussed below.

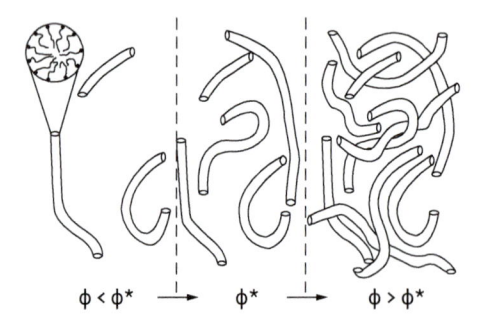

Fig. 10.3: Schematic representation of overlap of thread-like micelles.

10.5.4.2 Thickeners [8]

Most shampoos contain a high molecular weight polymer such as HEC, xanthan gum and some hydrophobically modified HEC or poly(ethylene oxide) (PEO) (associative thickeners). The concentration of the polymer required to produce a certain viscosity at low shear rates depends on its molecular weight M and structure. With HEC several grades are commercially available, e.g. the Natrosol range with M varying between 70 000 and 250 000. The concentration of HEC required (0.5–2 %) to reach a given optimum viscosity decreases with increasing M. With hydrophobically modified HEC (Natrosol Plus) a lower concentration can be used when compared with the unmodified HEC. Hydrophobically modified PEO (HEUR) is also available. Carbomers (crosslinked polyacrylic acids) and other acrylate crosspolymers such as Cabopol 934 and 941 can produce gels when neutralized using ethanolamine (forming microgel particles by swelling due to double layer effects). Unfortunately, they have low tolerance to electrolytes (due to compression of the double layers) and hence they are seldom used in shampoos. Alternatives to Carbomers are the modified acrylate derivatives, such as acrylates/steareth-20/methacrylate copolymer that is supplied as a latex. It is added to the shampoo and then neutralized to the appropriate pH. Care should be taken with this polymer to avoid high electrolyte concentrations and low pH values that may cause its precipitation.

10.5.5 Silicone oil emulsions in shampoos

Silicone oil offers a suitable replacement to sebum that is removed during shampooing. This needs to be formulated as small oil droplets which is not an easy task to obtain. The main advantage of silicone oil is its ability to spread and deposit uniformly on the hair surface, thus providing lubricant, lustre and softness to the hair. This stems from the low surface tension of silicone oils (< 20 mN m^{-1}) thus giving a negative work of spreading W_s. The latter is given by the balance of the solid/liquid interfacial tension, γ_{SL}, the liquid/vapour interfacial tension, γ_{LV}, and the solid/vapour interfacial tensions, γ_{SV},

$$W_s = \gamma_{SL} - \gamma_{LV} - \gamma_{SV}. \tag{10.9}$$

For W_s to become negative, both γ_{SL} and γ_{LV} have to be reduced while keeping γ_{SV} high. The main problem of incorporation of a silicone oil in the shampoo is its dispersion to small droplets and to cause these small droplets to coalesce on the hair surface.

10.6 Use of associative thickeners as rheology modifiers in shampoos [9]

Associative thickeners are hydrophobically modified polymer molecules whereby alkyl chains (C_{12}–C_{16}) are either randomly grafted on a hydrophilic polymer molecule such as hydroxyethyl cellulose (HEC) or simply grafted at both ends of the hydrophilic chain. An example of hydrophobically modified HEC is Natrosol Plus (Hercules) which contains 3–4 C_{16} randomly grafted onto hydroxyethyl cellulose [9]. An example of polymer that contains two alkyl chains at both ends of the molecule is HEUR (Rohm and Haas) that is made of polyethylene oxide (PEO) that is capped at both ends with linear C_{18} hydrocarbon chain. These hydrophobically modified polymers form gels when dissolved in water. Gel formation can occur at relatively lower polymer concentrations when compared with the unmodified molecule. The most likely explanation of gel formation is due to hydrophobic bonding (association) between the alkyl chains in the molecule. This effectively causes an apparent increase in molecular weight. These associative structures are similar to micelles, except the aggregation numbers are much smaller.

Another example of associative thickeners is a blend of PEG-150 distearate and PEG-2-hydroxyethyl cocamide (Promidium LTS, Croda, UK) which was used in a shampoo formulation consisting of 7 % sodium laureth sulphate (with 2 mol ethylene oxide), 3 % cocamidopropylbetaine (CAPB) and 1 % preservative (Germaben II). A comparison was made with the same shampoo thickened by addition of 1.6 % NaCl, using viscoelastic measurements. The latter were investigated using dynamic (oscillatory) measurements, using a Bohlin CVO rheometer (Malvern Instruments, UK). All measurements were carried out at 25 °C, using a cone-plate geometry.

Fig. 10.4 gives typical stress sweep results obtained at 1Hz for the surfactant base thickened with 1.6 % NaCl (Fig. 10.4 (a)) and 1.75 % Promidium LTS (Fig. 10.4 (b)). In both cases G' and G'' remained constant up to a critical stress, above which both G' and G'' start to decrease with decreasing applied stress. The region below the critical stress at which G' and G'' remain constant with increasing stress is denoted as the linear viscoelastic region. It should be mentioned that the surfactant system based on NaCl gives a lower critical stress when compared with the system thickened with Promidium LTS. This reflects the difference in "gel" structure between the two systems. It is likely that the system thickened with Promidium TLS gives a more coherent region (with a longer linear viscoelastic region) when compared with the system based on NaCl.

It can be also seen from the results in Fig. 10.4 that the surfactant base thickened with Promidium LTS (Fig. 10.4 (b)) is far more viscous than elastic ($G'' \gg G'$) when compared with the same base thickened with NaCl (Fig. 10.4 (a)) where $G' > G''$. In fact, regardless of the quantity of Promidium LTS used (and hence the final viscosity of the formula) the thickened surfactant always remains viscous dominant, even at extremely high viscosity. In the case of surfactant thickened with NaCl, at some critical concentration (in this case close to 1.6 %) the base becomes elastic dominant. This can be seen even at reasonably low viscosity.

Once the linear viscoelastic region was known it was possible to measure the effect of frequency on these surfactant bases. As an example typical frequency sweeps for surfactant bases thickened with 2.5 % NaCl and 2.5 % Promidium LTS are given in Fig. 10.5. It can be seen from Fig. 10.5 that the crossover point (at which $G' = G''$) occurs at much higher frequency for the surfactant base thickened with Promidium LTS when compared to the same base thickened with salt. This implies that the relaxation time for the base thickened with Promidium LTS is much smaller than the values for the salt thickened system. A plot of relaxation time versus both NaCl and Promidium LTS concentration is given in Fig. 10.6.

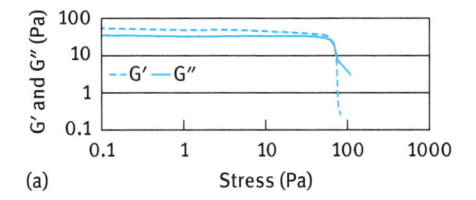

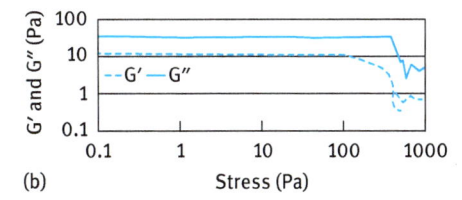

Fig. 10.4: Typical stress sweep results (1Hz) for surfactant blends thickened with 1.6 % NaCl (a) and Promidium LTS (b).

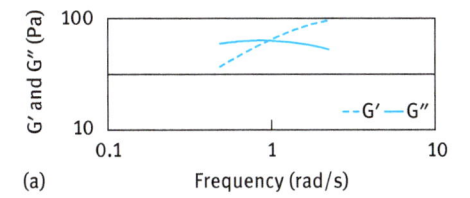

Fig. 10.5: Typical frequency sweeps for surfactant base thickened with 2.5 % NaCl (top) and 2.5 % Promidium LTS (bottom).

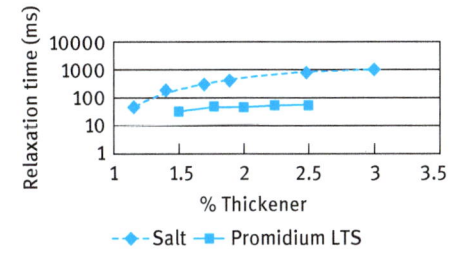

Fig. 10.6: Relaxation time versus NaCl and Promidium LTS concentration.

At the high frequencies (corresponding to short timescales) the response is more elastic than viscous (G' > G") for surfactants thickened with both NaCl and Promidium LTS. The high frequency modulus values are significantly higher for the bases thickened with Promidium LTS when compared to those thickened with NaCl. However, G" is beginning to plateau at 1.4 % NaCl while Promidium LTS G" values are continuing to rise (over the whole concentration range). Again, this implies that independent of Promidium LTS concentration, and the structure this gives, the surfactant bases thickened with this associative thickener remain viscous in behaviour. Those thickened with salt become predominantly elastic [9].

Figs. 10.7 and 10.8 show variation of G' and G" at a high frequency oscillation that is above the G'/G" crossover point. This is shown as a function of both NaCl concentration and Promidium LTS concentration. For NaCl the frequency is 10 rad s^{-1}, for LTS the frequency is 50 rad s^{-1}.

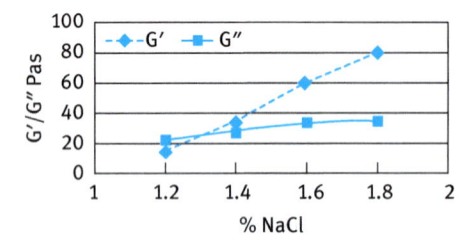

Fig. 10.7: Variation of G' and G" (at a frequency higher than the crossover point) for surfactants thickened with NaCl.

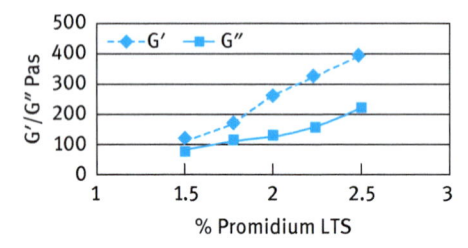

Fig. 10.8: Variation of G' and G" (at a frequency > G'/G" crossover point) for surfactant base thickened with Promidium LTS.

The increase in the viscosity or elasticity of surfactant blends thickened with NaCl is due to the change in micellar structure from spherical to rod-shaped micelles by considering the critical packing parameter, P, given by equation (10.8). However, such structures give a more elastic than viscous response even at low frequency (long timescales), i.e. 1 Hz. This could be comparable to the timescales used in pouring, pumping and spreading during in-use application. The crossover point for such electrolyte thickened systems occurs at much lower frequencies giving long relaxation times. With increasing NaCl concentration the relaxation time increases reaching very high values. For example, at 2.5 % NaCl, the crossover point occurs at 1 rad s^{-1} (0.16 Hz) giving a relaxation time of 1 s.

As mentioned above, an alternative and more elegant way of thickening shampoos is to use associative thickeners. The associative thickener studied above consists of a hydrophilic chain of 150 ethylene oxide units (PEG 150), with two stearate chains attached, one at each end of the hydrophilic chain. This produces "micelle-like" structures [9]. These structures are seen to have much shorter relaxation times when compared to surfactants thickened with salt (Fig. 10.6). The relaxation times are more than one order of magnitude lower than the bases thickened with salt. The crossover point for the formulations thickened using Promidium LTS occurs at much higher frequency when compared to those thickened by addition of NaCl. This means that at low frequency the system thickened with Promidium LTS is more viscous than elastic, independent of stress amplitude (Fig. 10.4). This will contribute a great degree to the sensory characteristics of the shampoo, both in terms of feel during application and also visually, i.e. a lack of stringiness and stickiness.

References

[1] Tadros, Th. F., "Cosmetics", in "Encyclopedia of Colloid and Interface Science", Th. F. Tadros (ed.), Springer, Germany (2013).
[2] Tadros, Th. F., "Applied Surfactants", Wiley-VCH, Germany (2005).
[3] Tadros, Th. F., "Introduction to Surfactants", De Gruyter, Germany (2014).
[4] Holmberg, K., Jonsson, B., Kronberg, B. and Lindman, B., "Surfactants and Polymers in Aqueous Solution", John Wiley & Sons, USA (2003).
[5] Penfield, K., IFSCC Magazine, **8** (2), 115 (2005).
[6] Tadros, Th. F., Advances in Colloid and Interface Science, **68**, 91 (1996).
[7] Israelachvili, J. N., "Intermolecular and Surface Forces", Academic Press of London (1985).
[8] Goddard, E. D. and Gruber, J. V. (eds.) "Principles of Polymer Science and Technology in Personal Care", Marcel Dekker, NY (1999).
[9] Tadros, Th. F. and Hously, S., in "Colloids in Cosmetics and Personal Care", Th. F. Tadros (ed.), Wiley-VCH, Germany (2008).

11 Formulation of hair conditioners in shampoos

11.1 Introduction

A hair conditioner is an ingredient or product that when applied to hair according to its recommended use, procedure and concentration improves the manageability, gloss and smooth touch of the hair [1]. When using shampoos containing anionic surfactants it leaves the hair difficult to comb while wet. It also results in a static charge build-up or fly away when the hair is combed dry. As will be discussed later, the isoelectric point of hair is approximately 3.67 and hence its surface will have a net negative charge at neutral pH. The anionic surfactants which are also negatively charged do not deposit (do not adsorb) on the hair and leave it in an unmanageable condition. Amphoteric surfactants that contain a positively charged nitrogen group are more substantive to hair and can impart some conditioning effect. Cationic surfactants such as stearyl benzyl dimethylammonium chloride, cetyltrimethylammonium chloride, distearyl dimethylammonium chloride or stearamidopropyldimethyl amine and diesterquats are also effective as hair conditioners. The main problem with using cationic surfactants is their strong interaction with the anionic surfactant molecules which may cause precipitation.

As we will see later, polymeric conditioners with their high molecular weight are deposited strictly on the fibre surface or can penetrate into the cuticule or even beyond it into the cortex. The most effective hair conditioners are the cationically modified polymers (e.g. Polyquaternium-10) that will be discussed later. These polymeric compounds are incorporated into shampoos with the major goal of improving the "condition" of hair, which includes its appearance and manageability. Properties such as combability, fly away, body and curl retention are affected by the deposition of polymers on the hair surface. Several other components can impart some conditioning effect, e.g. fatty alcohols, fatty acids, monoglycerides, lecithin, silicones, hydrolysed proteins, polyvinylpyrrolidone, gelatin, pectin, etc.

To understand the role of the conditioner it is essential to know the structure and properties of human hair with particular reference to its surface properties.

11.2 Morphology of hair

A schematic representation of a human hair fibre is shown in Fig. 11.1. This complex morphology [2] consists of four components, namely the cortex, the medulla, the cell membrane and the cuticle.

The major part of the interior of the fibre mass is the cortex that consists of elongated, spindle-shelled cells aligned in the direction of the fibre axis. The second component of hair morphology located in the centre of some thicker fibres and consisting

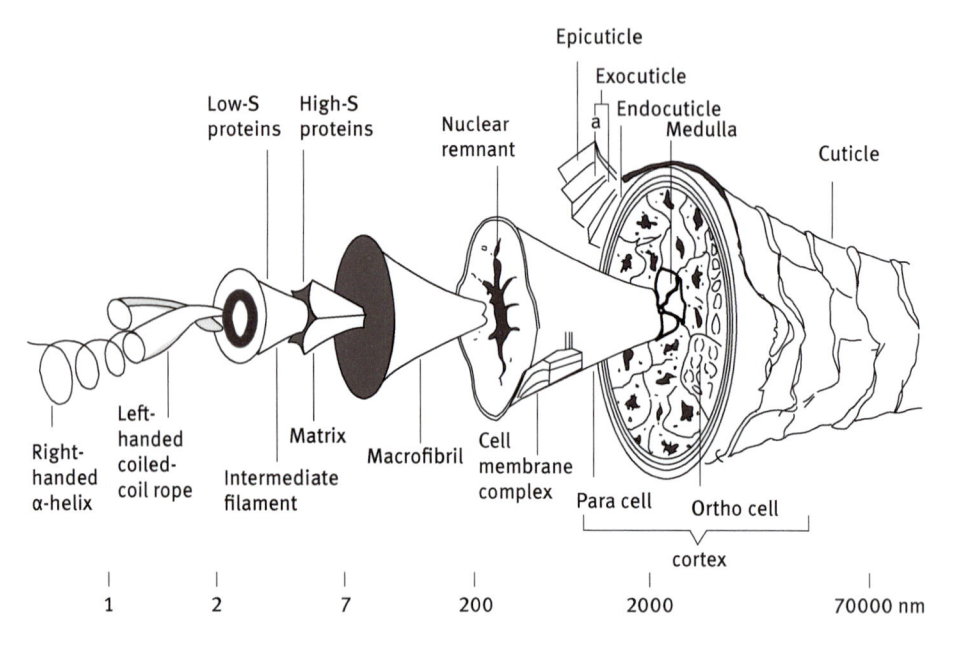

Fig. 11.1: Schematic representation of a human hair fibre.

of a loosely packed cellular structure is called the medulla (see Fig. 11.1). The third component fulfils the vital function of cementing the various cells of the cortex together, thus making a fibre out of a conglomerate of cells. This intercellular cement together with the cell membranes forms the cell membrane complex (illustrated in Fig. 11.2), is assumed to be the location of transport paths into the fibre. This intercellular transport is especially important for the incorporation of polymeric molecules into the cuticle, or even for diffusion into the cortex.

On the outside of the hair, a thick covering of several layers of overlapping cuticle cells provides protection against mechanical and environmental stresses. While at the root end up to 10 layers of cuticle cells are stacked over each other, the thickness of the cuticle layer decreases with increasing distance from the scalp as mechanical and environmental stresses cause ablation of cuticle fragments until occasionally the cuticle envelop has been totally worn away at the tip of long fibres. The cuticle cell itself is a multilayered structure as schematically illustrated in Fig. 11.3. The most important part of the cuticle from the point of view of surfactant and polymer deposition is its outermost surface, namely the epicuticle, which is about 2.5 nm thick [3]. It consists of 25 % lipids and 75 % protein, the latter having an ordered possibly β-pleated sheet structure with 12 % cystine. The cystine groups are acylated by fatty acids which form the hydrophobic surface region. A schematic representation of the epicuticle is shown in Fig. 11.4.

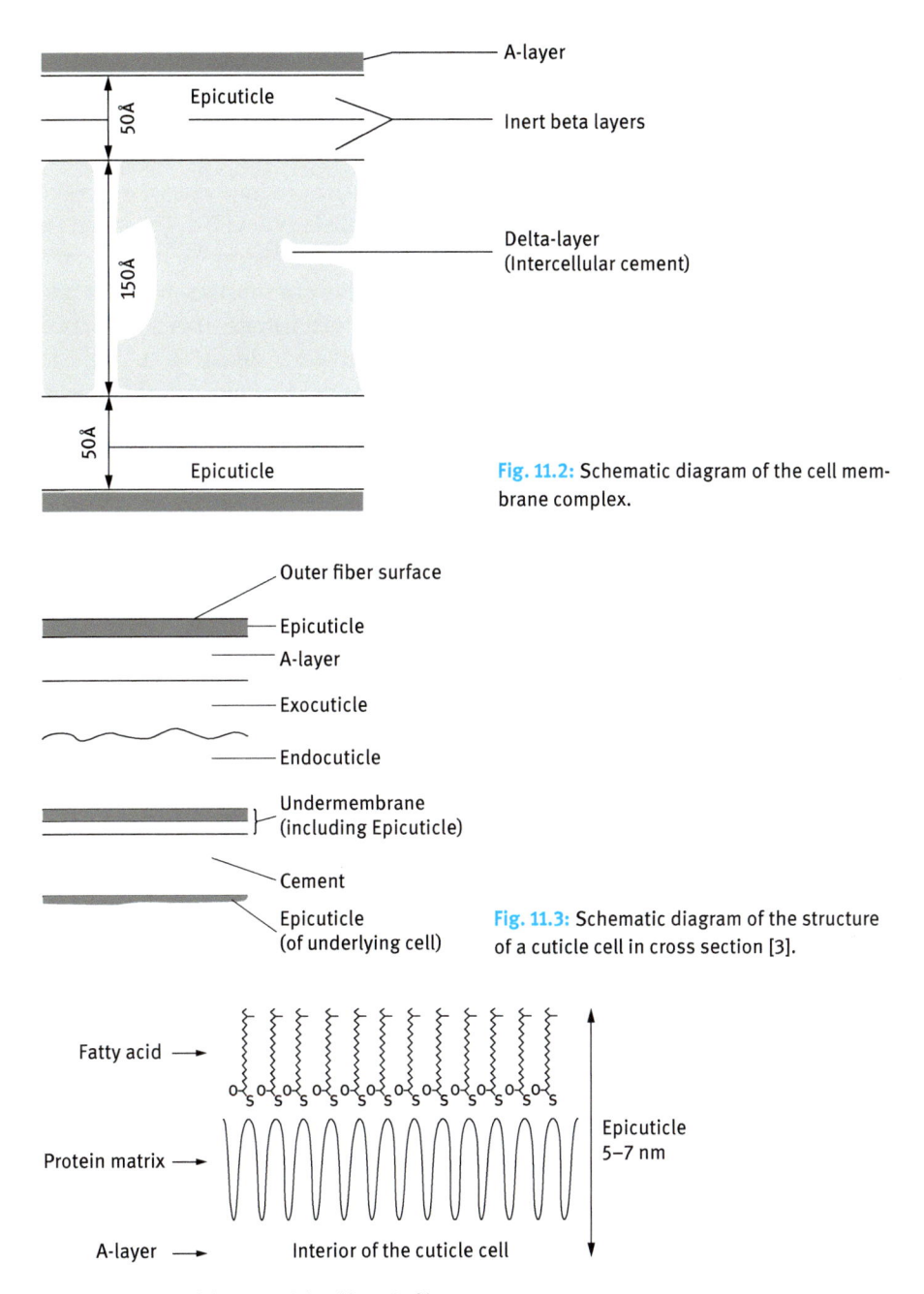

Fig. 11.2: Schematic diagram of the cell membrane complex.

Fig. 11.3: Schematic diagram of the structure of a cuticle cell in cross section [3].

Fig. 11.4: Model of the epicuticle of keratin fibres.

11.3 Surface properties of hair

11.3.1 Wettability investigations

The surface energy of the intact human hair is determined by the outermost layer of the epicuticle which consists of covalently bound, long chain fatty acids [2]. Thus, the low energy hydrophobic surface is not uniformly wetted by a high energy liquid like water. The most convenient method for assessing the wettability of a substrate by water is to measure the contact angle θ of a drop or air bubble on the substrate. This is illustrated in Fig. 11.5 which shows a schematic representation of a sessile drop (Fig. 11.5 (a)) and an air bubble (Fig. 11.5 (b)) resting on a flat surface for a wettable surface (with θ < 90°, Fig. 11.5) and a non-wettable surface (with θ > 90°, Fig. 11.5).

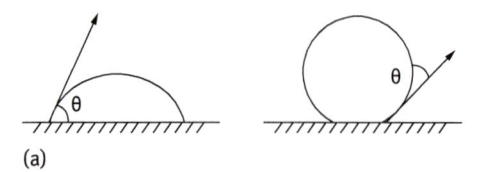

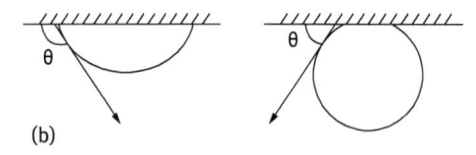

Fig. 11.5: Schematic representation of the sessile drop (a) and air bubble (b) resting on a surface.

The equilibrium aspect of wetting can be assessed using Young's equation by considering the balance of forces at the wetting line as illustrated in Fig. 11.6. Three interfacial tensions can be identified: γ_{SV}, γ_{SL} and γ_{LV} γ_{SL} (where S refers to solid, V to vapour and L to liquid).

$$\gamma_{SV} = \gamma_{SL} + \gamma_{LV} \cos\theta \tag{11.1}$$

$$\gamma_{LV} \cos\theta = \gamma_{SV} - \gamma_{SL}. \tag{11.2}$$

Clearly a hair is not a flat surface and it can be approximated as a cylinder. To measure the contact angle on the hair surface, the Wilhelmy plate method represented in Fig. 11.7 can be applied.

In the first case, the force on the plate F is given by the equation,

$$F = (\gamma_{LV} \cos\theta)p, \tag{11.3}$$

where p is the plate perimeter.

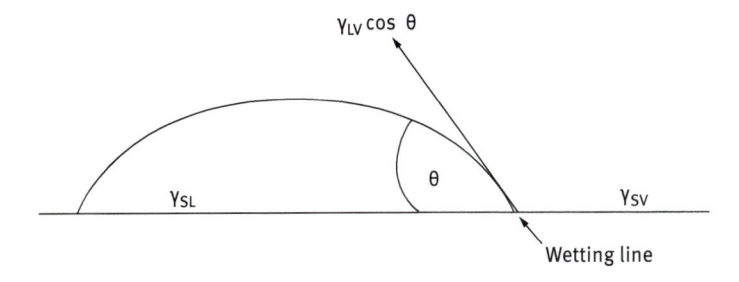

Fig. 11.6: Schematic representation of balance of forces at the wetting line.

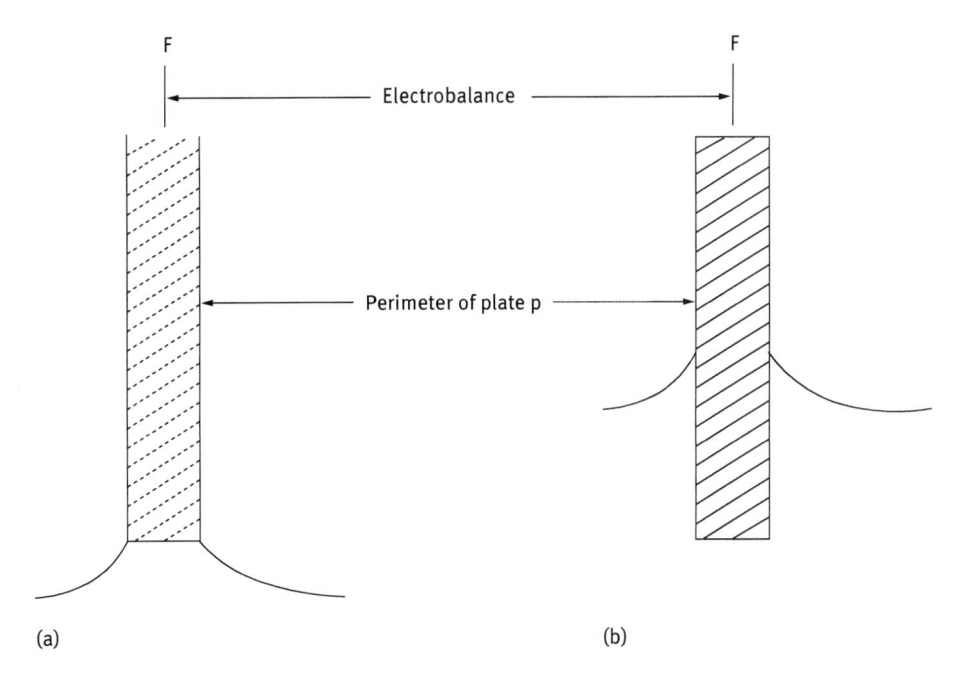

Fig. 11.7: Scheme for the Wilhelmy plate technique for measuring the contact angle: Left zero net depth; right finite depth.

In the second case, the force is given by,

$$F = (\gamma_{LV} \cos \theta)p - \Delta\rho g V. \qquad (11.4)$$

$\Delta\rho$ is the density difference between the plate and the liquid and V is the volume of liquid displaced.

A schematic representation for the set-up for measuring the wettability of a hair fibre [2] is shown in Fig. 11.8. An untreated intact hair fibre gives a contact angle θ greater than 90° and hence it produces a negative meniscus resulting in a negative

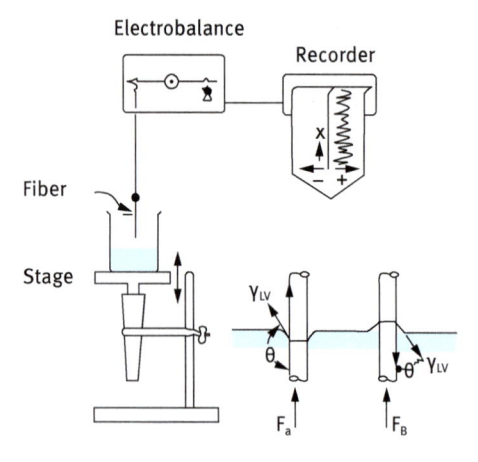

Fig. 11.8: Set-up for measuring the wettability of a hair fibre [2].

wetting force w that is given by,

$$F_w = w + F_b, \qquad (11.5)$$

Where F_w is the recorded force and F_b is the buoyancy force.

On deposition of a hydrophilic polymer on the surface of the hair, the contact angle becomes smaller than 90° giving a positive meniscus and a positive wetting force. If the wetted perimeter P at the line of contact between liquid and fibre is known, the wettability W can be calculated,

$$W = w/P = \gamma_{LV} \cos\theta, \qquad (11.6)$$

where γ_{LV} is the surface tension of the wetting liquid.

The perimeter of the fibre is calculated from the approximate relation,

$$P = 2\pi \left(\frac{A^2 + B^2}{2} \right)^{1/2}, \qquad (11.7)$$

where A and B are the major and minor half-axes of the fibre and can be determined by laser micrometry.

Another parameter than can be used to characterize the surface is the work of adhesion A,

$$A = \gamma_{LV}(\cos\theta + 1) = W + \gamma_{LV}. \qquad (11.8)$$

The deposition, uniformity and substantivity of the hair conditioner can be characterized by scanning the wettability along the length of the fibre before and after treatment. Typical results are illustrated in Fig. 11.9 for quaternized cellulose derivative (Polymer JR-400) that is commonly used in conditioner formulations [2]. The wetting force of untreated fibre shows minor irregularities due to the scale structure and surface heterogeneity of the fibre. First immersion in the JR solution shows a spotty deposition of the polymer. The second immersion in water shows a significant reduction in the wetting peak indicating a loss of the hydrophilic polymer from the surface. No further desorption of the polymer occurs after the third immersion in water. Interaction with the anionic surfactant such as sodium lauryl sulphate or PEG ether sulphate affects the polymer deposition and the fibre wettability.

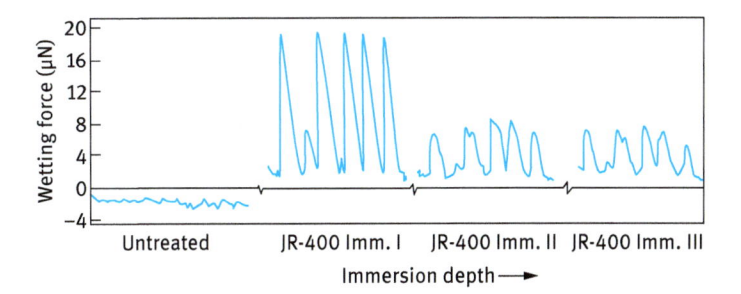

Fig. 11.9: Advanced wetting force curves in successive immersion of untreated hair fibre and the same hair fibre treated with 1 % JR-400.

The interaction between anionic surfactants and the polymer cation can affect the polymer deposition. For example, with sodium lauryl sulphate below the critical micelle concentration (cmc), high levels of deposition with low substantivity are observed. However, above the cmc the wettability decreases below that of the untreated fibre. This could be due to the interaction between the surfactant micelles and the cationic polymer forming a surfactant-polymer complex with reorientations that produce a hydrophobic surface.

11.3.2 Electrokinetic studies

The surface properties of hair can be investigated using streaming potential measurements [2] which can be applied to measure the zeta potential as a function of pH as well as the permeability of the plug (which can give information on swelling or shrinking of the fibre). A plug of hair is packed in a cell that contains two electrodes at its ends. The liquid under investigation is allowed to flow through the plug and the pressure drop P is measured. The potential difference at the electrodes is measured using an electrometer and the conductivity of the flowing liquid is simultaneously measured. This allows one to obtain the zeta potential and the permeability of the plug. Using this technique, the zeta potential–pH curves showed an isoelectric point for untreated hair of 3.7, indicating that in most practical conditions the hair surface is negatively charged (pH > 5).

11.4 Role of surfactants and polymers in hair conditioners

When using anionic surfactants alone in shampoos repulsion between the negatively charged hair and the anionic surfactant occurs, preventing deposition of the molecules on the hair surface. The electrostatic charges present on the hair surface result in difficult combing when the hair is wet [1]. In addition, when the hair is dry, the electrostatic build up on the surface of hair also makes the hair unmanageable, causing "fly away" or frizziness [1]. These problems can be reduced in part by incorporation of amphoteric surfactants which can deposit on the hair surface, thus reducing the negative charges. However, these molecules are not very effective in conditioning the hair and various more effective cationically charged molecules have been suggested for hair conditioning. One of the earliest conditioners tried are cationic surfactants which deposit on the hair by electrostatic attraction between the negative charge on the hair surface and the cationic charge of the surfactant. However, when added to a shampoo based on anionic surfactant, interaction between the molecules occurs, resulting in associative phase-separated complexes that are incompatible with the nonionic formulation. Efforts have been made to minimize these interactions but in general the resulting systems provide poor conditioning from a shampoo. The use of soluble cationic surfactants that form soluble ionic complexes that remain compatible in the formulation but do not deposit well on the hair surface. The use of cationic surfactants that are compatible in the formulation, but form insoluble complexes on dilution also did not result in good conditioning. The breakthrough in hair conditioners came from the development of cationically modified water soluble polymers.

The earliest studies used the cationic polymer polyethyleneimine (PEI) which could be radiolabelled (^{14}C) allowing one to accurately measuring the uptake of the polymer by hair. Although this polymer was later withdrawn from hair conditioners (due to its toxicity) it can be considered as an initial model for an adsorbing polycation [4–7]. Two homopolymers of PEI with molar mass 600 and 60 000 were used in these studies. As an illustration Fig. 11.10 shows the sorption of ^{14}C-labelled PEI 600 (expressed in % based on the weight of hair) from a 5 % aqueous solution as a function of contact time.

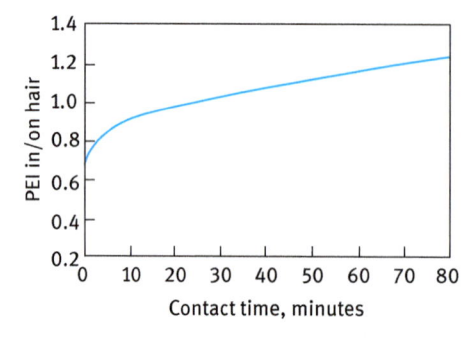

Fig. 11.10: Sorption of ^{14}C-labelled PEI 600 (expressed in % based on the weight of hair) from a 5 % aqueous solution as a function of contact time.

11.4 Role of surfactants and polymers in hair conditioners

The results in Fig. 11.10 show that sorption occurs almost immediately once the hair comes into contact with the PEI solution. This sorption increases with increasing contact time reaching more than 1 % after 60 minutes. Similar results were obtained with the higher molar mass PEI and the sorption is compatible for the two polymers. Bleaching of hair increases the uptake of the polymer, in particular with the higher molar mass PEI. After 1 hour, the sorption of PEI 60 000 increases from 1.2 to 3.4 % on bleaching the hair. Reducing the concentration of PEI causes a decrease in the sorption amount (an 80 % reduction in concentration decreases the sorption amount by 50 %). The sorption was highest at pH 7 and it decreases when the pH is increased to 10. Reduction of the pH to 2 significantly reduces the sorption amount since at this pH the hair becomes positively charged.

Polyquaternium-10, which is a cationically modified hydroxyethyl cellulose (HEC), with the cationic groups being hydroxypropyltrimethylammonium is commonly used as a hair conditioner in shampoos [4–7]. The grade of Polyquaternium-10 that is commonly used in shampoos has a number average molecular weight of 400 000 and about 1300 cationic sites. Several other cationically modified HEC have been developed such as Polymer JR with three molecular weight grades of 250 000 (JR-30M), 400 000 (JR-400) and 600 000 (JR-125). These polymers have the generic formula represented in Fig. 11.11.

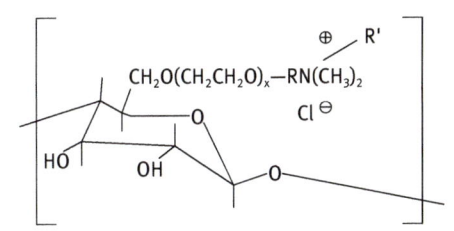

Fig. 11.11: Generic formula of Polymer JR.

The cross-sectional area of these polymers is considerably larger than that of PEI. The adsorption of HEC and JR polymers on hair was studied by Goddard [6] using radiolabelled polymers. In all experiments, the concentration of polymer was kept constant at 0.1 % and the amount sorbed (mg g^{-1}) was measured as a function of time for several days. The sorption of HEC reached equilibrium in 5 minutes, whereas the sorption of charged JR polymers did not reach its equilibrium value even after 2 days. The results are shown in Figs. 11.12 and 11.13 which show the variation of the amount sorbed (mg g^{-1}) with time.

The amount of adsorption of HEC on hair (0.05 mg g^{-1}) corresponds to the value expected for a close-packed monolayer of the cellulose (in flat orientation) giving an area per HEC residue of ≈ 0.85 nm^2. The adsorption of JR polymers is higher than the corresponding amount for flat orientation. It has been suggested that the polycation diffuses in the keratinous substrate. The sorption of the polymer on bleached hair was much higher (Fig. 11.14) which shows an order of magnitude higher adsorption when

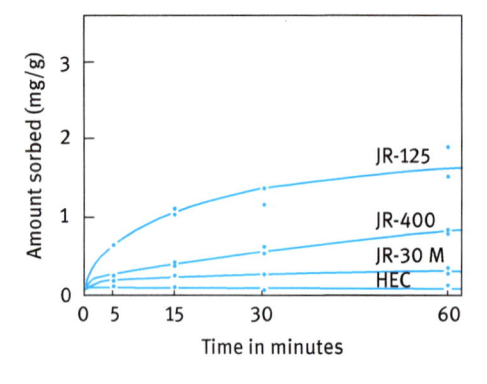

Fig. 11.12: Sorption of ^{14}C-labelled polymer (from 0.1% solution) by virgin hair; short time experiment [6].

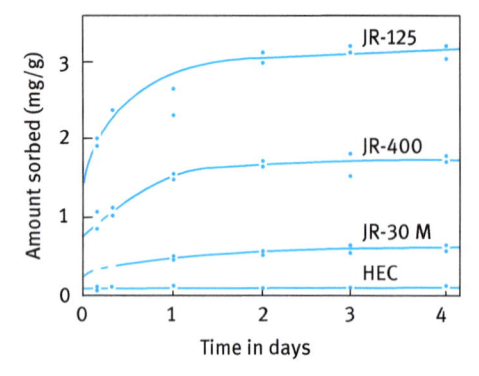

Fig. 11.13: Sorption of ^{14}C-labelled polymer (from 0.1% solution) by virgin hair; long time experiment [6].

compared with that of unbleached hair (Fig. 11.13). This indicates the more damaged and porous nature of the bleached fibre.

The electrostatic attraction between the cationic groups on the JR polymers and the negative charges on the surface of hair seems to be the driving force for the adsorption process. Evidence for this was obtained by studying adsorption in the presence of added electrolytes, 0.1 and 1% NaCl, which reduced the adsorption by approximately three- and 10-fold respectively (Fig. 11.15).

For a given molarity of electrolyte the reduction in sorption increases with increasing electrolyte valency as shown in Fig. 11.16 (in accordance with the Schultze–Hardy rule).

If electrostatic attraction between the polycation and the negatively charged hair is the driving force for adsorption, one would expect a large effect of pH which determines the charge on the hair surface. The effect of pH on the sorption of JR-125 (expressed as the amount of sorption σ in g g^{-1}) is shown in Fig. 11.17. Initial work showed little variation of the adsorption of Polymer JR-125 on bleached hair within the pH range 4–10. However, later work showed a catastrophic reduction in the sorption of this polymer on virgin hair when the pH was reduced below the isoelectric point (pH 3.7). Under these conditions, the amount of uptake approximated that displayed

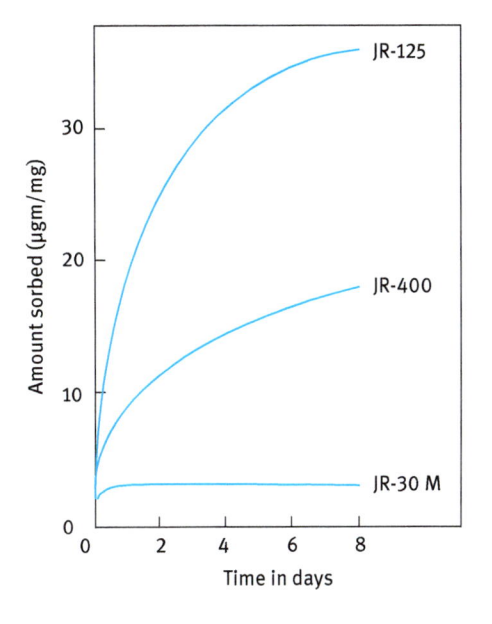

Fig. 11.14: Sorption of different grades of Polymer JR on bleached hair from 0.1% solution [6].

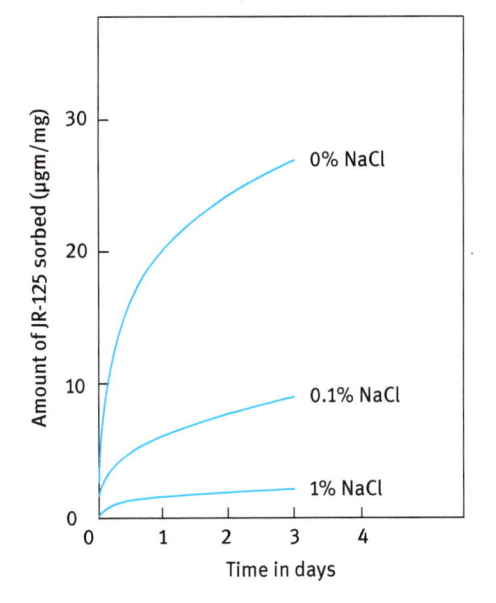

Fig. 11.15: Effect of addition of NaCl on the sorption of JR-125 by bleached hair; 0.1% polymer solution [6].

by the uncharged HEC molecule. This result provides further evidence that electrostatic forces govern the adsorption of the polyelectrolyte [6].

As mentioned above, the adsorption of the cationically modified polymers on hair in the presence of anionic surfactants is complicated by the polyelectrolyte–surfactant interaction. Results for the interaction between JR-400 or RETEN (polycation of acrylamide/β-methacryloyloxyethyl trimethylammonium chloride) and sodium dodecyl

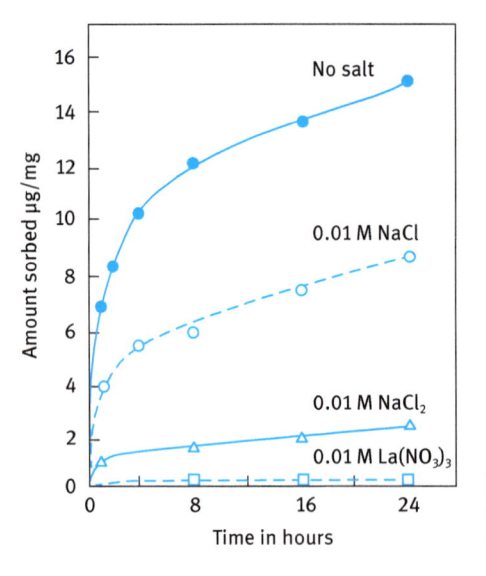

Fig. 11.16: Effect of different electrolytes on the sorption of Polymer JR-125 by bleached hair; 0.1 % polymer solution [6].

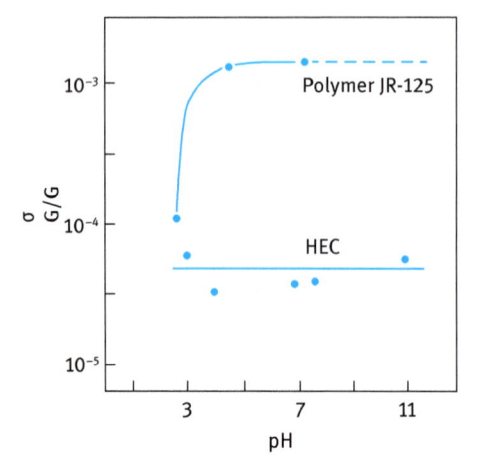

Fig. 11.17: Sorption of JR-125 onto virgin hair as a function of solution pH; 0.1 % polymer solution [6].

sulphate (SDS) were obtained by Goddard [4] using surface tension and viscosity measurements. Fig. 11.18 shows the viscosity results where the relative viscosity is plotted as a function of SDS concentration at a constant JR-400 or RETEN concentration of 1 %. With Polymer JR-400, the relative viscosity showed a rapid increase in the immediate precipitation zone. In the precipitation zone a network is invoked in which surfactant molecules bound to one polycation molecule associate with similarly linked surfactant molecules on the other polymer chains. At high SDS concentration, the solution viscosity falls since the properties are now dominated by surfactant micelles. In contrast with RETEN at 1 % concentration, a change in viscosity with added SDS in the

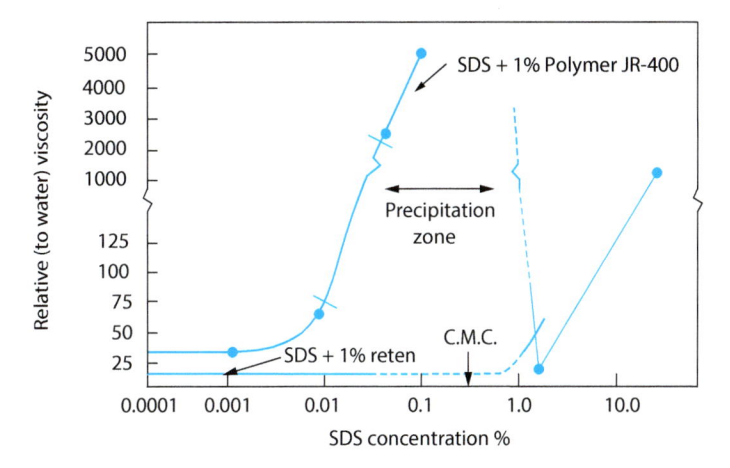

Fig. 11.18: Relative viscosity of 1% Polymer JR-400 and 1% RETEN as a function of SDS concentrations [4].

precipitation zone and only a modest increase in viscosity is observed at 1% SDS concentration.

The above interaction between the anionic surfactants and polycation has a major influence on the uptake of Polymer JR. This interaction leads to considerable reduction of polyelectrolyte deposition on hair. Nonionic surfactants such as Tergitol 15-S-9 show substantially "unimpeded" deposition of the polycation, whereas amphoteric surfactants (based on imidazoline) showed substantial deposition of the polymer. In contrast, the cationic surfactant cetyltrimethyl ammonium bromide (CTAB) virtually eliminated the polymer adsorption. This is due to the faster diffusion of CTA^+ which neutralized the negative charges on the hair.

References

[1] Tadros, Th. F., "Cosmetics", in "Encyclopedia of Colloid and Interface Science", Th. F. Tadros (ed.), Springer, Germany (2013).
[2] Weigman, H. D. and Kamath, Y., "Evaluation Methods for Conditioned Hair", in "Principles of Polymer Science and Technology in Cosmetics and Personal Care", D. E. Goddard and J. V. Gruber (eds.), Marcel Dekker, NY (1999) Chapter 12.
[3] Robbins, C. R., "Chemical and Physical Behaviour of Human Hair", 3rd ed., Springer-Verlag, NY (1994).
[4] Goddard, D. E., "Polymer/Surfactant Interaction, Methods and Mechanisms" in "Principles of Polymer Science and Technology in Cosmetics and Personal Care", D. E. Goddard and J. V. Gruber (eds.), Marcel Dekker, NY (1999) Chapter 4.
[5] Goddard, D. E., "Polymer/Surfactant Interaction, Applied Systems" in "Principles of Polymer Science and Technology in Cosmetics and Personal Care", D. E. Goddard and J. V. Gruber (eds.), Marcel Dekker, NY (1999) Chapter 5.

[6] Goddard, D. E., "Measuring and Interpreting Polycation Adsorption" in "Principles of Polymer Science and Technology in Cosmetics and Personal Care", D. E. Goddard and J. V. Gruber (eds.), Marcel Dekker, NY (1999) Chapter 10.

[7] Goddard, D. E., "The Adsorptivity of Charged and Uncharged Cellulose Ethers" in "Principles of Polymer Science and Technology in Cosmetics and Personal Care", D. E. Goddard and J. V. Gruber (eds.), Marcel Dekker, NY (1999) Chapter 11.

12 Formulation of sunscreens for UV protection

12.1 Introduction

The increase in skin cancers has heightened public awareness to the damaging effects of the sun and many skin preparations are now available to help protect the skin from UV radiation [1, 2]. The actives employed in these preparations are of two basic types: organics, which can absorb UV radiation of specific wavelengths due to their chemical structure, and inorganics, which both absorb and scatter UV radiation. Inorganics have several benefits over organics in that they are capable of absorbing over a broad spectrum of wavelengths and they are mild and non-irritant. Both of these advantages are becoming increasingly important as the demand for *daily* UV protection against both UVB (wavelength 290–320 m) and UVA (wavelength 320–400 nm) radiation increases. Since UVB is much more effective than UVA at causing biological damage, solar UVB contributes about 80 % towards a sunburn reaction, with solar UVA contributing the remaining 20 %. In people with white skin living in the tropics (30° N to 30° S), sun protection is necessary all year, whereas those living in temperate altitudes (40° to 60°), sun awareness is generally limited to the 6-month period encompassing the summer solstice. Several harmful effects can be quoted on prolonged exposure to UV radiation. For UVB the main effects are DNA damage, immunosuppression, sunburn and skin cancer. For UVA the main effects are generation of active oxygen species, photodermatoses, premature skin ageing, skin wrinkles and skin cancer.

The ability of fine particle inorganics to absorb radiation depends upon their refractive index. For inorganic semiconductors such as titanium dioxide and zinc oxide this is a complex number indicating their ability to absorb light. The band gap in these materials is such that UV light up to around 405 nm can be absorbed. They can also scatter light due to their particulate nature and their high refractive indices make them particularly effective scatterers. Both scattering and absorption depend critically on particle size [1–3]. Particles of around 250 nm for example are very effective at scattering visible light and TiO_2 of this particle size is the most widely used white pigment. At smaller particle sizes absorption and scattering maxima shift to the UV region and at 30–50 nm UV attenuation is maximized.

The use of TiO_2 as a UV attenuator in cosmetics was, until recently, largely limited to baby sun protection products due to its poor aesthetic properties (viz; scattering of visible wavelengths results in whitening). Recent advances in particle size control and coatings have enabled formulators to use fine particle titanium dioxide and zinc oxide in daily skin care formulations without compromising the cosmetic elegance [3]

The benefits of a pre-dispersion of inorganic sunscreens are widely acknowledged. However it requires an understanding of the nature of colloidal stabilization in order to optimize this pre-dispersion (for both UV attenuation and stability) and exceed the performance of powder based formulations. Dispersion rheology and its dependence on interparticle interactions is a key factor in this optimization. Optimization of sunscreen actives however does not end there; an appreciation of the end application is crucial to maintaining performance. Formulators need to incorporate the particulate actives into an emulsion, mousse or gel with due regard to aesthetics (skin feel and transparency), stability and rheology.

In this chapter, I will demonstrate how the application of colloid and interface science principles give a sound basis on which to carry out true optimization of consumer acceptable sunscreen formulations based upon particulate TiO_2. I will show that both dispersion stability and dispersion rheology depend upon adsorbed amount Γ and steric layer thickness δ (which in turn depends on oligomer molecular weight M_n and solvency χ). In order to optimize formulation, the adsorption strength χ^s must also be considered. The nature of interaction between particles, dispersant, emulsifiers and thickeners must be considered with regard to competitive adsorption and/or interfacial stability if a formulation is to deliver its required protection when spread on the skin.

12.2 Mechanism of absorbance and scattering by TiO_2 and ZnO

As mentioned in the introduction, TiO_2 and ZnO absorb and scatter UV light. They provide a broad spectrum and they are inert and safe to use. Larger particles scatter visible light and they cause whitening. The scattering and absorption depend on the refractive index (which depends on the chemical nature), the wavelength of light and the particle size and shape distribution. The total attenuation is maximized in UVB for 30–50 nm particles. A schematic representation of the scattering of light is given in Fig. 12.1 whereas Fig. 12.2 shows the effect of particle size on UVA and UVB absorption.

The performance of any sunscreen formulation is defined by a number referred to as the sun protection factor (SPF). The basic principle of calculation of the SPF [4] is based on the fact that the inverse of the UV transmission through an absorbing layer, $1/T$, is the factor by which the intensity of the UV light is reduced. Thus, at a certain wavelength λ, $1/T(\lambda)$ is regarded as a monochromatic protection factor (MPF). Since the spectral range relevant for the in vivo SPF is between 290 and 400 nm (see Fig. 12.2), the monochromatic protection factors have to be averaged over this range. This average must be weighted using the intensity of a standard sun, $S_s(\lambda)$ and the erythemal action spectrum, $S_{er}(\lambda)$, leading to the following definition of SPF [4],

$$\text{SPF} = \frac{\sum_{290}^{400} S_{er}(\lambda) S_s(\lambda)}{\sum_{290}^{400} S_{er}(\lambda) S_s(\lambda) T(\lambda)}. \tag{12.1}$$

Data for $S_s(\lambda)$ and $S_{er}(\lambda)$ are available in the literature; the product of $S_s(\lambda) S_{er}(\lambda)$ is called the erythemal efficiency. $T(\lambda)$ has to be determined for the respective sunscreen; this can be done either via transmission measurements with special UV spectrometers using substrates and a rough surface or via the calculation of the transmission.

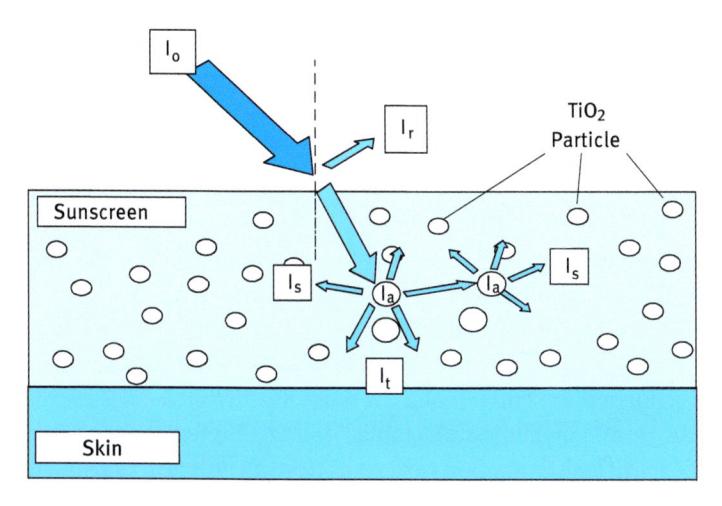

Fig. 12.1: Schematic representation of scattering of light by TiO$_2$ particles.

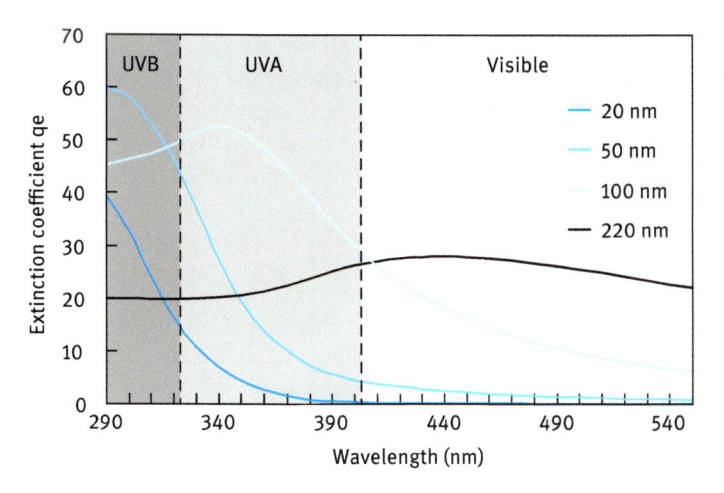

Fig. 12.2: Effect of particle size on UVA and UVB absorption.

12.3 Preparation of well-dispersed particles

To keep the particles well dispersed (as single particles) high steric repulsion is required to overcome strong van der Waals attraction. The mechanism of steric stabilization has been described in detail in Chapter 4 and only a summary is given here.

Small particles tend to aggregate as a result of the universal van der Waals attraction unless this attraction is screened by an effective repulsion between the particles. The van der Waals attraction energy $G_A(h)$ at close approach depends upon the distance, h, between particles of radius, R, and is characterized by the effective Hamaker constant, A,

$$G_A(h) = -\frac{AR}{12h}. \tag{12.2}$$

The effective Hamaker constant A is given by the following equation,

$$A = \left(A_{11}^{1/2} - A_{22}^{1/2}\right)^2. \tag{12.3}$$

A_{11} is the Hamaker constant of the particles and A_{22} is that for the medium. For TiO_2 A_{11} is exceptionally high so that in nonaqueous media with relatively low A_{22} the effective Hamaker constant A is high and despite the small size of the particles a dispersant is always needed to achieve colloidal stabilization. This is usually obtained using adsorbed layers of polymers or surfactants. The most effective molecules are the A–B, A–B–A block or BA_n graft polymeric surfactants [5] where B refers to the anchor chain. For a hydrophilic particle this may be a carboxylic acid, an amine or phosphate group or other larger hydrogen bonding type block such as polyethylene oxide. The A chains are referred to as the stabilizing chains which should be highly soluble in the medium and strongly solvated by its molecules. For nonaqueous dispersions the A chains could be polypropylene oxide, a long chain alkane, oil soluble polyester or polyhydroxystearic acid (PHS). A schematic representation of the adsorbed layers and the resultant interaction energy-distance curve is shown in Fig. 12.3.

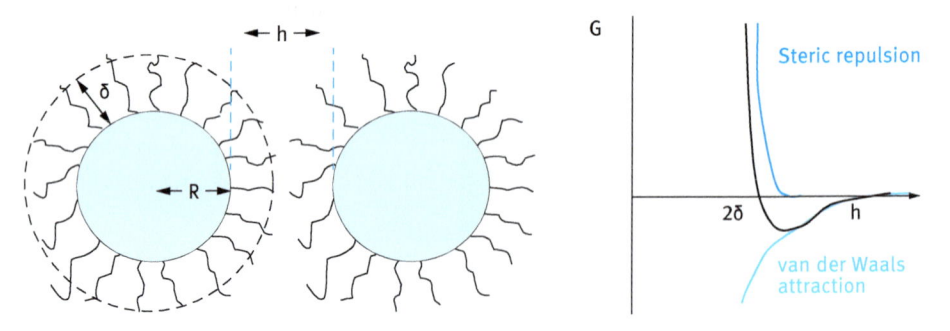

Fig. 12.3: Schematic representation of adsorbed polymer layers and resultant interaction energy G on close approach at distance h < 2R.

When two particles with an adsorbed layer of hydrodynamic thickness δ approach to a separation distance h that is smaller than 2δ, repulsion occurs as a result of two main effects: (i) unfavourable mixing of the A chains when these are in good solvent condition; (ii) reduction in configurational entropy on significant overlap.

Napper [6] derived a form for the so-called steric potential G(h) which arises as polymer layers begin to overlap (equation (12.4)),

$$G(h) = 2\pi k T R^2 \Gamma^2 N_A \left(\frac{v_p^2}{V_s}\right)\left(\frac{1}{2} - \chi\right)\left(1 - \frac{h}{2\delta}\right)^2 + G_{elastic}, \tag{12.4}$$

where k is the Boltzmann constant, T is temperature, R is the particle radius, Γ is the adsorbed amount, N_A the Avogadro constant, v is the specific partial volume of the polymer, V_s the molar volume of the solvent, χ is the Flory–Huggins parameter and δ is the maximum extent of the adsorbed layer.

It is useful to consider the terms in equation (12.4): (i) The adsorbed amount Γ; the higher the value the greater the interaction/repulsion. (ii) Solvent conditions as determined by the value of χ; two very distinct cases emerge. Maximum interaction occurs on overlap of the stabilizing layers when the chains are in good solvent conditions, i.e. $\chi < 0.5$. Osmotic forces cause solvent to move into the highly concentrated overlap zone forcing the particles apart. If $\chi = 0.5$, a theta solvent, the steric potential goes to zero and for poor solvent conditions ($\chi > 0.5$) the steric potential becomes negative and the chains will attract, enhancing flocculation. (iii) Adsorbed layer thickness δ. The steric interaction starts at $h = 2\delta$ as the chains begin to overlap and increases as the square of the distance. Here it is not the size of the steric potential that is important but the distance h at which it begins. (iv) The final interaction potential is the superposition of the steric potential and the van der Waals attraction as shown in Fig. 12.1.

The adsorbed layer thickness depends critically on the solvation of the polymer chain and it is therefore important to gain at least a qualitative view as to the relative solubilities of a polymer in different oils employed in dispersion. In this study solubility parameters were employed to provide that comparison. Generally the affinity between two materials is considered to be high when the chemical and physical properties of the two materials resemble each other. For example, nonpolar materials can be easily dispersed in nonpolar solvents but hardly dissolved in polar solvents and vice versa.

One of the most useful concepts for assessing solvation of any polymer by the medium is to use the Hildebrand's solubility parameter δ^2 which is related to the heat of vaporization ΔH by the following equation [7],

$$\delta^2 = \frac{\Delta H - RT}{V_M}, \qquad (12.5)$$

where V_M is the molar volume of the solvent.

Hansen [8] first divided Hildebrand's solubility parameter into three terms as follows:

$$\delta^2 = \delta_d^2 + \delta_p^2 + \delta_h^2, \qquad (12.6)$$

where δ_d, δ_p and δ_h correspond to London dispersion effects, polar effects and hydrogen bonding effects, respectively.

Hansen and Beerbower [9] developed this approach further and proposed a stepwise approach such that theoretical solubility parameters can be calculated for any solvent or polymer based upon its component groups. In this way we can arrive at theoretical solubility parameters for dispersants and oils. In principle, solvents with a similar solubility parameter to the polymer should also be a good solvent for it (low χ).

For sterically stabilized dispersions, the resulting energy–distance curve (Fig. 12.3) often shows a shallow minimum G_{min} at particle-particle separation distance h comparable to twice the adsorbed layer thickness δ. The depth of this minimum depends upon the particle size R, Hamaker constant A and adsorbed layer thickness δ. At constant R and A, G_{min} decreases with increasing δ/R [10]. This is illustrated in Fig. 12.4.

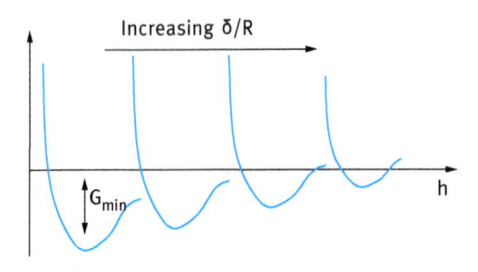

Fig. 12.4: Schematic representation of energy–distance curves at increasing δ/R ratios.

When δ becomes smaller than 5 nm, G_{min} may become deep enough to cause weak flocculation. This is particularly the case with concentrated dispersions since the entropy loss on flocculation becomes very small and a small G_{min} would be sufficient to cause weak flocculation ($\Delta G_{flocc} < 0$). This can be explained by considering the free energy of flocculation [10],

$$\Delta G_{flocc} = \Delta H_{flocc} - T\Delta S_{flocc}. \tag{12.7}$$

Since for concentrated dispersions ΔS_{flocc} is very small, then ΔG_{flocc} depends only on the value of ΔH_{flocc}. This in turn depends on G_{min}, which is negative. In other words, ΔG_{flocc} becomes negative causing weak flocculation. This will result in a three-dimensional coherent structure with a measurable yields stress [11]. This weak gel can be easily redispersed by gentle shaking or mixing. However, the gel will prevent any separation of the dispersion on storage. So, we can see that the interaction energies also determine the dispersion rheology.

At high solids content and for dispersions with larger δ/R, viscosity is also increased by steric repulsion. With a dispersion consisting of very small particles, as is the case with UV attenuating TiO_2, significant rheological effects can be observed even at moderate volume fraction of the dispersion. This is due to the much higher effective volume fraction of the dispersion compared with the core volume fraction due to the adsorbed layer.

Let us for example consider a 50 w/w% TiO_2 dispersion with a particle radius of 20 nm with a 3000 molecular weight stabilizer giving an adsorbed layer thickness of ≈ 10 nm. The effective volume fraction is given by [2]:

$$\phi_{eff} = \phi \left[1 + \frac{\delta}{R}\right]^3$$
$$= \phi [1 + 10/20]^3 \quad (12.8)$$
$$\approx 3\phi.$$

The effective volume fraction can be three times that of the core particle volume fraction. For a 50 % solid (w/w) TiO$_2$ dispersion the core volume fraction $\phi \approx 0.25$ (taking an average density of 3 g cm^{-3} for the TiO$_2$ particles) which means that ϕ_{eff} is about 0.75 which is sufficient to fill the whole dispersion space producing a highly viscous material. It is important therefore to choose the minimum δ for stabilization.

In the case of steric stabilization as employed in these oil dispersions the important success criteria for well stabilized but handleable dispersions are [2]: (i) complete coverage of the surface – high Γ adsorbed amount); (ii) strong adsorption (or "anchoring") of the chains to the surface; (iii) effective stabilizing chain, chain well solvated, $\chi < 0.5$ and adequate (but not too large) steric barrier δ. However, a colloidally stable dispersion does not guarantee a stable and optimized final formulation. TiO$_2$ particles are always surface modified in a variety of ways in order to improve dispersability and compatibility with other ingredients. It is important that we understand the impact these surface treatments may have upon the dispersion and more importantly upon the final formulation. As will be disused below, TiO$_2$ is actually formulated into a suspoemulsion, i.e. a suspension in an emulsion. Many additional ingredients are added to ensure cosmetic elegance and function. The emulsifiers used are structurally and functionally not very different to the dispersants used to optimize the fine particle inorganics. Competitive adsorption may occur with some partial desorption of a stabilizer from one or other of the available interfaces. Thus one requires strong adsorption (which should be irreversible) of the polymer to the particle surface.

12.4 Experimental results for sterically stabilized TiO$_2$ dispersions in nonaqueous media

Dispersions of surface modified TiO$_2$ (Tab. 12.1) in alkyl benzoate and hexamethyltetracosane (squalane) were prepared at various solids loadings using a polymeric/oligomeric polyhydroxystearic acid (PHS) surfactant of molecular weight 2500 (PHS2500) and 1000 (PHS1000) [2]. For comparison, results were also obtained using a low molecular weight (monomeric) dispersant, namely isostearic acid, ISA. The titania particles had been coated with alumina and/or silica. The electron micrograph in Fig. 12.5 shows the typical size and shape of these rutile particles. The surface area and particle size of the three powders used are summarized in Tab. 12.1.

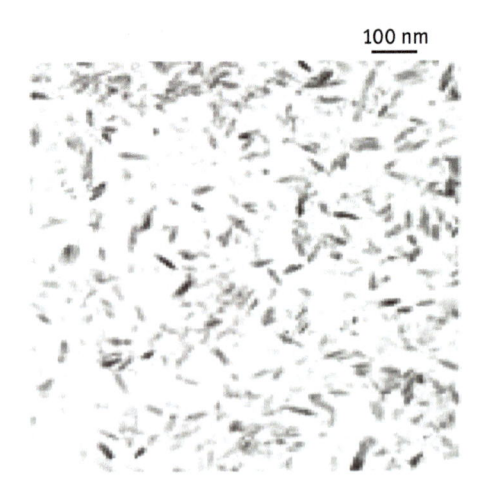

Fig. 12.5: Transmission electron micrograph of titanium dioxide particles.

Tab. 12.1: Surface modified TiO$_2$ powders.

Powder	Coating	Surface Area* / m^2 g^{-1}	Particle size** / nm
A	Alumina/silica	95	40–60
B	Alumina/stearic acid	70	30–40
C	Silica/stearic acid	65	30–40

* BET N2, ** equivalent sphere diameter, X-ray disc centrifuge

The dispersions of the surface modified TiO$_2$ powder, dried at 110 °C, were prepared by milling (using a horizontal bead mill) in polymer solutions of different concentrations for 15 minutes and were then allowed to equilibrate for more than 16 hours at room temperature before making the measurements.

The adsorption isotherms were obtained by preparing dispersions of 30 w/w% TiO$_2$ at different polymer concentration (C_0, mg l^{-1}). The particles and adsorbed dispersant were removed by centrifugation at 20 000 rpm ($\approx$ 48 000 g) for 4 hours, leaving a clear supernatant. The concentration of the polymer in the supernatant was determined by acid value titration. The adsorption isotherms were calculated by mass balance to determine the amount of polymer adsorbed at the particle surface (Γ, mg m^{-2}) of a known mass of particulate material (mg) relative to that equilibrated in solution (C_e, mg l^{-1}).

$$\Gamma = \frac{(C_0 - C_e)}{mA_s}. \tag{12.9}$$

The surface area of the particles (A_s, m^2 g^{-1}) was determined by BET nitrogen adsorption method. Dispersions of various solids loadings were obtained by milling at progressively increasing TiO$_2$ concentration at an optimum dispersant/solids ratio [2]. The dispersion stability was evaluated by viscosity measurement and by attenuation of UV/vis radiation. The viscosity of the dispersions was measured by subjecting the dispersions to an increasing shear stress, from 0.03 Pa to 200 Pa over 3 minutes at 250 °C using a Bohlin CVO rheometer. It was found that the dispersions exhibited shear thinning behaviour and the zero shear viscosity, identified from the plateau region at low shear stress (where viscosity was apparently independent of the applied shear stress), was used to provide an indication of the equilibrium energy of interaction that had developed between the particles.

UV-vis attenuation was determined by measuring transmittance of radiation between 250 nm and 550 nm. Samples were prepared by dilution with a 1 w/v% solution of dispersant in cyclohexane to approximately 20 mg l^{-1} and placed in a 1 cm pathlength cuvette in a UV-vis spectrophotometer. The sample solution extinction ε, (l g^{-1} cm^{-1}) was calculated from Beer's Law (equation (12.10)):

$$\varepsilon = \frac{A}{cl}, \tag{12.10}$$

where A is absorbance, c is concentration of attenuating species (g l^{-1}), l is the pathlength (cm).

The dispersions of powders B and C were finally incorporated into typical water-in-oil sunscreen formulations at 5% solids with an additional 2% of organic active (butyl methoxy dibenzoyl methane) and assessed for efficacy, SPF (sun protection factor) as well as stability (visual observation, viscosity). SPF measurements were made on an Optometrics SPF-290 analyser fitted with an integrating sphere, using the method of Diffey and Robson [12].

Fig. 12.6 shows the adsorption isotherms of ISA, PHS1000 and PHS2500 on TiO$_2$ in alkylbenzoate (Fig. 12.6 (a)) and in squalane (Fig. 12.6 (b)).

The adsorption of the low molecular weight ISA from alkylbenzoate is of low affinity (Langmuir type) indicating reversible adsorption (possibly physisorption). In contrast, the adsorption isotherms for PHS100 and PHS2500 are of the high affinity type indicating irreversible adsorption and possible chemisorption due to acid-base inter-

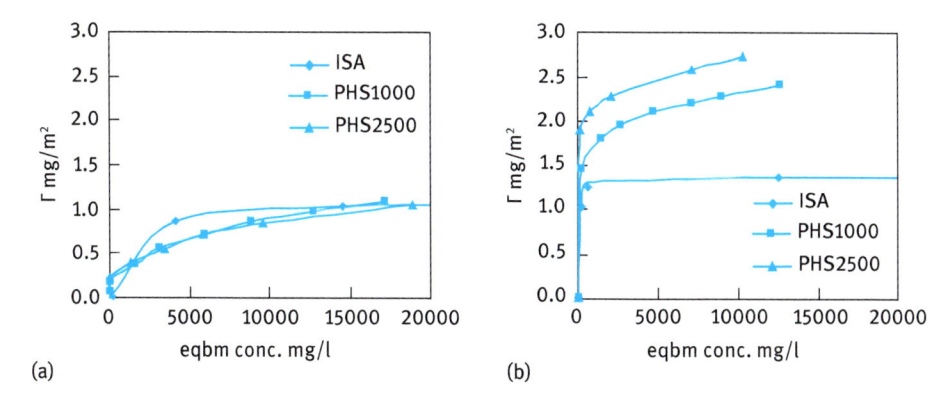

Fig. 12.6: Adsorption isotherms in (a) alkylbenzoate and (b) squalane.

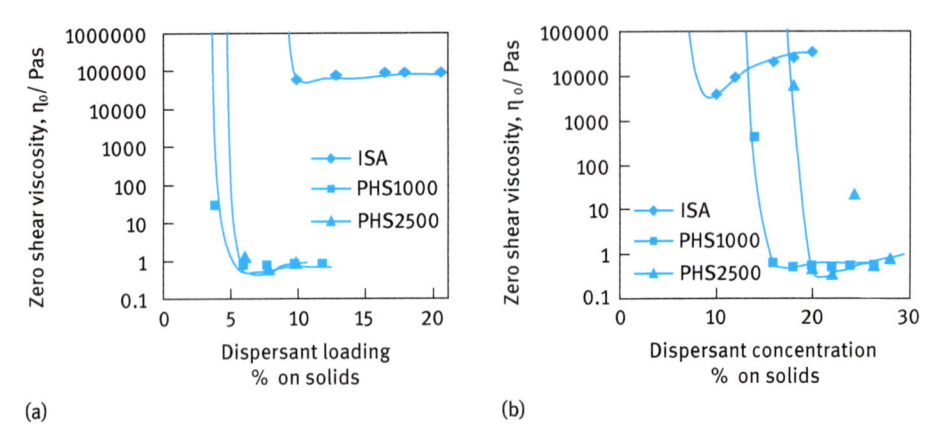

Fig. 12.7: Dispersant demand curve in alkylbenzoate (left) and squalane (right).

action. From squalane, all adsorption isotherms show high affinity type and they show higher adsorption values when compared with the results using alkylbenzoate. This reflects the difference in solvency of the dispersant by the medium as will be discussed below.

Fig. 12.7 shows the variation of zero shear viscosity with dispersant loading % on solid for a 40 % dispersion. It can be seen that the zero shear viscosity decreases very rapidly with increasing dispersant loading and eventually the viscosity reaches a minimum at an optimum loading that depends on the solvent used as well as the nature of the dispersant.

With the molecular dispersant ISA, the minimum viscosity that could be reached at high dispersant loading was very high (several orders of magnitude more than the optimized dispersions) indicating poor dispersion of the powder in both solvents. Even reducing the solids content of TiO_2 to 30 % did not result in a low viscosity dispersion. With PHS1000 and PHS2500, a low minimum viscosity could be reached at 8–10 % dispersant loading in alkylbenzoate and 18–20 % dispersant loading in squalane. In the latter case the dispersant loading required for reaching a viscosity minimum is higher for the higher molecular weight PHS.

The quality of the dispersion was assessed using UV-vis attenuation measurements. At very low dispersant concentration a high solids dispersion can be achieved by simple mixing but the particles are aggregated as demonstrated by the UV-vis curves (Fig. 12.8 (a)). These large aggregates are not effective as UV attenuators. As the PHS dispersant level is increased, UV attenuation is improved and above 8 wt|% dispersant on particulate mass, optimized attenuation properties (high UV, low visible attenuation) are achieved (for the PHS1000 in alkylbenzoate). However, milling is also required to break down the aggregates into their constituent nanoparticles and a simple mixture which is unmilled has poor UV attenuation even at 14 % dispersant loading.

The UV-vis curves obtained when monomeric isostearic acid was incorporated as a dispersant (Fig. 12.8 (b)) indicate that these molecules do not provide a sufficient barrier to aggregation, resulting in relatively poor attenuation properties (low UV, high visible attenuation).

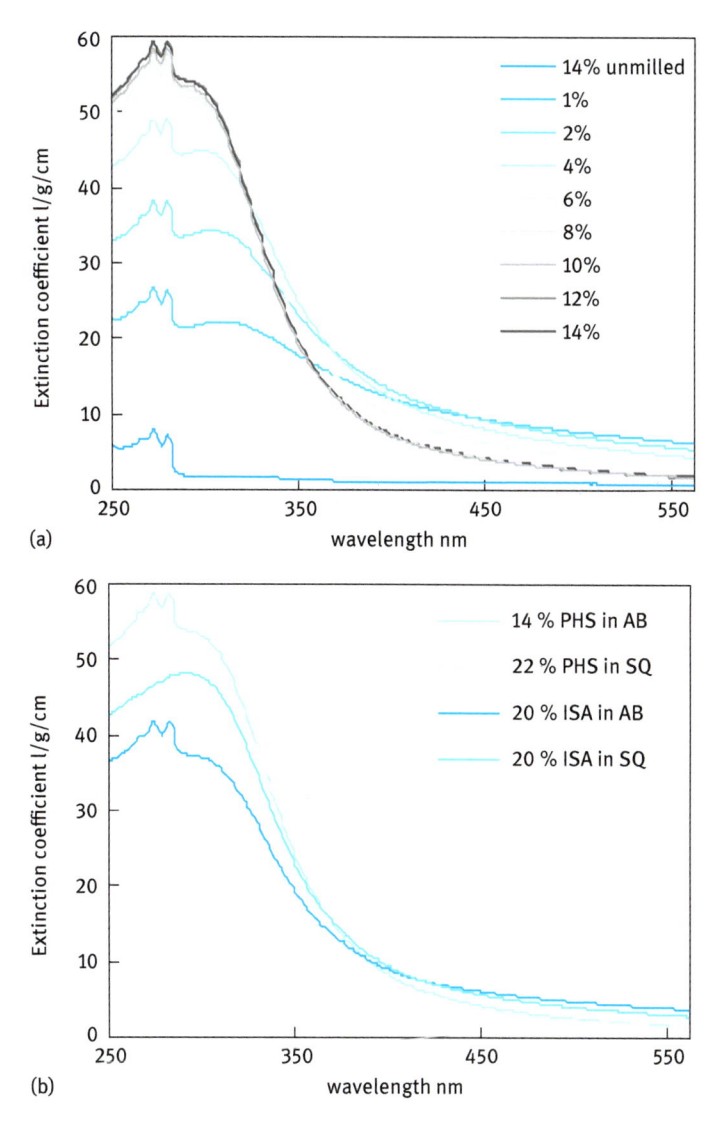

Fig. 12.8: (a) UV-vis attenuation for milled dispersions with 1–14 % PHS1000 dispersant and unmilled at 14 % dispersant on solids. (b) UV-vis attenuation for dispersions in squalane (SQ) and in alkylbenzoate (AB) using 20 % isostearic acid (ISA) as dispersant compared to optimized PHS1000 dispersions in the same oils.

The steric layer thickness δ could be varied by altering the dispersion medium and hence the solvency of the polymer chain. This had a significant effect upon dispersion rheology. Solids loading curves (Figs. 12.9 (a) and (b)) demonstrate the differences in effective volume fraction due to the adsorbed layer (equation (12.8)).

In the poorer solvent case (squalane) the effective volume fraction and adsorbed layer thickness showed a strong dependence upon molecular weight with solids loading becoming severely limited above 35 % for the higher molecular weight whereas ≈ 50 % could be reached for the lower molecular weight polymer. In alkylbenzoate no strong dependence was seen with both systems achieving more than 45 % solids. Solids weight fraction above 50 % resulted in very high viscosity dispersions in both solvents.

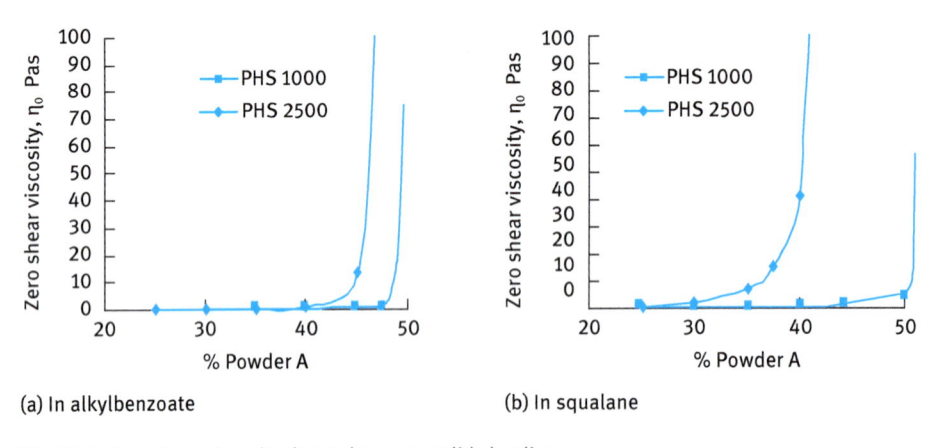

(a) In alkylbenzoate

(b) In squalane

Fig. 12.9: Zero shear viscosity dependence on solids loading.

The same procedure described above enabled optimized dispersion of equivalent particles with alumina and silica inorganic coatings (powders B and C). Both particles additionally had the same level of organic (stearate) modification. These optimized dispersions were incorporated into water-in-oil formulations and their stability/efficacy monitored by visual observation and SPF measurements (Tab. 12.2).

Tab. 12.2: Sunscreen emulsion formulations from dispersions of powders B and C.

Emulsion	Visual observation	SPF	Emulsifier level
Powder B emulsion 1	Good homogenous emulsion	29	2.0 %
Powder C emulsion 1	Separation, inhomogeneous	11	2.0 %
Powder C emulsion 2	Good homogeneous emulsion	24	3.5 %

12.4 Steric stabilization — 243

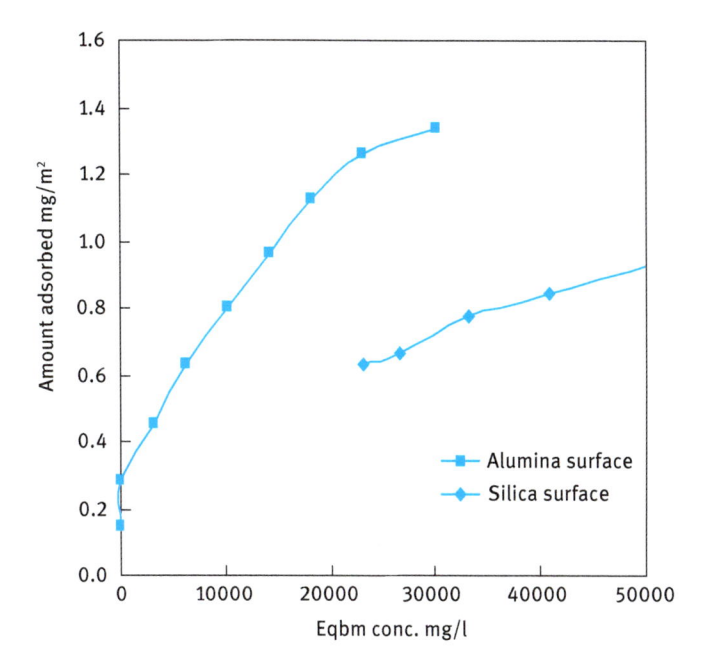

Fig. 12.10: Adsorption isotherms for PHS2500 on powder B (alumina surface) and powder C (silica surface).

The formulation was destabilized by the addition of the powder C dispersion and poor efficacy was achieved despite an optimized dispersion before formulation. When emulsifier concentration was increased from 2 to 3.5 % (emulsion 2) the formulation became stable and efficacy was restored.

The anchor of the chain to the surface (described qualitatively through χ^s) is very specific and this could be illustrated by silica-coated particles which showed lower adsorption of the PHS (Fig. 12.10).

In addition, when a quantity of emulsifier was added to an optimized dispersion of powder C (silica surface) the acid value of the equilibrium solution was seen to rise indicating some displacement of the PHS2500 by the emulsifier.

The dispersant demand curves (Fig. 12.7 (a) and (b)) and solids loading curves (Fig. 12.9 (a) and (b)) show that one can reach a stable dispersion using PHS1000 or PHS2500 both in alkylbenzoate and in squalane. These can be understood in terms of the stabilization produced when using these polymeric dispersants. Addition of sufficient dispersant enables coverage of the surface and results in a steric barrier (Fig. 12.3) preventing aggregation due to van der Waals attraction. Both molecular weight oligomers were able to achieve stable dispersions. The much smaller molecular weight "monomer", isostearic acid is however insufficient to provide this steric barrier and dispersions were aggregated, leading to high viscosities, even at 30 % solids. UV-vis curves confirm that these dispersions are not fully dispersed since their full

UV potential is not realized (Fig. 12.8). Even at 20 % isostearic acid the dispersions are seen to give a lower E_{max} and increased scattering at visible wavelengths indicating a partially aggregated system.

The differences between alkylbenzoate and squalane observed in the optimum dispersant concentration required for maximum stability can be understood by examining the adsorption isotherms in Fig. 12.6 (a) and (b). The nature of the steric barrier depends on the solvency of the medium for the chain, and is characterized by the Flory–Huggins interaction parameter χ. Information on the value of χ for the two solvents can be obtained from solubility parameter calculations (equation (12.3)). The results of these calculations are given in Tab. 12.3 for PHS, alkylbenzoate and squalane.

Tab. 12.3: Hansen and Beerbower solubility parameters for the polymer and both solvents.

	δ_T	δ_d	δ_p	δ_h	$\Delta\delta_T$
PHS	19.00	18.13	0.86	5.60	
alkyl benzoate	17.01	19.13	1.73	4.12	1.99
squalane	12.9	15.88	0	0	6.1

It can be seen that both PHS and alkylbenzoate have polar and hydrogen bonding contributions to the solubility parameter δ_T. In contrast, squalane which is nonpolar has only a dispersion component to δ_T. The difference in the total solubility parameter $\Delta\delta_T$ value is much smaller for alkylbenzoate when compared with squalane. Thus one can expect that alkylbenzoate is a better solvent for PHS when compared with squalane. This explains the higher adsorption amounts of the dispersants in squalane when compared with alkyl benzoate (Fig. 12.6). The PHS finds adsorption at the particle surface energetically more favourable than remaining in solution. The adsorption values at the plateau for PHS in squalane (> 2 mg m^{-2} for PHS1000 and > 2.5 mg m^{-2} for PHS2500) is more than twice the value obtained in alkylbenzoate (1 mg m^{-2} for both PHS1000 and PHS2500).

It should be mentioned, however, that both alkylbenzoate and squalane will have χ values less than 0.5, i.e., good solvent conditions and a positive steric potential. This is consistent with the high dispersion stability produced in both solvents. However, the relative difference in solvency for PHS between alkylbenzoate and squalane is expected to have a significant effect on the conformation of the adsorbed layer. In squalane, a poorer solvent for PHS, the polymer chain is denser when compared with the polymer layer in alkylbenzoate. In the latter case a diffuse layer that is typical for polymers in good solvents is produced. This is illustrated in Fig. 12.11 (a) which shows a higher hydrodynamic layer thickness for the higher molecular weight PHS2500. A schematic representation of the adsorbed layers in squalane is shown in Fig. 12.11 (b) which also shows a higher thickness for the higher molecular weight PHS2500.

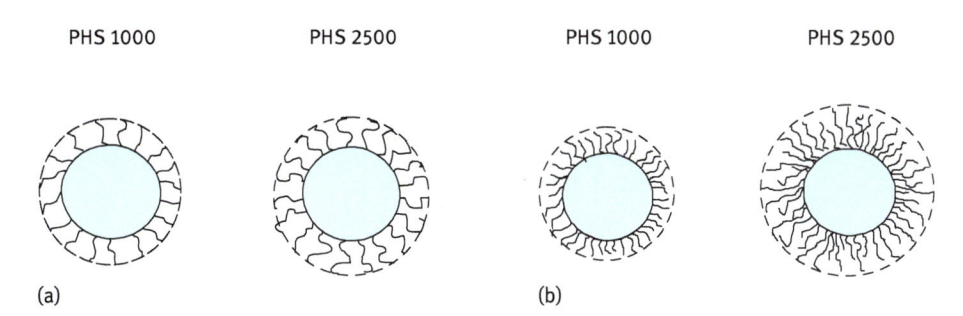

Fig. 12.11: (a) Well solvated polymer results in diffuse adsorbed layers (alkylbenzoate). (b) Polymers are not well solvated and form dense adsorbed layers (squalane).

In squalane the dispersant adopts a close packed conformation with little solvation and high amounts are required to reach full surface coverage ($\Gamma > 2$ mg m^{-2}). It seems also that in squalane the amount of adsorption depends much more on the molecular weight of PHS than in the case of alkylbenzoate. It is likely that with the high molecular weight PHS2500 in squalane the adsorbed layer thickness can reach higher values when compared with the results in alkylbenzoate. This larger layer thickness increases the effective volume fraction and this restricts the total solids that can be dispersed. This is clearly shown from the results of Fig. 12.9 which shows a rapid increase in zero shear viscosity at a solids loading > 35 %. With the lower molecular weight PHS1000, with smaller adsorbed layer thickness, the effective volume fraction is lower and high solids loading ($\approx$ 50 %) can be reached. The solids loading that can be reached in alkylbenzoate when using PHS2500 is higher ($\approx$ 40 %) than that obtained in squalane. This implies that the adsorbed layer thickness of PHS2500 is smaller in alkylbenzene when compared with the value in squalane as schematically shown in Fig. 12.11. The solids loading with PHS1000 in alkylbenzene is similar to that in squalane, indicating a similar adsorbed layer thickness in both cases.

The solids loading curves demonstrate that with an extended layer such as that obtained with the higher molecular weight (PHS2500) the maximum solids loading becomes severely limited as the effective volume fraction (equation (12.5)) is increased.

In squalane the monomeric dispersant, isostearic acid shows high affinity adsorption isotherm with a plateau adsorption of 1 mg m^{-2} but this provides an insufficient steric barrier (δ/R too small, Fig. 12.4) to ensure colloidal stability.

12.5 Competitive interactions in sunscreen formulations

Most sunscreen formulations consist of an oil-in-water (O/W) emulsion in which the particles are incorporated. These active particles can be in either the oil phase, or the water phase, or both as is illustrated in Fig. 12.12. For a sunscreen formulation based

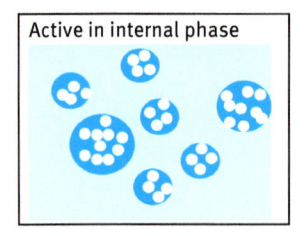

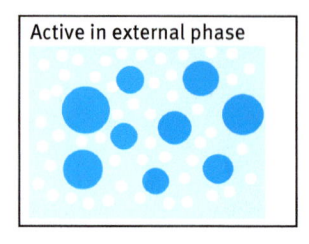

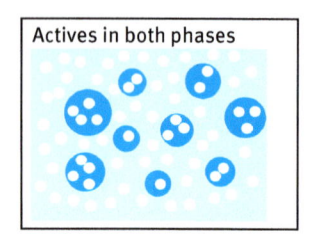

Fig. 12.12: Schematic representation of the location of active particles in sunscreen formulations.

on a W/O emulsion, the added nonaqueous sunscreen dispersion mostly stays in the oil continuous phase.

On addition of the sunscreen dispersion to an emulsion to produce the final formulation, one has to consider the competitive adsorption of the dispersant/emulsifier system. In this case the strength of adsorption of the dispersant to the surface modified TiO$_2$ particles must be considered. As shown in Fig. 12.10 the silica coated particles (C) show lower PHS2500 adsorption compared to the alumina coated particles (B). However, the dispersant demand for the two powders to obtain a colloidally stable dispersion was similar in both cases (12–14 % PHS2500). This appears at first sight to indicate similar stabilities. However, when added to a water-in-oil emulsion prepared using an A–B–A block copolymer of PHS–PEO–PHS as emulsifier, the system based on the silica coated particles (C) became unstable showing separation and coalescence of the water droplets. The SPF performance also dropped drastically from 29 to 11. In contrast, the system based on alumina coated particles (B) remained stable showing no separation as illustrated in Tab. 12.2. These results are consistent with the stronger adsorption (higher χ^s) of PHS2500 on the alumina coated particles. With the silica coated particles, it is likely that the PHS–PEO–PHS block copolymer becomes adsorbed on the particles thus depleting the emulsion interface from the polymeric emulsifier and this is the cause of coalescence. It is well known that molecules based on PEO can adsorb on silica surfaces [13]. By addition of more emulsifier (increasing its concentration from 2 to 3.5 %) the formulation remained stable as is illustrated in Tab. 12.2. This final set of results demonstrates how a change in surface coating can alter the adsorption strength which can have consequences for the final formulation. The same optimization process used for powder A enabled stable dispersions to be formed from powders B and C. Dispersant demand curves showed optimized dispersion rheology at similar added dispersant levels of 12–14 % PHS2500. To the dispersion scientist these appeared to be stable TiO$_2$ dispersions. However, when the optimized dispersions were formulated into the external phase of a water-in-oil emulsion differences were observed and alterations in formulation were required to ensure emulsion stability and performance.

References

[1] Tadros, Th. F., "Cosmetics", in "Encyclopedia of Colloid and Interface Science", Th. F. Tadros (ed.), Springer, Germany (2013).
[2] Kessel, L. M., Naden, B. J., Tooley, I. R. and Tadros, Th. F., "Application of Colloid and Interface Science Principles for Optimisation of Sunscreen Dispersions", in "Colloids in Cosmetics and Personal Care", Th. F. Tadros (ed.), Wiley-VCH, Germany (2008).
[3] Robb, J. L., Simpson, L. A. and Tunstall, D. F., Scattering and absorption of UV radiation by sunscreens containing fine particle and pigmentary titanium dioxide, Drug. Cosmet. Ind., March, 32–39 (1994).
[4] Herzog, B., "Models for the Calculation of Sun Protection Factor and Parameters Characterising the UVA Protection Ability of Cosmetic Sunscreens", in "Colloids in Cosmetics and Personal Care", Th. F. Tadros (ed.), Wiley-VCH, Germany (2008).
[5] Fleer, G. J., Cohen-Stuart, M. A., Scheutjens, J. M. H. M., Cosgrove, T. and Vincent, B., "Polymers at Interfaces", Chapman and Hall, London (1993).
[6] Napper, D. H., "Polymeric Stabilization of Colloidal Dispersions", Academic Press, London (1983).
[7] Hildebrand, J. H., "Solubility of Non-Electrolytes", 2nd Ed., Reinhold, NY (1936).
[8] Hansen, C. M., J. Paint Technol., **39**, 104–117, 505 (1967).
[9] Hansen, C. M. and Beerbower, A., in "Handbook of Solubility Parameters and Other Cohesion Parameters", A. F. M. Barton (ed.), CRC Press, Boca Raton, Florida, (1983).
[10] Tadros, Th. F., Izquierdo, P., Esquena, J. and Solans, C., Formation and stability of nano-emulsions, Adv. Colloid Interface Sci., **108/109**, 303–318 (2004).
[11] Kessell, L. M., Naden, B. J. and Tadros, Th. F., "Attractive and Repulsive Gels From Inorganic Sunscreen Actives", Proceedings of the IFSCC 23rd Congress, October 2004.
[12] Diffey, B. L. and Robson, J., J. Soc. Cosmet. Chem **40**, 127–133 (1989).
[13] Shar, J. A., Obey, T. M. and Cosgrove, T., Colloids and Surfaces **A 150**, 15–23 (1999).

13 Formulation of colour cosmetics

13.1 Introduction

Pigments are the primary ingredient of any colour cosmetic and the way in which these particulate materials are distributed within the product will determine many aspects of product quality including functional activity (colour, opacity, UV protection) but also stability, rheology and skin feel [1, 2]. Several colour pigments are used in cosmetic formulations ranging from inorganic pigments (such as red iron oxide) to organic pigments of various types. The formulation of these pigments in colour cosmetics requires a great deal of skill since the pigment particles are dispersed in an emulsion (oil-in-water or water-in-oil). The pigment particles may be dispersed in the continuous medium in which case one should avoid flocculation with the oil or water droplets. In some cases, the pigment may be dispersed in an oil which is then emulsified in an aqueous medium. Several other ingredients are added such as humectants, thickeners, preservatives, etc. and the interaction between the various components can be very complex.

The particulate distribution depends on many factors such as particle size and shape, surface characteristics, processing and other formulation ingredients but ultimately is determined by the interparticle interactions. A thorough understanding of these interactions and how to modify them can help to speed up product design and solve formulation problems.

In this chapter, I will start with a section describing the fundamental principles of preparation of pigment dispersion. These consist of three main topics, namely wetting of the powder, its dispersion (or wet milling including comminution) and stabilization against aggregation. A schematic representation of this process is shown in Fig. 13.1 [3]. This will be followed by a section on the principles of dispersion stability for both aqueous and nonaqueous media. The use of rheology in assessing the performance of a dispersant will be included. The application of these fundamental principles for colour cosmetic formulation will be discussed. Finally, the interaction with other formulation ingredients when these particulates are incorporated in an emulsion (forming a suspoemulsion) will be discussed. Particular attention will be given to the process of competitive adsorption of the dispersant and emulsifier.

In this chapter I will try to demonstrate that optimization of colour cosmetics can be achieved through a fundamental understanding of colloid and interface science. I will show that the dispersion stability and rheology of particulate formulations depend on interparticle interactions which in turn depend on the adsorption and conformation of the dispersant at the solid/liquid interface. Dispersants offer the possibility of being able to control the interactions between particles such that consistency is improved. Unfortunately, it is not possible to design a universal dispersant due to specificity of anchor groups and solvent-steric interactions. Colour formulators should be encouraged to understand the mechanism of stabilizing pigment particles and how to improve it. In order to optimize performance of the final cosmetic colour formulation one must consider the interactions between particles, dispersant, emulsifiers and thickeners and strive to reduce the competitive interactions through proper choice of the modified surface as well as the dispersant to optimize adsorption strength.

Agglomerates (particles connected by their corners)

Aggregates (particles joined at their faces)

Liquid + Dispersing agent

Wet milling

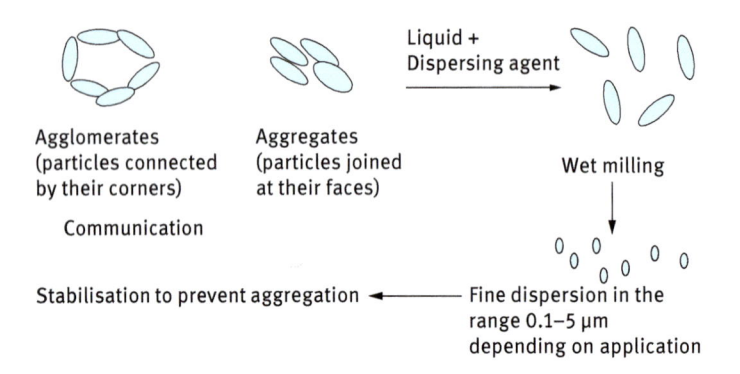

Communication

Stabilisation to prevent aggregation ← Fine dispersion in the range 0.1–5 µm depending on application

Fig. 13.1: Schematic representation of the dispersion process.

13.2 Fundamental principles for preparation of a stable colour cosmetic dispersion

13.2.1 Powder wetting

Wetting of powders of colour cosmetics is an important prerequisite for dispersion of that powder in liquids. It is essential to wet both the external and internal surfaces of the powder aggregates and agglomerates as schematically represented in Fig. 13.1. In all these processes one has to consider both the equilibrium and dynamic aspects of the wetting process [4]. The equilibrium aspects of wetting can be studied at a fundamental level using interfacial thermodynamics. Under equilibrium, a drop of a liquid on a substrate produces a contact angle θ, which is the angle formed between planes tangent to the surfaces of solid and liquid at the wetting perimeter. This is illustrated in Fig. 13.2 which shows the profile of a liquid drop on a flat solid substrate. An equilibrium between vapour, liquid and solid is established with a contact angle θ (that is smaller than 90°).

The wetting perimeter is frequently referred to as the three-phase line (solid/liquid/vapour); the most common name is the wetting line. Most equilibrium wetting studies centre around measurements of the contact angle. The smaller the angle, the better the liquid is said to wet the solid [4].

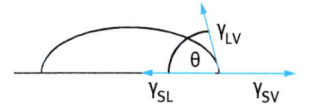

Fig. 13.2: Schematic representation of contact angle and wetting line.

The dynamic process of wetting is usually described in terms of a moving wetting line which results in contact angles that change with the wetting velocity. The same name is sometimes given to contact angles that change with time [4].

Wetting of a porous substrate may also be considered a dynamic phenomenon. The liquid penetrates through the pores and gives different contact angles depending on the complexity of the porous structure. Studying the wetting of porous substrates is very difficult. The same applies to wetting of agglomerates and aggregates of powders. However, even measurements of apparent contact angles can be very useful for comparing one porous substrate with another and one powder with another [3].

The liquid drop takes the shape that minimizes the free energy of the system. Consider a simple system of a liquid drop (L) on a solid surface (S) in equilibrium with the vapour of the liquid (V) as was illustrated in Fig. 13.2. The sum ($\gamma_{SV} A_{SV} + \gamma_{SL} A_{SL} + \gamma_{LV} A_{LV}$) should be a minimum at equilibrium and this leads to the Young's equation [4],

$$\gamma_{SV} = \gamma_{SL} + \gamma_{LV} \cos\theta, \qquad (13.1)$$

where θ is the equilibrium contact angle. The angle which a drop assumes on a solid surface is the result of the balance between the cohesion force in the liquid and the adhesion force between the liquid and solid, i.e.,

$$\gamma_{LV} \cos\theta = \gamma_{SV} - \gamma_{SL} \qquad (13.2)$$

or,

$$\cos\theta = \frac{\gamma_{SV} - \gamma_{SL}}{\gamma_{LV}}. \qquad (13.3)$$

There is no direct way by which γ_{SV} or γ_{SL} can be measured. The difference between γ_{SV} and γ_{SL} can be obtained from contact angle measurements. This difference is referred to as the "Wetting Tension" or "Adhesion Tension" [3, 4],

$$\text{Adhesion Tension} = \gamma_{SV} - \gamma_{SL} = \gamma_{LV} \cos\theta. \qquad (13.4)$$

Another useful parameter for describing wetting of liquids on solid substrates is the work of adhesion, W_a. Consider a liquid drop with surface tension γ_{LV} and a solid surface with surface tension γ_{SV}. When the liquid drop adheres to the solid surface it forms a surface tension γ_{SL}. This is schematically illustrated in Fig. 13.3. The work of adhesion is simply the difference between the surface tensions of the liquid/vapour and solid/vapour and that of the solid/liquid [3],

$$W_a = \gamma_{SV} + \gamma_{LV} - \gamma_{SL}. \qquad (13.5)$$

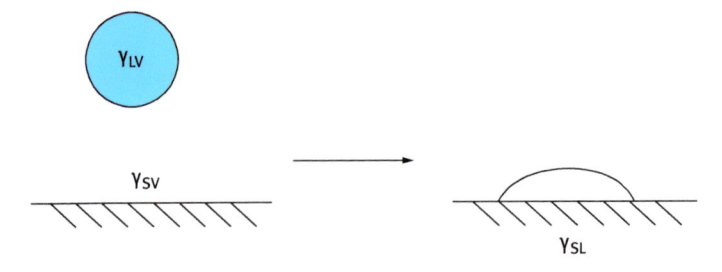

Fig. 13.3: Representation of adhesion of a drop on a solid substrate.

Using Young's equation,

$$W_a = \gamma_{LV}(\cos\theta + 1). \tag{13.6}$$

The work of cohesion W_c is the work of adhesion when the two phases are the same. Consider a liquid cylinder with unit cross-sectional area. When this liquid is subdivided into two cylinders, two new surfaces are formed. The two new areas will have a surface tension of $2\gamma_{LV}$ and the work of cohesion is simply,

$$W_c = 2\gamma_{LV}. \tag{13.7}$$

Thus, the work of cohesion is simply equal to twice the liquid surface tension. An important conclusion may be drawn if one considers the work of adhesion given by equation (13.6) and the work of cohesion given by equation (13.7): When $W_c = W_a$, $\theta = 0°$. This is the condition for complete wetting. When $W_c = 2W_a$, $\theta = 90°$ and the liquid forms a discrete drop on the substrate surface. Thus, the competition between the cohesion of the liquid to itself and its adhesion to a solid gives an angle of contact that is constant and specific to a given system at equilibrium [3]. This shows the importance of Young's equation in defining wetting.

The spreading of liquids on substrates is also an important industrial phenomenon. A useful concept introduced by Harkins [3, 4] is the spreading coefficient which is simply the work in destroying a unit area of solid/liquid and liquid/vapour interface to produce an area of solid/air interface. The spreading coefficient is simply determined from the contact angle θ and the liquid/vapour surface tension γ_{LV},

$$S = \gamma_{LV}(\cos\theta - 1). \tag{13.8}$$

For spontaneous spreading S has to be zero or positive. If S is negative only limited spreading is obtained.

The energy required to achieve dispersion wetting, W_d, is given by the product of the external area of the powder, A, and the difference between γ_{SL} and γ_{SV},

$$W_d = A(\gamma_{SL} - \gamma_{SV}). \tag{13.9}$$

13.2 Fundamental principles for preparation of a stable colour cosmetic dispersion — 253

Using Young's equation,

$$W_d = -A\gamma_{LV} \cos \theta. \tag{13.10}$$

Thus, wetting of the external surface of the powder depends on the liquid surface tension and contact angle [3]. If $\theta < 90°$, $\cos \theta$ is positive and the work of dispersion is negative, i.e. wetting is spontaneous.

For agglomerates (represented in Fig. 13.1) which are found in all powders, wetting of the internal surface between the particles in the structure requires liquid penetration through the pores. Assuming the pores to behave as simple capillaries of radius r, the capillary pressure Δp is given by the following equation [3],

$$\Delta p = \frac{2\gamma_{LV} \cos \theta}{r}. \tag{13.11}$$

For liquid penetration to occur, Δp must be positive and hence θ should be less than 90°. The maximum capillary pressure is obtained when $\theta = 0$ and Δp is proportional to γ_{LV} which means that a high γ_{LV} is required. Thus to achieve wetting of the internal surface a compromise is needed since the contact angle only decreases as γ_{LV} decreases. One needs to make θ as close to 0 as possible while not having a too low liquid surface tension [3].

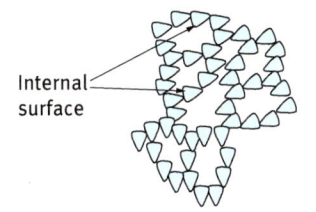

Fig. 13.4: Schematic representation of an agglomerate.

The most important parameter that determines wetting of the powder is the dynamic surface tension, $\gamma_{dynamic}$ (i.e. the value at short times). $\gamma_{dynamic}$ depends both on the diffusion coefficient of the surfactant molecule as well as its concentration [3]. Since wetting agents are added in sufficient amounts ($\gamma_{dynamic}$ is lowered sufficiently) spontaneous wetting is the rule rather than the exception.

Wetting of the internal surface requires penetration of the liquid into channels between and inside the agglomerates. The process is similar to forcing a liquid through fine capillaries. To force a liquid through a capillary with radius r, a pressure Δp is required that was given by equation (13.11).

To assess the wettability of the internal surface, one must consider the rate of penetration of the liquid through the pores of the agglomerates [3]. Assuming the pores to be represented by horizontal capillaries with radius r, neglecting the effect of gravity, the depth of penetration l in time t is given by the Rideal–Washburn equation,

$$l^2 = \left[\frac{r\gamma_{LV} \cos \theta}{2\eta} \right] t. \tag{13.12}$$

To enhance the rate of penetration, γ_{LV} has to be made as high as possible, θ as low as possible and η as low as possible. For a packed bed of particles, r may be replaced by r/k^2, where r is the effective radius of the bed and k is the tortuosity factor, which takes into account the complex path formed by the channels between the particles [3], i.e.,

$$l^2 = \left(\frac{r\gamma_{LV} \cos\theta}{2\eta k^2}\right) t. \qquad (13.13)$$

Thus a plot of l^2 versus t gives a straight line and from the slope of the line one can obtain θ.

The Rideal–Washburn equation can be applied to obtain the contact angle of liquids (and surfactant solutions) in powder beds [3]. k should first be obtained using a liquid that produces zero contact angle.

13.2.2 Powder dispersion and milling (comminution)

The dispersion of the powder is achieved by using high speed stirrers such as the Ultra-Turrax or Silverson mixers. This results in dispersion of the wetted powder aggregate or agglomerate into single units [3]. The primary dispersion (sometimes referred to as the mill base) may then be subjected to a bead milling process to produce nanoparticles which are essential for some colour cosmetic applications. Subdivision of the primary particles into much smaller units in the nano-size range (10–100 nm) requires application of intense energy. In some cases high pressure homogenizers (such as the Microfluidizer, USA) may be sufficient to produce nanoparticles. This is particularly the case with many organic pigments. In some cases, the high pressure homogenizer is combined with application of ultrasound to produce the nanoparticles.

Milling or comminution (the generic term for size reduction) is a complex process and there is little fundamental information on its mechanism. For the breakdown of single crystals or particles into smaller units, mechanical energy is required. This energy in a bead mill is supplied by impaction of the glass or ceramic beads with the particles. As a result permanent deformation of the particles and crack initiation results. This will eventually lead to the fracture of particles into smaller units. Since the milling conditions are random, some particles receive impacts far in excess of those required for fracture whereas others receive impacts that are insufficient for the fracture process. This makes the milling operation grossly inefficient and only a small fraction of the applied energy is used in comminution. The rest of the energy is dissipated as heat, vibration, sound, inter-particulate friction, etc.

The role of surfactants and dispersants on the grinding efficiency is far from being understood. In most cases the choice of surfactants and dispersant is made by trial and error until a system is found that gives the maximum grinding efficiency [3]. Rehbinder and his collaborators investigated the role of surfactants in the grinding process [3]. As a result of surfactant adsorption at the solid/liquid interface, the surface energy at the boundary is reduced and this facilitates the process of deformation or destruction. The adsorption of surfactants at the solid/liquid interface in cracks facilitates their propagation. This mechanism is referred to as the Rehbinder effect.

13.2.3 Stabilization of the dispersion against aggregation

For stabilization of the dispersion against aggregation (flocculation) one needs to create a repulsive barrier that can overcome the van der Waals attraction [5, 6]. The process of stabilization of dispersions in cosmetics has been described in detail in Chapter 4 and only a summary is given in this chapter. As discussed in Chapter 4, all particles experience attractive forces on close approach. The strength of this van der Waals attraction $V_A(h)$ depends upon the distance h between particles of radius R and is characterized by the Hamaker constant, A. The Hamaker constant expresses the attraction between particles (in a vacuum). A depends upon the dielectric and physical properties of the material and for some materials such as TiO_2, iron oxides and alumina this is exceptionally high so that (in nonaqueous media at least) despite their small size a dispersant is always needed to achieve colloidal stabilization. In order to achieve stability one must provide a balancing repulsive force to reduce interparticle attraction. This can be done in two main ways by electrostatic or steric repulsion as illustrated in Fig. 13.5 (a) and (b) (or a combination of the two, Fig. 13.5 (c)). A polyelectrolyte dispersant such as sodium polyacrylate is required to achieve high solids content. This produces a more uniform charge on the surface and some steric repulsion due to the high molecular weight of the dispersant. Under these conditions the dispersion becomes stable over a wider range of pH at moderate electrolyte concentration. This is electrosteric stabilization (Fig. 13.2 (c) shows a shallow minimum at long separation distances, a maximum (of the DLVO type) at intermediate h and a sharp increase in repulsion at shorter separation distances. This combination of electrostatic and steric repulsion can be very effective for stabilization of the suspension [3].

Electrostatic stabilization can be achieved if the particles contain ionizable groups on their surface such as inorganic oxides, which means that in aqueous media they can therefore develop a surface charge depending upon pH, which affords an electrostatic stabilization to the dispersion. On close approach the particles experience a repulsive potential overcoming the van der Waals attraction which prevents aggregation [3]. This stabilization is due to the interaction between the electric double layers surrounding the particles as illustrated in Fig. 13.6.

Double layer repulsion depends upon the pH and electrolyte concentration and can be predicted from zeta potential measurements (Fig. 13.6). Surface charge can also be produced by adsorption of ionic surfactants. This balance of electrostatic repulsion with van der Waals attraction is described in the well-known theory of colloid stability by Deryaguin–Landau–Verwey–Overbeek (DLVO theory) [7, 8]. Fig. 13.4 (a) shows two attractive minima at long and short separation distances; V_{sec} that is shallow and of

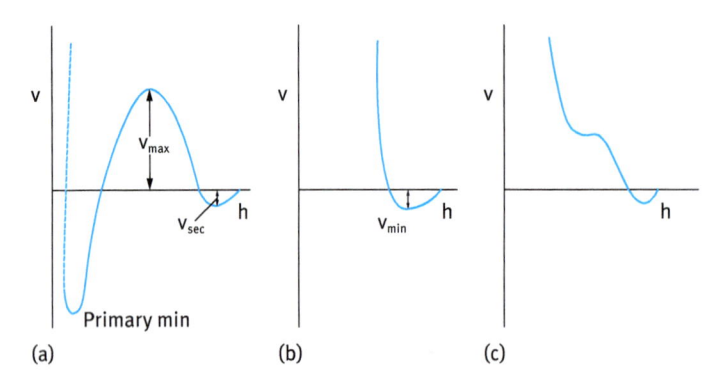

Fig. 13.5: Energy–distance curves for three stabilization mechanisms: (a) electrostatic; (b) steric; (c) electrosteric.

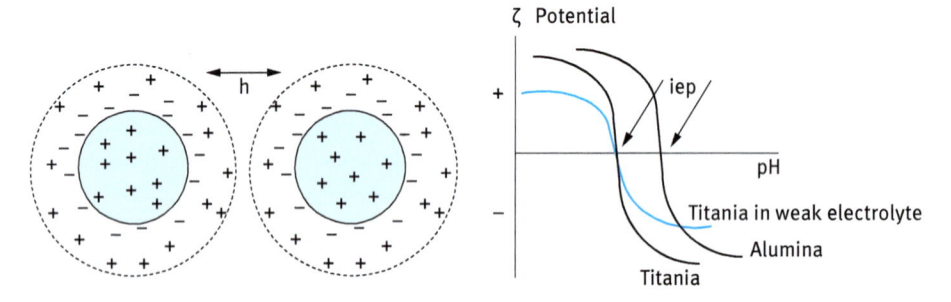

Fig. 13.6: Schematic representation of double layer repulsion (left) and variation of ζ potential with pH for titania and alumina (right).

few kT units and $V_{primary}$ that is deep and exceeds several 100 kT units. These two minima are separated by an energy maximum V_{max} that can be greater than 25 kT thus preventing flocculation of the particles into the deep primary minimum.

When the pH of the dispersion is well above or below the isoelectric point or the electrolyte concentration is less than 10^{-2} mol dm^{-3} 1 : 1 electrolyte, electrostatic repulsion is often sufficient to produce a dispersion without the need for added dispersant. However, in practice this condition often cannot be reached since at high solids content the ionic concentration from the counter- and co-ions of the double layer is high and the surface charge is not uniform. Therefore a polyelectrolyte dispersant such as sodium polyacrylate is required to achieve this high solids content. This produces a more uniform charge on the surface and some steric repulsion due to the high molecular weight of the dispersant. Under these conditions the dispersion becomes stable over a wider range of pH at moderate electrolyte concentration. This is electrosteric stabilization (Fig. 13.5 (c) shows a shallow minimum at long separation distances, a maximum (of the DLVO type) at intermediate h and a sharp increase in repulsion at shorter separation distances. This combination of electrostatic and steric repulsion can be very effective for stabilization of the suspension.

Steric stabilization is usually obtained using adsorbed layers of polymers or surfactants. The most effective molecules are the A–B or A–B–A block or BA_n graft polymeric surfactants [9] where B refers to the anchor chain. This anchor should be strongly adsorbed to the particle surface. For a hydrophilic particle this may be a carboxylic acid, an amine or phosphate group or other larger hydrogen bonding type block such as polyethylene oxide. The A chains are referred to as the stabilizing chains which should be highly soluble in the medium and strongly solvated by its molecules. A schematic representation of the adsorbed layers is shown in Fig. 13.7. When two particles with an adsorbed layer of hydrodynamic thickness δ approach to a separation distance h that is smaller than 2δ, repulsion occurs (Fig. 13.5 (b)) as a result of two main effects: (i) unfavourable mixing of the A chains when these are in good solvent condition; (ii) reduction in configurational entropy on significant overlap.

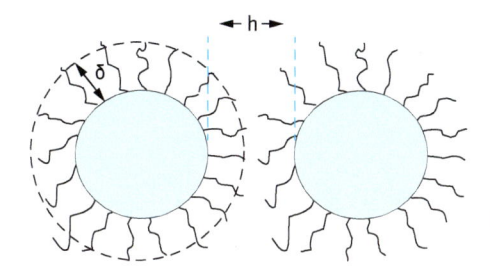

Fig. 13.7: Schematic representation of steric layers.

The efficiency of steric stabilization depends on both the architecture and the physical properties of the stabilizing molecule. Steric stabilizers should have an adsorbing anchor with a high affinity for the particles and/or insoluble in the medium. The stabilizer should be soluble in the medium and highly solvated by its molecules. For aqueous or highly polar oil systems the stabilizer block can be ionic or hydrophilic such as polyalkylene glycols and for oils it should resemble the oil in character. For silicone oils silicone stabilizers are best, other oils could use a long chain alkane, fatty ester or polymers such as poly(methylmethacrylate) (PMMA) or polypropylene oxide.

Various types of surface–anchor interactions are responsible for the adsorption of a dispersant to the particle surface: Ionic or acid/base interactions; sulphonic acid, carboxylic acid or phosphate with a basic surface e.g. alumina; amine or quat with acidic surface e.g. silica. H bonding; surface esters, ketones, ethers, hydroxyls; multiple anchors – polyamines and polyols (h-bond donor or acceptor) or polyethers (h-bond acceptor). Polarizing groups, e.g. polyurethanes can also provide sufficient adsorption energies and in nonspecific cases lyophobic bonding (van der Waals) driven by insolubility (e.g. PMMA). It is also possible to use chemical bonding e.g. by reactive silanes.

For relatively reactive surfaces, specific ion pairs may interact giving particularly good adsorption to a powder surface. An ion pair may even be formed in situ particularly if in low dielectric media. Some surfaces are actually heterogeneous and can have both basic and acidic sites, especially near the IEP. Hydrogen bonding is weak but is particularly important for polymerics which can have multiple anchoring.

The adsorption strength is measured in terms of the segment/surface energy of adsorption χ^s. The total adsorption energy is given by the product of the number of attachment points n by χ^s. For polymers the total value of $n\chi^s$ can be sufficiently high for strong and irreversible adsorption even though the value of χ^s may be small (less than 1 kT, where k is the Boltzmann constant and T is the absolute temperature). However, this situation may not be adequate particularly in the presence of an appreciable concentration of wetter and/or in the presence of other surfactants used as adjuvants. If the χ^s of the individual wetter and/or other surfactant molecules is higher than the χ^s of one segment of the B chain of the dispersant, these small molecules can displace the polymeric dispersant particularly at high wetter and/or other surfactant molecules and this could result in flocculation of the suspension. It is, therefore, essential to make sure that the χ^s per segment of the B chain is higher than that of the wetter and/or surfactant adsorption and that the wetter concentration is not excessive.

In order to optimize steric repulsion one may consider the steric potential as expressed by Napper [9],

$$V(h) = 2\pi kTR\Gamma^2 N_A \left[\frac{V_p^2}{V_s}\right][0.5 - \chi]\left(1 - \frac{h}{2\delta}\right)^2 + V_{elastic}, \qquad (13.14)$$

where k is the Boltzmann constant, T is temperature, R is the particle radius, Γ is the adsorbed amount, N_A the Avogadro constant, V_p is the specific partial volume of the polymer, V_s the molar volume of the solvent, χ is the Flory–Huggins parameter and δ is the maximum extent of the adsorbed layer. $V_{elastic}$ takes account of the compression of polymer chains on close approach.

It is instructive to examine the terms in equation (13.14): (i) The adsorbed amount Γ; higher adsorbed amounts will result in more interaction/repulsion. (ii) Solvent conditions as determined by χ, the Flory–Huggins chain-solvent interaction parameter; two very distinct cases emerge. We see maximum interaction on overlap of the stabilizing layers when the chain is in good solvent conditions ($\chi < 0.5$). Osmotic forces cause solvent to move into the highly concentrated overlap zone forcing the particles apart. If $\chi = 0.5$, a theta solvent, the steric potential goes to zero and for poor solvent conditions ($\chi > 0.5$) the steric potential becomes negative and the chains will attract, enhancing flocculation. Thus, a poorly solvated dispersant can enhance flocculation/aggregation. (iii) Adsorbed layer thickness δ. Steric interaction starts at $h = 2\delta$ as the chains begin to overlap and increases as the square of the distance. Here it is not the size of the steric potential that is important but the distance h at which it begins. (iv) The final interaction potential is the superposition of the steric potential and the van der Waals attraction as shown in Fig. 13.5 (b).

13.2 Fundamental principles for preparation of a stable colour cosmetic dispersion — 259

For sterically stabilized dispersions, the resulting energy-distance curve often shows a shallow minimum V_{min} at particle-particle separation distance h comparable to twice the adsorbed layer thickness δ. For a given material the depth of this minimum depends upon the particle size R, and adsorbed layer thickness δ. So V_{min} decreases with increasing δ/R as is illustrated in Fig. 13.8. This is because as we increase the layer thickness the van der Waals attraction is weakening so the superposition of attraction and repulsion will have a smaller minimum. For very small steric layers V_{min} may become deep enough to cause weak flocculation resulting in a weak attractive gel. So we can see how the interaction energies can also determine the dispersion rheology.

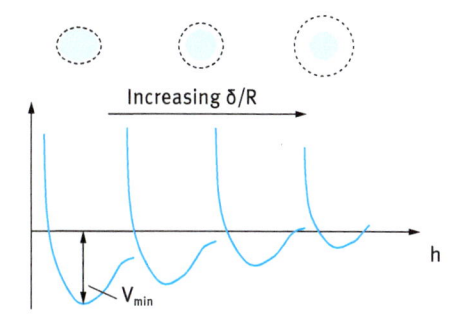

Fig. 13.8: Variation of V_{min} with δ/R.

On the other hand if the layer thickness is too large, the viscosity is also increased due to repulsion. This is due to the much higher effective volume fraction ϕ_{eff} of the dispersion compared to the core volume fraction [2]. We can calculate the effective volume fraction of particles plus dispersant layer by geometry and we see it depends on the thickness of that adsorbed layer as illustrated in Fig. 13.9. The effective volume fraction increases with relative increase of the dispersant layer thickness. Even at 10 % volume fraction we can soon reach maximum packing ($\phi = 0.67$) with an adsorbed layer comparable to the particle radius. In this case overlap of the steric layers will result in significant viscosity increases. Such considerations help to explain why solids loading can be severely limited especially with small particles. In practice solids loading curves can be used to characterize the system and will take the form of those illustrated in Fig. 13.10.

Higher solids loading might be achieved with thinner adsorbed layers but may also result in interparticle attraction resulting in particle aggregation. Clearly a compromise is needed; choosing an appropriate steric stabilizer for the particle size of the pigment.

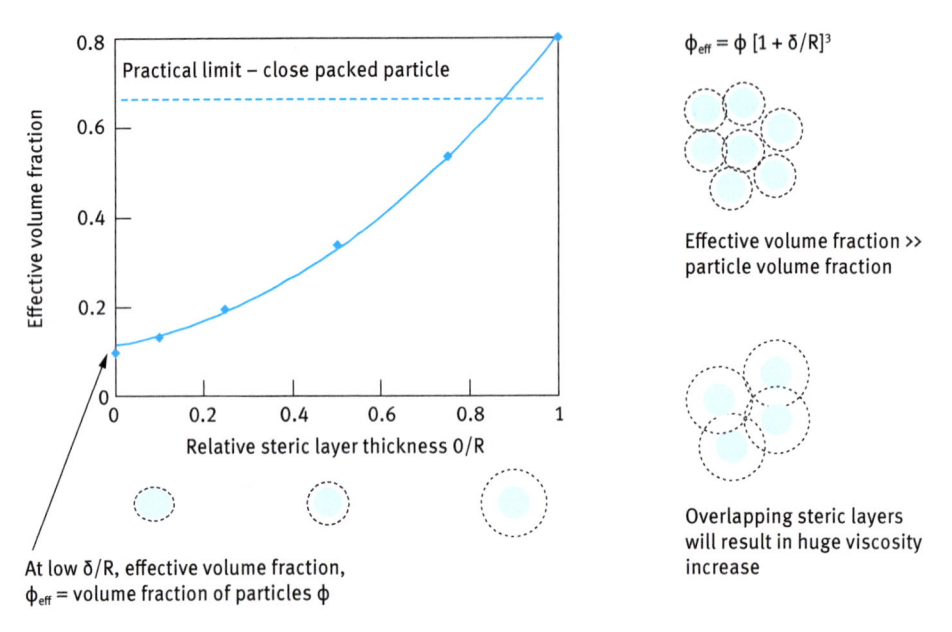

Fig. 13.9: Schematic representation of the effective volume fraction.

$\phi_{eff} = \phi [1 + \delta/R]^3$

Effective volume fraction >> particle volume fraction

Overlapping steric layers will result in huge viscosity increase

At low δ/R, effective volume fraction, ϕ_{eff} = volume fraction of particles ϕ

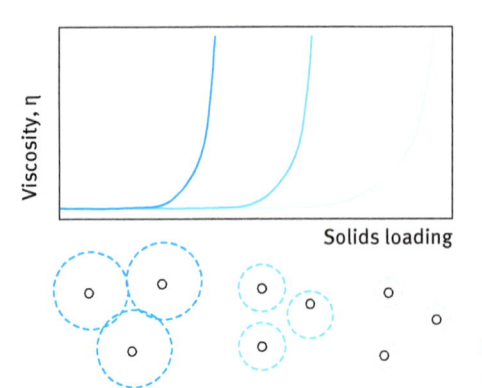

Fig. 13.10: Dependence of solids loading on adsorbed layer thickness.

13.3 Classes of dispersing agents

One of the most commonly used dispersants for aqueous media are nonionic surfactants. The most common nonionic surfactants are the alcohol ethoxylates R–O–(CH$_2$–CH$_2$–O)$_n$–H, e.g. C$_{13/15}$(EO)$_n$ with n being 7, 9, 11 or 20. These nonionic surfactants are not the most effective dispersants since the adsorption by the C$_{13/15}$ chain is not very strong. To enhance the adsorption on hydrophobic surfaces a polypropylene oxide (PPO) chain is introduced in the molecule giving R–O–(PPO)$_m$–(PEO)$_n$–H.

The above nonionic surfactants can also be used for stabilization of polar solids in nonaqueous media. In this case the PEO chain adsorbs on the particle surface leaving the alkyl chains in the nonaqueous solvent. Provided these alkyl chains are sufficiently long and strongly solvated by the molecules of the medium, they can provide sufficient steric repulsion to prevent flocculation.

A better dispersant for polar solids in nonaqueous media is poly(hydroxystearic acid) (PHS) with molecular weight in the region of 1000–2000 Daltons. The carboxylic group adsorbs strongly on the particle surface leaving the extended chain in the nonaqueous solvent. With most hydrocarbon solvents the PHS chain is strongly solvated by its molecules and an adsorbed layer thickness in the region of 5–10 nm can be produced. This layer thickness prevents any flocculation and the suspension can remain fluid up to high solids content [2].

The most effective dispersants are those of the A–B, A–B–A block and BA$_n$ types. A schematic representation of the architecture of block and graft copolymers is shown in Fig. 13.11.

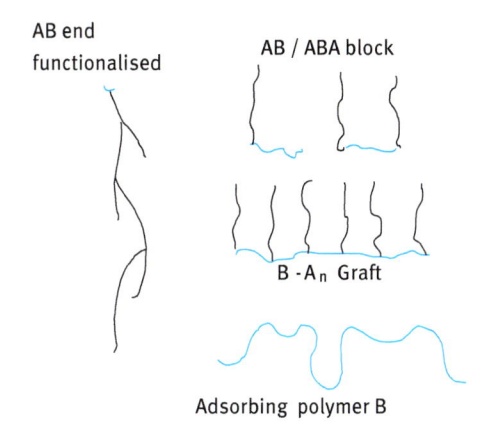

Fig. 13.11: Schematic representation of the architecture of block and graft copolymers.

B, the "anchor chain" (red), is chosen to be highly insoluble in the medium and has a strong affinity to the surface. Examples of B chains for hydrophobic solids are polystyrene (PS), polymethylmethacrylate (PMMA), poly(propylene oxide) (PPO) or alkyl chains provided these have several attachments to the surface. The A stabilizing (blue) chain has to be soluble in the medium and strongly solvated by its molecules. The A chain/solvent interaction should be strong giving a Flory–Huggins χ-parameter < 0.5 under all conditions. Examples of A chains for aqueous media are polyethylene oxide (PEO), polyvinyl alcohol (PVA) and polysaccharides (e.g. polyfructose). For nonaqueous media, the A chains can be polyhydroxystearic acid (PHS).

One of the most commonly used A–B–A block copolymers for aqueous dispersions are those based on PEO (A) and PPO (B). Several molecules of PEO–PPO–PEO are available with various proportions of PEO and PPO. The commercial name is followed by a letter: L (Liquid), P (Paste) and F (Flake). This is followed by two numbers that represent the composition – The first digits represent the PPO molar mass and the last digit represents the % PEO. F68 (PPO molecular mass 1508–1800 + 80 % or 140 mol EO). L62 (PPO molecular mass 1508–1800 + 20 % or 15 mol EO). In many cases two molecules with high and low EO content are used together to enhance the dispersing power.

An example of BA_n graft copolymers is based on polymethylmethacrylate (PMMA) backbone (with some polymethacrylic acid) on which several PEO chains (with average molecular weight of 750) are grafted. It is a very effective dispersant particularly for high solids content suspensions. The graft copolymer is strongly adsorbed on hydrophobic surfaces with several attachment points along the PMMA backbone and a strong steric barrier is obtained by the highly hydrated PEO chains in aqueous solutions.

Another effective graft copolymer is hydrophobically modified inulin, a linear polyfructose chain A (with degree of polymerization > 23) on which several alkyl chains have been grafted. The polymeric surfactant adsorbs with multipoint attachment with several alkyl chains.

13.4 Assessment of dispersants

13.4.1 Adsorption isotherms

These are by far the most quantitative method for assessment and selection of a dispersant [3]. A good dispersant should give a high affinity isotherm as illustrated in Fig. 13.12. The adsorbed amount Γ is recorded as a function of the equilibrium solution concentration, i.e. that left in solution after adsorption.

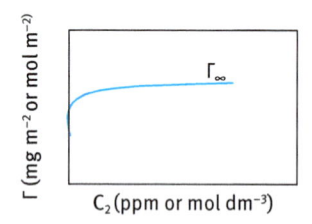

Fig. 13.12: High affinity isotherm.

In general, the value of Γ_∞ is reached at lower C_2 for polymeric surfactant adsorption when compared with small molecules. The high affinity isotherm obtained with polymeric surfactants implies that the first added molecules are virtually completely adsorbed and such a process is irreversible. The irreversibility of adsorption is checked by carrying out a desorption experiment. The suspension at the plateau value is centrifuged and the supernatant liquid is replaced by pure carrier medium. After redispersion, the suspension is centrifuged again and the concentration of the polymeric surfactant in the supernatant liquid is analytically determined. For lack of desorption, this concentration will be very small indicating that the polymer remains on the particle surface.

13.4.2 Measurement of dispersion and particle size distribution

An effective dispersant should result in complete dispersion of the powder into single particles [3]. In addition, on wet milling (comminution) smaller particle distribution should be obtained (this could be assessed by light diffraction, e.g. using the Malvern Master Sizer). The efficiency of dispersion and reduction of particle size can be understood from the behaviour of the dispersant. Strong adsorption and an effective repulsive barrier prevent any aggregation during the dispersion process. It is necessary in this case to include the wetter (which should be kept at the optimum concentration). Adsorption of the dispersant at the solid/liquid interface results in lowering of γ_{SL} and this reduces the energy required for breaking the particles into smaller units. In addition by adsorption in crystal defects, crack propagation occurs (the Rehbinder effect) and this results in production of smaller particles [3].

13.4.3 Rheological measurements

Although "Brookfield" viscometers are still widely used in the industry they should be used with caution in the assessment of dispersion stability; high shear viscosity (as measured by "Brookfield") can be misleading as a predictor of sedimentation velocity. Low or zero shear viscosity measured on a rheometer is the best indicator [10]. Fig. 13.13 demonstrates how at high shear, dispersion B has the highest viscosity and might be expected to give the best resistance to sedimentation. However, dispersion A has the highest low shear viscosity and will hence have the best sedimentation stability.

Concentrated dispersions are viscoelastic; that is they have both viscous and elastic characteristics. Oscillatory rheometry can therefore give us much more information

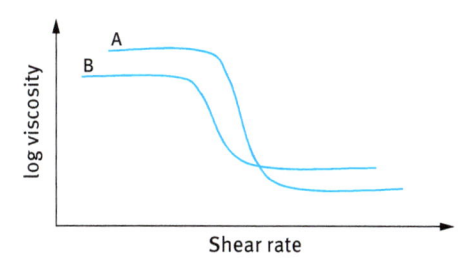

Fig. 13.13: Schematic flow curve for particulate dispersions.

about the interparticle interactions than viscometry [10]. For example, an elastic modulus which dominates the shear sweep can confer significant stability to a formulation with dispersed solids.

Rheological techniques are often the most informative techniques for assessment and selection of a dispersant [3, 10]. The best procedure is to follow the variation of relative viscosity η_r with the volume fraction ϕ of the dispersion. For this purpose a concentrated suspension (say 50 w/w%) is prepared by milling using the optimum dispersant concentration. This suspension is further concentrated by centrifugation and the sedimented suspension is diluted with the supernatant liquid to obtain volume fractions ϕ in the range 0.1–0.7. The relative viscosity η_r is measured for each suspension using the flow curves. η_r is then plotted as a function of ϕ and the results are compared with the theoretical values calculated using the Dougherty–Krieger equation [11] as discussed below.

Dougherty and Krieger [11] derived an equation for the variation of the relative viscosity η_r with the volume fraction ϕ of suspensions assumed to behave like hard spheres,

$$\eta_r = \left[1 - \frac{\phi}{\phi_p}\right]^{-[\eta]\phi_p}, \quad (13.15)$$

where $[\eta]$ is the intrinsic viscosity that is equal to 2.5 for hard spheres and ϕ_p is the maximum packing fraction that is ≈ 0.6–0.7. The maximum packing fraction ϕ_p is obtained by plotting $1/(\eta_r)^{1/2}$ versus ϕ and in most cases a straight line is obtained which is then extrapolated to $1/(\eta_r)^{1/2} = 0$ and this gives ϕ_p.

η_r–ϕ curves are established from the experimental data using the flow curves. The theoretical η_r–ϕ curves obtained from the Dougherty–Krieger equation are also established using a value of 2.5 for the intrinsic viscosity $[\eta]$ and ϕ_p calculated using the above extrapolation procedure. As an illustration Fig. 13.14 shows a schematic representation for results for an aqueous suspension of hydrophobic particles that are dispersed using a graft copolymer of polymethylmethacrylate (PMMA) backbone on which several polyethylene oxide (PEO) chains have been grafted [3].

Both the experimental and theoretical η_r–ϕ curves show an initial slow increase of η_r with increasing ϕ but at a critical ϕ value η_r shows a rapid increase with any further increase in ϕ. It can be seen from Fig. 13.14 that the experimental η_r values

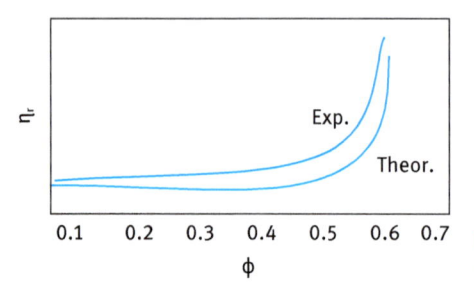

Fig. 13.14: Variation of η_r with ϕ for suspensions stabilized with a graft copolymer.

show a rapid increase above a high ϕ value (> 0.6). The theoretical η_r–ϕ curve (using equation (13.15)) shows an increase in η_r at a ϕ value close to the experimental results [12]. This shows a highly deflocculated (sterically stabilized) suspension. Any flocculation will cause a shift in the η_r–ϕ curve to lower values of ϕ. These η_r–ϕ curves can be used for assessment and selection of dispersants. The higher the value of ϕ at which the viscosity shows a rapid increase, the more effective the dispersant is. Strong adsorption of the graft polymeric surfactant and the high hydration of the PEO chains ensure such high stability. In addition, such polymeric surfactant is not likely to be displaced by the wetter surfactant molecules provided these are not added at high concentrations. It is essential to use the minimum wetter concentration that is sufficient for complete wetting of the powder.

13.5 Application of the above fundamental principles to colour cosmetics

Pigments are in fact the primary ingredient of any modern colour cosmetic. Pigments need to be incorporated first into slurries and for most colour chemists the primary objective is to reduce the viscosity and improve ease of use of these slurries. It is important to remember that both attractive and repulsive interactions result in viscosity increase. The aim is thus to reduce particle-particle interactions. It is not just in the processing where optimization is required; the particle distribution in the final cosmetic formulation will determine its functional activity (colour, opacity, UV protection), stability, rheology and skin feel. The particle distribution depends upon a number of characteristics such as particle size and shape, surface characteristics, processing, compatibilities but is ultimately also determined by interparticle interactions.

There are two main consequences of instability in particulate dispersions; flocculation or agglomeration and sedimentation. For colour cosmetics, insufficient deagglomeration (all pigments are agglomerated as supplied) can manifest itself as poor colour consistency or streaking with colour being liable to change on application. Sedimentation effects can appear as colour flotation or plate-out. Sedimentation is determined by gravity and is not necessarily a sign of colloidal instability. It simply needs to be controlled. Sedimentation velocity tends to increase with particle size (thus aggregation is bad) but is reduced by increased fluid viscosity.

Dispersion stability may manifest itself in different ways and for the formulator one must have lower viscosity during manufacture. Fig. 13.15 demonstrates the potential benefits (for viscosity dependence on pigment concentration) when a suitable dispersant is added. This can be liberating in removing formulation restrictions and more practically in reducing processing times and cost. Higher pigment concentrations can be achieved giving increased functionality. The colour strength often improves with milling time but again can be stepped up by the incorporation of suitable dispersants.

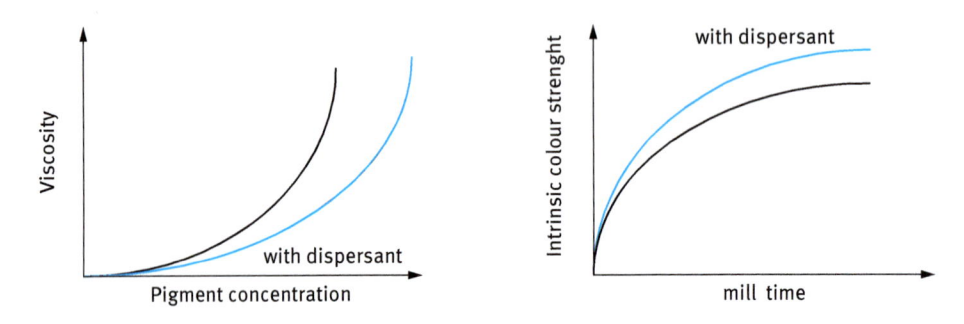

Fig. 13.15: Effect of dispersant on viscosity and intrinsic colour strength.

With improved product quality, one can expect improvements in stability, consistency and function.

Product quality is the key to product differentiation in the market and it is highly desirable therefore to reduce flooding and floating caused by flocculation of differing pigments. The control and reproducibility of gloss/shine and brightness and the ability to control rheology and skin feel, particularly at high solids loadings are all within reach here.

Finally, the optimization of functionality can often depend strongly on the state of dispersion. As mentioned in Chapter 12, opacity and UV attenuation of TiO_2 for example is strongly dependent on particle size [13] as illustrated in Fig. 13.16. A titanium dioxide pigment, designed to provide opacity in a formulation will not realize its maximum hiding power unless it is dispersed and remains dispersed in its constituent particles of 200–300 nm. A UV attenuating grade of TiO_2 on the other hand must be

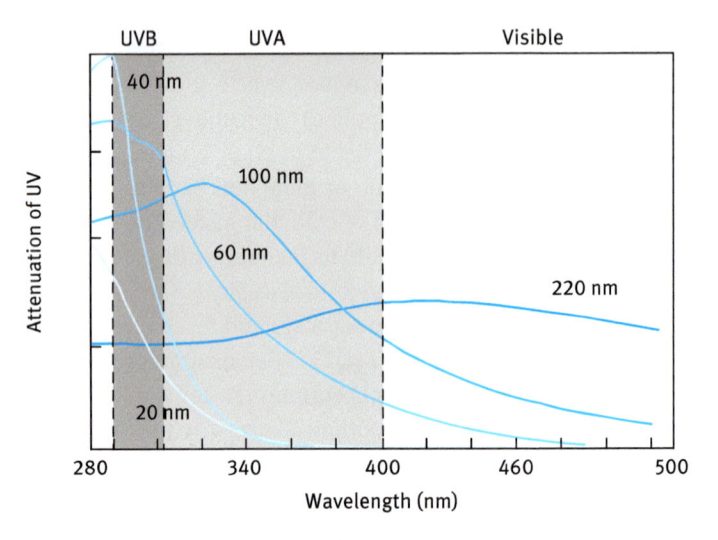

Fig. 13.16: Attenuation of UV versus wavelength for TiO_2 dispersion.

dispersed down to its primary particle size of 50–100 nm in order to be optimally functional as a sunscreen agent. Both powders as supplied (in order to be handleable), however, have similar agglomerate sizes of several microns.

13.6 Principles of preparation of colour cosmetics

As mentioned above, the first task is to obtain complete wetting of the powder. Both external and internal surfaces of the agglomerates must be adequately wetted by using a suitable surfactant. For aqueous dispersions, the above mentioned wetting agents such as Aerosol OT or alcohol ethoxylates are generally efficient. For hydrophilic pigments in oil one can use coated particles (with hydrophobic coating) or sodium stearate which strongly binds to the hydroxyl surface. A schematic representation for binding of stearate to hydrophilic TiO_2 is shown in Fig. 13.17, thus rendering it easily wetted and dispersed in oils. This figure also shows the effect of addition of an alcohol ethoxylate to this coated TiO_2 which can then be dispersed in an aqueous medium.

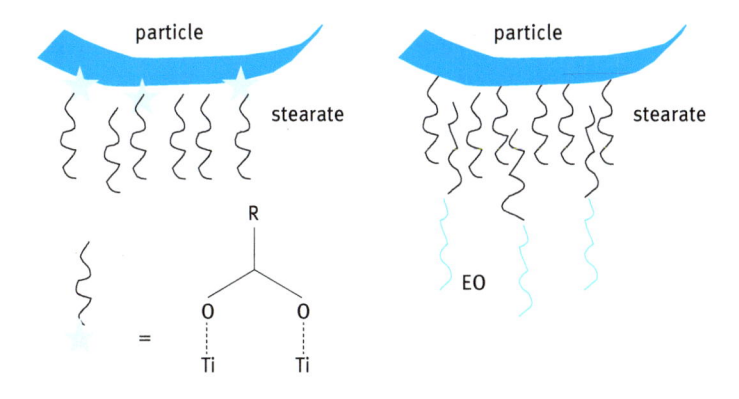

Eg specific attachment via carboxylate group to inorganic surface

Fig. 13.17: Schematic representation of specific interaction of stearate to TiO_2 (left) and effect of addition of alcohol ethoxylate (right).

This process is followed by complete dispersion and/or comminution and adequate stabilization of the resulting single particles as illustrated in Fig. 13.18.

The next step is to control the process of dispersion and comminution. Simple mixing of inorganic powders can produce a fluid dispersion even at high solids. However this is not necessarily an indication of a "well dispersed" material and indeed a particle size analysis (and for UV attenuators, spectral analysis) demonstrates that

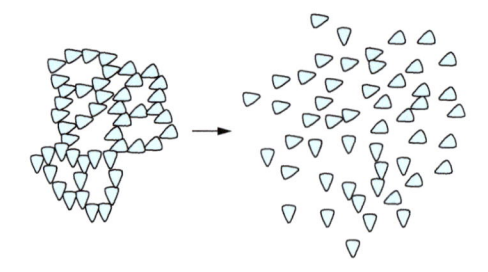

Fig. 13.18: Schematic representation of the dispersion process.

particle dispersion is not optimized. Particulate powders are supplied in an aggregated state. However, they must be milled down to their individual units in order to provide their designed function. This process must allow transport of the dispersant to the particle surface and adsorption there. Finally, the dispersion must remain stable to dilution or addition of further formulation components. The presence of a suitable dispersant/stabilizer at the right level can be critical in achieving a usable and stable dispersion and preventing re-aggregation on standing. In practice, the formulation chemist may use some simple laboratory tools to assess dispersion quality and arrive at an appropriate dispersion recipe. Having assessed wetting as previously described one will often plot a dispersant demand curve in order to establish the optimum dispersant loading. The pigment is processed (milling or grinding) in the presence of the carrier oil and wetting agent with varying levels of dispersant. The state of dispersion can be effectively monitored by rheology and/or some functional measurement (e.g. colour strength, UV attenuation). Fig. 13.19 shows the results for some fine particle TiO_2 in isopropyl isostearate as dispersing fluid and polyhydroxystearic acid as dispersant [14].

Dispersions were produced at 30 wt% solids so that they could be prepared on a bead mill at all dispersant loadings and their UV attenuation properties compared. Zero shear viscosities give an indication of interparticle interactions and were found to be at a minimum at around 5% dispersant. UV attenuation was used as an indicator of particle size.

The unmilled dispersions [2] appeared very fluid, but UV measurement revealed poor attenuation properties implying that the particles are still aggregated. The solid particles quickly settled out of suspension to form a sediment in the bottom of the beaker. An improvement of UV attenuation properties, along with an increase in viscosity, was observed upon milling. The aggregates are broken down into their constituent particles in the mill [2], but in the absence of dispersant they quickly reaggregate by van der Waals attraction in a more open structure. This caused the mill to block. Further improvements in UV properties were observed when the dispersion was milled in the presence of the dispersant [2] but viscosity was still high. Addition of sufficient dispersant allows the particles to disperse to single particulates [3] which are well stabilized and the viscosity drops. This is an optimized dispersion. UV properties are well developed. On addition of further dispersant the particles gain an extended

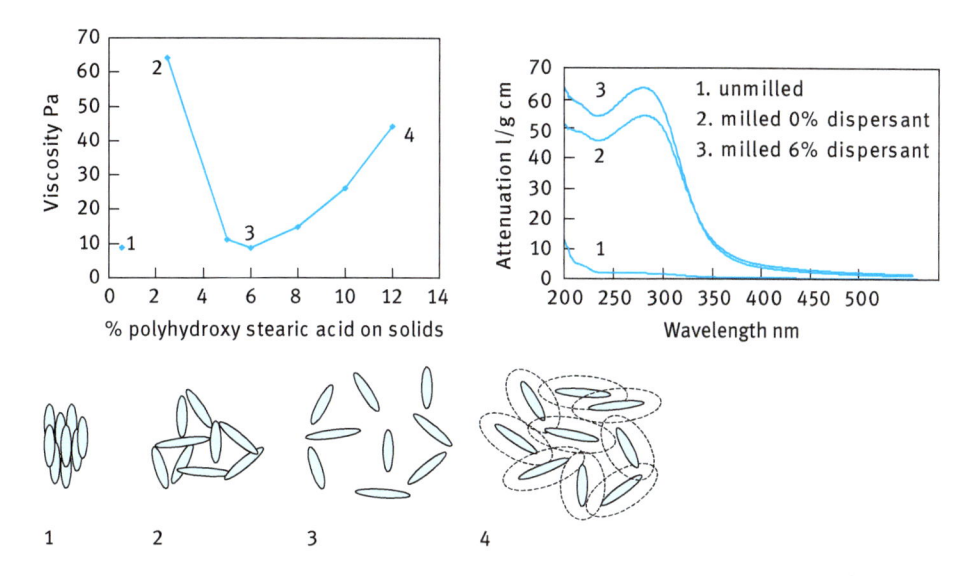

Fig. 13.19: Zero shear viscosity (dispersant demand curve), UV attenuation curves and a schematic of the milling process.

stabilization layer [2] causing potential overlap of stabilization layers which is sufficient to produce a weak repulsive gel. The viscosity again rises and the dispersion has a measurable yield value. UV properties are still well developed but the solids loading becomes very limited.

The dispersant demand curves described in Chapter 12, particle size monitoring in addition to solids loading curves (Fig. 13.15) are very useful tools in optimizing a pigment dispersion in practice. Further examples were given in Chapter 12.

13.7 Competitive interactions in colour cosmetic formulations

The colour cosmetic pigments are added to oil-in-water (O/W) or water-in-oil (W/O) emulsions. The resulting system is referred to as a suspoemulsion [2]. The particles can be in the internal or external phases or both as is illustrated in Fig. 12.12 of Chapter 12 for sunscreen formulations. An understanding of competitive interactions is important in optimizing formulation stability and performance. Several possible instabilities might arise in the final formulations: (i) Heteroflocuation from particles and droplets of differing charge sign. (ii) Electrolyte intolerance of electrostatically stabilized pigments. (iii) Competitive adsorption/desorption of a weakly anchored stabilizer which can lead to homoflocculation of the pigment particles and/or emulsion droplet coalescence. (iv) Interaction between thickeners and charge stabilized pigments.

Several steps can be taken to improve the stability of colour cosmetic formulations which are in fact very similar to those for optimal steric stabilization [2]: (i) Use of a strongly adsorbed ("anchored") dispersant, e.g. by multipoint attachment of a block or graft copolymer. (ii) Use of a polymeric stabilizer for the emulsion (also with multipoint attachment). (iii) Preparation of the suspension and emulsion separately and allowing enough time for complete adsorption (equilibrium). (iv) Using low shear when mixing the suspension and emulsion. (v) Use of rheology modifiers that reduce the interaction between the pigment particles and emulsion droplets. (vi) Increasing dispersant and emulsifier concentrations to ensure that the lifetime of any bare patches produced during collision is very short. (vii) Use same polymeric surfactant molecule for emulsifier and dispersant. (viii) Reducing emulsion droplet size.

References

[1] Tadros, Th. F., "Cosmetics", in "Encyclopedia of Colloid and Interface Science", Th. F. Tadros (ed.), Springer, Germany (2013).
[2] Kessel, L. M. and Tadros, Th. F., "Interparticle Interactions in Color Cosmetics", in "Colloids in Cosmetics and Personal Care", Th. F. Tadros (ed.), Wiley-VCH, Germany (2008).
[3] Tadros, Th. F., "Dispersion of Powders in Liquids and Stabilisation of Suspensions", Wiley-VCH, Germany (2012).
[4] Blake, T. B., in "Surfactants", Th. F. Tadros (ed.), Academic Press, London, (1984).
[5] Tadros, Th. F., "Applied Surfactants", Wiley-VCH, Germany (2005).
[6] Visser, J., Advances Colloid Interface Sci , **3**, 331 (1972) Hamaker constants.
[7] Deryaguin, B. V. and Landau, L., Acta Physicochem. USSR, **14**, 633 (1941).
[8] Verwey, E. J. W. and Overbeek, J. Th. G., "Theory of Stability of Lyophobic Colloids", Elsevier, Amsterdam (1948).
[9] Napper, D. H., "Polymeric Stabilisation of Colloidal Dispersions", Academic Press, London (1983).
[10] Tadros, Th. F., "Rheology of Dispersions", Wiley-VCH, Germany (2010).
[11] Krieger, I. M., Advances Colloid and Interface Sci., **3**, 111 (1972).
[12] Tadros, Th. F., Advances Colloid Interface Sci., **104**, 191 (2003).
[13] Robb, J. L., Simpson, L. A. and Tunstall, D. F., Scattering and absorption of UV radiation by sunscreens containing fine particle and pigmentary titanium dioxide, Drug. Cosmet. Ind., March, 32–39 (1994).
[14] Kessel, L. M., Naden, B. J. and Tadros, Th. F., "Attractive and Repulsive Gels", IFSCC Congress, Orlando, Florida (2004).

14 Industrial examples of cosmetic and personal care formulations

A useful text that gives many examples of commercial cosmetic formulations has been published by Polo [1] to which the reader should refer for any detailed information. Below only a summary of some personal care and cosmetic formulations is given, illustrating some of the principles applied. As far as possible, a qualitative description of the role of the ingredients used in these cosmetic formulations is given. For more fundamental information, the reader should refer to the previous chapters in this book.

14.1 Shaving formulations

Three main types of shaving preparations may be distinguished: (i) wet shaving formulations; (ii) dry shaving formulations and (iii) after shave preparations. The main requirements for wet shaving preparations are to soften the beard, to lubricate the passage of the razor over the face and to support the beard hair. The hair of a typical beard is very coarse and difficult to cut and hence it is important to soften the hair for easier shaving and this requires the application of soap and water. The soap makes the hair hydrophilic and hence it becomes easy to wet by water which also may cause swelling of the hair. Most soaps used in shaving preparations are sodium or potassium salts of long chain fatty acids (sodium or potassium stearate or palmitate). Sometimes, the fatty acid is neutralized with triethanolamine. Other surfactants such as ether sulphates and sodium lauryl sulphate are included in the formulation to produce stable foam. Humectants such as glycerol may also be included to hold the moisture and prevent drying of the lather during shaving.

The most commonly used shaving formulations are those of the aerosol type, whereby hydrocarbon propellants (e.g. butane) are used to dispense the foam. The amount of propellant is critical for foam characteristics. More recently, several companies introduced the concept of post-foaming gel, whereby the product is discharged in the form of a clear gel which can be easily spread on the face and the foam is then produced by vaporization of low boiling hydrocarbons such as isopentene. Due to the high viscosity of the gel, the latter is packed in a bag separated from the propellant used to expel the gel.

The above aerosol type formulations are complex, consisting of an O/W emulsion (whereby the propellant forms most of the oil phase) with the continuous phase consisting of soap/surfactant mixtures. The aerosol shaving foam which was introduced first is relatively more simple, whereby a pressurized can is used to release the soap/surfactant mixture in the form of a foam. The sudden release of pressure results in the formation of fine foam bubbles throughout the emerging liquid phase. Two main factors should be considered. Firstly, that foam stability should be maintained during the shaving process. In this case, one has to consider the intermolecular forces that operate in a foam film. The lifetime of a foam film is determined by the disjoining pressure [2] that operates across the liquid lamellae. By using the right combination of soap and surfactants, one can optimize the foam characteristics. The second important property of the foam is its feel on the skin. This is determined by the amount of propellant used in the formulation. If the propellant level is too low, the foam will appear "watery". In contrast, a high amount of propellant will produce a "rubbery" dry foam. The humectant added also plays an important role in the skin feel of the foam. Again, an optimum concentration is required to prevent drying out of the foam during shaving. However, if the humectant level is too high it may cause problems by pulling out moisture from the hair thus making it more difficult to shave.

From the above discussion, it becomes clear that to formulate a shaving foam, the chemist has to consider a large number of physicochemical factors, such as the interaction between the soap and surfactant, the quality of the emulsion produced and the bulk properties of the foam produced. It is no surprise that most shaving foams consist of complex recipes and understanding the role of each component at a molecular level is far from being achieved at present.

As mentioned above, aerosol shaving foam has been replaced with the more popular aerosol post-forming gel. The latter is more difficult to produce, since one has to produce a clear gel with the right rheological characteristics for discharge from the aerosol container and good spreading on the surface of the skin. The foam should then be produced by vaporization of a low boiling liquid such as isobutene or isopentene.

The first problem that must be addressed is the gel characteristics, which are produced by a combination of soap/surfactant mixtures and some polymer (that acts as a "thickener"), e.g. polyvinyl pyrrolidone. The interaction between the surfactants and the polymer should be considered to arrive at the optimum composition [2]. The heat on the skin causes the isopentene to evaporate forming a rich thick gel. One can incorporate skin conditioners and lubricating agent in the gel to obtain good skin feel. Again, most aerosol post-forming gels consist of complex recipes and the interactions between the various components are difficult to understand at a molecular level. A fundamental colloid and interface science investigation is essential to arrive at the optimum composition. In addition, the rheology of the gel, in particular its viscoelastic properties must be considered in detail [3]. Measurements of the viscoelasticity of these gels are difficult, since the foam is produced during such measurements.

One of the main properties that should also be considered in these shaving foams and post-forming gels is the lubricity of the formulation. Skin friction can be reduced by incorporating some oils, e.g. silicone, and gums. When shaving, the first stroke by the razor causes no problem since the shave foam or gel is present in sufficient quantities to ensure lubricity of the skin. However, the second stroke in shaving will produce a very high frictional force and hence one should ensure that a residual amount of a lubricant should be preset on the skin after the first stroke.

Another type of wet shaving preparation is the non-aerosol type, which is now much less popular than the aerosol type. Two types may be distinguished, namely the brushless and lather shaving creams. These formulations are still marketed, although they are much less popular than the aerosol type systems. The brushless shaving cream is an O/W emulsion with high concentrations of oil and soap. The thick film of lubricant oil provides emolliency and protection to the skin surface. This reduces razor drag during shaving. The main disadvantage of these creams is the difficulty of rinsing them from the razor and the formulation may leave a "greasy" feeling on the skin. Due to the high oil content of the formulation, the hair softening action is less effective when compared to the aerosol type.

The lather shaving cream is a concentrated dispersion of alkali metal soap in a glycerol-water mixture. This formulation suffers from adequate physical stability, particularly if the manufacturing process is not carefully optimized. Phase separation of the formulation may occur at elevated temperatures.

Dry shaving is a process using electric shavers. In contrast to wet shaving, when using an electric razor the hair should remain dry and stiff. This requires removal of the moisture film and sebum from the face. This may be achieved by using a lotion based on an alcohol solution. A lubricant such as fatty acid ester or isopropyl myristate may be added to the lotion. Alternatively, a dry talc stick may be used that can absorb the moisture and sebum from the face.

Another important formulation that is used after shaving is that used to reduce skin irritation and to provide a pleasant feel. This can be achieved by providing emolliency accompanied by a cooling effect. In some cases an antiseptic agent is added to keep the skin free from bacterial infection. Most of these after shave formulations are aqueous based gels which should be non-greasy and easy to rub into the skin.

14.2 Bar soaps

These are one of the oldest toiletries products that have been used over centuries. The earliest formulations were based on simple fatty acid salts, such as sodium or potassium palmitate. However, these simple soaps suffer from the problem of calcium soap precipitation in hard water. For that reason, most soap bars contain other surfactants such as cocomonoglyceride sulphate or sodium cocoglyceryl ether sulphonate that prevent precipitation with calcium ions. Other surfactants used in soap bars include sodium cocyl isethinate, sodium dodecyl benzene sulphonate and sodium stearyl sulphate.

Several other functional ingredients are included in soap bar formulations, e.g. antibacterials, deodorants, lather enhancers, anti-irritancy materials, vitamins, etc. Other soap bar additives include antioxidants, chelating agents, opacifying agents (e.g. titanium dioxide), optical brighteners, binders, plasticizers (for ease of manufacture), anticracking agents, pearlescent pigments, etc. Fragrants are also added to impart pleasant smell to the soap bar.

14.3 Liquid hand soaps

Liquid hand soaps are concentrated surfactant solutions which can be simply applied from a plastic squeeze bottle or a simple pump container. The formulation consists of a mixture of various surfactants such as alpha olefin sulphonates, lauryl sulphates or lauryl ether sulphates. Foam boosters such as cocoamides are added to the formulation. A moisturizing agent such as glycerine is also added. A polymer such as poyquaternium-7 is added to hold the moisturizers and to impart a good skin feel. More recently, some manufacturers have been using alkyl polyglucosides in their formulations. The formulation may also contain other ingredients such as proteins, mineral oil, silicones, lanolin, etc. In many cases a fragrant is added to impart a pleasant smell to the liquid soap.

One of the major properties of liquid soaps that needs to be addressed is its rheology, which affects its dispensing properties and spreading on the skin. Most liquid soap formulations have high viscosities to give them a "rich" feel, but some shear thinning properties are required for ease of dispensation and spreading on the surface of the skin. These rheological characteristics are achieved by two main effects: (i) Addition of electrolyte that causes a change of the micellar structure from spherical (with low viscosity) to rod-shaped micelles (thread-like micelles) that form a three-dimensional gel network structure by overlap of the threads that is viscoelastic in nature. This is schematically shown in Fig. 14.1. (ii) The rheology of liquid soap formulations can also be controlled by addition of "thickeners" (high molecular weight polymers such as xanthan gum) which produce viscoelastic structures by polymer chain overlap as illustrated in Fig. 14.2.

Many liquid soap formulations also contain some preservatives and bactericides such as sodium benzoate and cresol type chemicals.

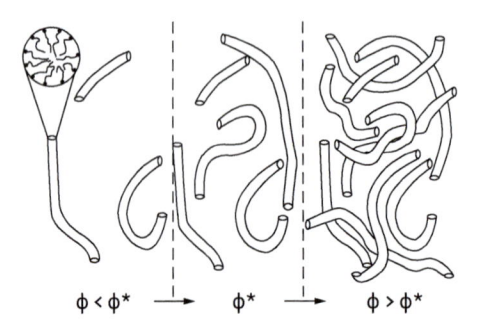

$\phi < \phi^*$ → ϕ^* → $\phi > \phi^*$

Fig. 14.1: Overlap of thread-like micelles.

(a) Dilute
C < C*

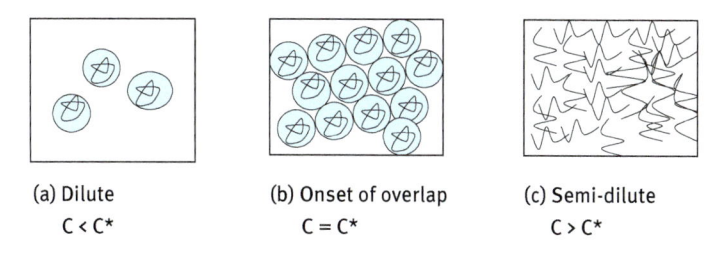

(b) Onset of overlap
C = C*

(c) Semi-dilute
C > C*

Fig. 14.2: Crossover between dilute and semi-dilute solutions.

14.4 Bath oils

Three types of bath oils may be distinguished: floating or spreading oil; dispersible, emulsifying or blooming oil and milky oil. The floating or spreading bath oils (usually mineral or vegetable oils or cosmetic esters such as isopropyl myristate) are the most effective for lubricating dry skin as well as carrying the fragrant. However, they suffer from "greasiness" and deposit formation around the bath tub. These problems are overcome by using self-emulsifying oils which are formulated with surfactant mixtures. When added to water they spontaneously emulsify forming small oil droplets that deposit on the skin surface. However, these self-emulsifying oils produce less emolliency compared with the floating oils. These bath oils usually contain a high level of fragrance since they are used in a large amount of water.

14.5 Foam (or bubble) baths

These can be produced in the form of liquids, creams, gels, powders, granules (beads). Their main function is to produce maximum foam into running water. The basic surfactants used in bubble bath formulations are anionic, nonionic or amphoteric together with some foam stabilizers, fragrants and suitable solubilizers. These formulations should be compatible with soap and they may contain other ingredients for enhancing skincare properties.

14.6 After bath preparations

These are formulations designed to counteract the damaging effects caused after bathing, e.g. skin drying caused by removal of natural fats and oils from the skin. Several formulations may be used, e.g. lotions and creams, liquid splashes, dry oil spray, dusting powders or talc, etc. Lotions and creams, which are the most commonly used formulations, are simply O/W emulsions with skin conditioners and emollients. The liquid splashes are hydroalcoholic products that contain some oil to provide skin conditioning. They can be applied as a liquid spread on the skin by hand or by spraying.

14.7 Skincare products

The skin forms an efficient permeability barrier with the following essential functions: (i) protection against physical injury, wear and tear and it may also protect against ultraviolet (UV) radiation; (ii) it protects against penetration of noxious foreign materials including water and micro-organisms; (iii) it controls loss of fluids, salts, hormones and other endogenous materials from within; (iv) it provides thermoregulation of the body by water evaporation (through sweat glands).

For the above reasons skincare products are essential materials for protection against skin damage. A skincare product should have two main ingredients, a moisturizer (humectant) that prevents water loss from the skin and an emollient (the oil phase in the formulation) that provides smoothing, spreading, degree of occlusion and moisturizing effect. The term emollient is sometimes used to encompass both humectant and oils.

The moisturizer should keep the skin humid and it should bind moisture in the formulation (reducing water activity) and protect it from drying out. The term water content implies the total amount of water in the formulation (both free and bound), whereas water activity is a measure of the free (available) water only. The water content of the deeper, living epidermic layers is of the order of 70 % (same as the water content in living cells). Several factors can be considered to account for drying of the skin. One should distinguish between the water content of the dermis, viable epidermis and the horny layer (stratum corneum). During dermis aging, the amount of mucopolysaccharides decreases leading to a decrease in the water content. This aging process is accelerated by UV radiation (in particular the deep penetrating UVA, see Chapter 12 on sunscreens). Chemical or physical changes of the epidermis during aging also lead to dry skin. As discussed in Chapter 1, the structured lipid/water bilayer system in the stratum corneum forms a barrier against water loss and protects the viable epidermis from the penetration of exogenous irritants. The skin barrier may be damaged by extraction of lipids by solvents or surfactants and water loss can also be caused by low relative humidity.

Dry skin, caused by a loss of the horny layer, can be cured by formulations containing extracts of lipids from horny layers of humans or animals. Due to loss of water from the lamellar liquid crystalline lipid bilayers of the horny layer, phase transition to crystalline structures may occur and this causes contraction of the intercellular regions. The dry skin becomes inflexible and inelastic and it may also crack.

For the above reasons, it is essential to use skincare formulations that contain moisturizers (e.g. glycerine) that draw and strongly bind water, thus trapping water on the skin surface. Formulations prepared with nonpolar oils (e.g. paraffin oil) also help in water retention. Occlusion of oil droplets on the skin surface reduces the rate of trans-epidermal water loss. Several emollients can be applied, e.g. petrolatum, mineral oils, vegetable oils, lanolin and its substitutes and silicone fluids. Apart from glycerine, which is the most widely used humectant, several other moisturizers can be used, e.g. sorbitol, propylene glycol and polyethylene glycols (with molecular weights in the range 200–600). As mentioned in Chapter 9, liposomes or vesicles, neosomes can also be used as skin moisturizers.

In general, emollients may be described as products that have softening and smoothing properties. They could be hydrophilic substances such as glycerine, sorbitol, etc. (mentioned above) and lipophilic oils such as paraffin oil, castor oil, triglycerides, etc. For the formulation of stable O/W or W/O emulsions for skincare application, the emulsifier system has to be chosen according to the polarity of the emollient. The polarity of an organic molecule may be described by its dielectric constant or dipole moment. Oil polarity can also be related to the interfacial tension of oil against water γ_{OW}. For example, a nonpolar substance such as isoparaffinic oil will give an interfacial tension in the region of 50 mN m^{-1}, whereas a polar oil such as cyclomethicone gives γ_{OW} in the region of 20 mN m^{-1}. The physicochemical nature of the oil phase determines its ability to spread on the skin, the degree of occlusivity and skin protection. The optimum emulsifier system also depends on the property of the oil (its HLB number) as discussed in detail in Chapter 5 on cosmetic emulsions.

The choice of an emollient for a skincare formulation is mostly based on sensorial evaluation using well trained panels. These sensorial attributes are classified into several categories: ease of spreading, skin feeling directly after application and 10 minutes later, softness, etc. A lubricity test is also conducted to establish a friction factor. Spreading of an emollient may also be evaluated by measurement of the spreading coefficient.

14.8 Haircare formulations

Haircare comprises two main operations: (i) Care and stimulation of the metabolically active scalp tissue and its appendages, the pilosibaceous units. This process is normally carried out by dermatologists or specialized hair salons. (ii) Protection and care of the lifeless hair shaft as it passes beyond the surface of the skin. The latter is the subject of cosmetic preparations, which should acquire one or more of the following functions: (i) Hair conditioning for ease of combing. This could also include formulations that can easily manage styling by combing and brushing and its capacity to stay in place for a while. The difficulty to manage hair is due to the static electric charge which may be eliminated by hair conditioning. (ii) Hair "body", i.e. the apparent volume of a hair assembly as judged by sight and touch.

Another important type of cosmetic formulations are those used for hair dyeing, i.e. changing the natural colour of the hair. This subject will also be briefly discussed in this section.

As mentioned in Chapter 11, hair is a complex multicomponent fibre with both hydrophilic and hydrophobic properties. It consists of 65–95 % by weight of protein and up to 32 % water, lipids, pigments and trace elements. The proteins are made of structured hard α-keratin embedded in an amorphous, proteinaceous matrix. Human hair is a modified epidermal structure taking its origin from small sacs called follicles that are located at the border line of dermis and hypodermis. A cross section of human hair shows three morphological regions, the medulla (inner core), the cortex that consists of fibrous proteins (α-keratin and amorphous protein) and an outer layer namely the cuticle. The major constituent of the cortex and cuticle of hair is protein or polypeptide (with several amino acid units). The keratin has an α-helix structure (molecular weight in the region of 40 000–70 000 Daltons, i.e. 363–636 amino acid units).

The surface of hair has both acidic and basic groups (i.e. amphoteric in nature). For unaltered human hair, the maximum acid combining capacity is approximately 0.75 mmol g^{-1} hydrochloric, phosphoric or ethyl sulphuric acid. This value corresponds to the number of dibasic amino acid residues, i.e. arginine, lysine or histidine. The maximum alkali combining capacity for unaltered hair is 0.44 mmol g^{-1} potassium hydroxide. This value corresponds to the number of acidic residues, i.e. aspartic and glutamic side chains. The isoelectric point (IEP) of hair keratin (i.e. the pH at which there is an equal number of positive, $-NH^+$ and negative, $-COO^-$ groups) is pH ≈ 6.0. However, for unaltered hair, the IEP is at pH = 3.67.

The above charges on human hair play an important role in the reaction of hair to cosmetic ingredients in a haircare formulation. Electrostatic interaction between anionic or cationic surfactants in any haircare formulation will occur with these charged groups. Another important factor in application of haircare products is the water content of the hair, which depends on the relative humidity (RH). At low RH (< 25 %), water is strongly bound to hydrophilic sites by hydrogen bonds (sometimes this is referred to as "immobile" water). At high RH (> 80 %), the binding energy for water molecules is lower because of the multimolecular water-water interactions (this is sometimes referred to as "mobile" or "free" water). With increasing RH, the hair swells; increasing relative humidity from 0 to 100 %, the hair diameter increases by ≈ 14 %. When water-soaked hair is put into a certain shape while drying, it will temporarily retain its shape. However, any change in RH may lead to the loss of setting.

Both surface and internal lipids exist in hair. The surface lipids are easily removed by shampooing with a formulation based on an anionic surfactant. Two successive steps are sufficient to remove the surface lipids. However, the internal lipids are difficult to remove by shampooing due to the slow penetration of surfactants.

Analysis of hair lipids reveals that they are very complex consisting of saturated and unsaturated, straight and branched fatty acids with chain length from 5 to 22 carbon atoms. The difference in composition of lipids between persons with "dry" and "oily" hair is only qualitative. Fine straight hair is more prone to "oiliness" than curly coarse hair.

From the above discussion, it is clear that hair treatment requires formulations for cleansing and conditioning of hair and this is mostly achieved by using shampoos. These are now widely used by most people and various commercial products are available with different claimed attributes. The primary function of a shampoo is to clean both hair and scalp of soils and dirt. Modern shampoos fulfil other purposes, such as conditioning, dandruff control and sun protection. The main requirements for a hair shampoo are: (i) safe ingredients (low toxicity, low sensitization and low eye irritation); (ii) low substantivity of the surfactants; (iii) absence of ingredients that can damage the hair.

The main interactions of the surfactants and conditioners in the shampoo occur in the first few µm of the hair surface. Conditioning shampoos (sometimes referred to as 2-in-1 shampoos) deposit the conditioning agent onto the hair surface. These conditioners neutralize the charge on the surface of the hair, thus decreasing hair friction and this makes the hair easier to comb. The adsorption of the ingredients in a hair shampoo (surfactants and polymers) occurs both by electrostatic and hydrophobic forces. The hair surface has a negative charge at the pH at which a shampoo is formulated. Any positively charged species such as a cationic surfactant or cationic polyelectrolyte will adsorb by electrostatic interaction between the negative groups on the hair surface and the positive head group of the surfactant. The adsorption of hydrophobic materials such as silicone or mineral oils occurs by hydrophobic interaction.

Several hair conditioners are used in shampoo formulations, e.g. cationic surfactants such as stearyl benzyl dimethyl ammonium chloride, cetyl trimethyl ammonium chloride, distearyl dimethyl ammonium chloride or stearamidopropyldimethyl amine. As mentioned above, these cationic surfactants cause dissipation of static charges on the hair surface, thus allowing ease of combing by decreasing the hair friction. Sometimes, long chain alcohols such as cetyl alcohol, stearyl alcohol and cetostearyl alcohol are added, which are claimed to have a synergistic effect on hair conditioning. Thickening agents, such as hydroxyethyl cellulose or xanthan gum are added, which act as rheology modifiers for the shampoo and may also enhance deposition to the hair surface. Most shampoos also contain lipophilic oils such as dimethicone or mineral oils, which are emulsified into the aqueous surfactant solution. Several other ingredients, such as fragrants, preservatives and proteins are also incorporated in the formulation. Thus, a formula of shampoo contains several ingredients and the interaction between the various components should be considered both for the long-term physical stability of the formulation and its efficiency in cleaning and conditioning the hair.

Another haircare formulation is that used for permanent-waving, straightening and depilation. The steps in hair waving involve reduction, shaping and hardening of the hair fibres. Reduction of cystine bonds (disulphide bonds) is the primary reaction in permanent waving, straightening and depilation of human hair. The most commonly used depilatory ingredient is calcium thioglycollate that is applied at pH = 11–12. Urea is added to increase the swelling of the hair fibres. In permanent waving, this reduction is followed by molecular shifting through stressing the hair on rollers and ended by neutralization with an oxidizing agent where cysteine bonds are reformed. Recently, superior "cold waves" have replaced the "hot waves" by using thioglycollic acid at pH = 9–9.5. Glycerylmonothio-glycolate is also used in hair waving. An alternative reducing agent is sulphite, which could be applied at pH = 6 and this followed by hydrogen peroxide neutralizer.

Another process that is also applied in the cosmetic industry is hair bleaching which has the main purpose of lightening the hair. Hydrogen peroxide is used as the primary oxidizing agent and salts of persulphate are added as "accelerators". The system is applied at pH = 9–11. The alkaline hydrogen peroxide produces disintegration of the melanin granules, which are the main source of hair colour, with subsequent destruction of the chromophore. Heavy metal complexants are added to reduce the rate of decomposition of the hydrogen peroxide. It should be mentioned that during hair bleaching, an attack of the hair keratin occurs producing cystic acid.

Another important formulation in the cosmetic industry is that used for hair dyeing. Three main steps may be involved in this process: bleaching, bleaching and colouring combined as well as dyeing with artificial colours. Hair dyes can be classified into several categories: permanent or oxidative dyes, semipermanent dyes and temporary dyes or colour rinses. The colouring agent for hair dyes may consist of an oxidative dye, an ionic dye, a metallic dye or a reactive dye. The permanent or oxidative dyes are the most commercially important systems and they consist of dye precursors such as p-phenylenediamine which is oxidized by hydrogen peroxide to a diimminium ion. The active intermediate condenses in the hair fibre with an electron-rich dye coupler such as resorcinol and with possibly electron-rich side chain groups of the hair, forming di-, tri- or polynuclear product that is oxidized into an indo dye.

Semipermanent dyes refer to formulations that dye the hair without the use of hydrogen peroxide to a colour that only persists for 4–6 shampooings. The objective of temporary hair dyes or colour rinses is to provide colour that is removed after the first shampooing process.

14.9 Sunscreens

As discussed in Chapter 12, the damaging effect of sunlight (in particular ultraviolet light) has been recognized for several decades and this has led to a significant demand for improved photoprotection by topical application of suncreening agents. Three main wavelength of ultraviolet (UV) radiation may be distinguished, referred to as UV-A (wavelength 320–400, sometimes subdivided into UV-A1 (340–360) and UV-A2

(320–340), UV-B (covering the wavelength 290–320) and UV-C (covering the wavelength range 200–290). UV-C is of little practical importance since it is absorbed by the ozone layer of the stratosphere. UV-B is energy rich and it produces intense short-range and long-range pathophysiological damage to the skin (sunburn). About 70 % is reflected by the horny layer (stratum corneum), 20 % penetrates into the deeper layers of the epidermis and 10 % reaches the dermis. UV-A is of lower energy, but its photobiological effects are cumulative causing long-term effects. UV-A penetrates deeply into the dermis and beyond, i.e., 20-30 % reaches the dermis. As it has a photoaugmenting effect on UV-B, it contributes about 8 % to UV-B erythema.

Several studies have shown that sunscreens are able not only to protect against UV-induced erythema in human and animal skin but also to inhibit photocarcinogenesis in animal skin. The increasing harmful effect of UV-A on UV-B has led to a quest for sunscreens that absorb in the UV-A range with the aim of reducing the direct dermal effects of UV-A which causes skin ageing and several other photosensitivity reactions. Sunscreens are given a sun protection factor (SPF) which is a measure of the ability of a sunscreen to protect against sunburn within the UV-B wavelength (290–320). The formulation of sunscreen with high SPF (> 50) has been the object of many in the cosmetic industry.

An ideal sunscreen formulation should protect against both UV-B and UV-A. Repeated exposure to UV-B accelerates skin ageing and can lead to skin cancer. UV-B can cause thickening of the horny layer (producing "thick" skin). UV-B can also cause damage to DNA and RNA. Individuals with fair skin cannot develop a protective tan and they must protect themselves from UV-B.

UV-A can cause also several effects: (i) Large amounts of UV-A radiation penetrate deep into the skin and reach the dermis causing damage to blood vessels, collagen and elastic fibres. (ii) Prolonged exposure to UV-A can cause skin inflammation and erythema. (iii) UV-A contributes to photoageing and skin cancer. It augments the biological effect of UV-B. (iv) UV-A can cause phytotoxicity and photoallergy and it may cause immediate pigment darkening (immediate tanning) which may be undesirable for some ethnic populations.

From the above discussion, it is clear that formulation of effective sunscreen agents is necessary with the following requirements: (i) Maximum absorption in the UV-B and/or UV-A. (ii) High effectiveness at low dosage. (iii) Non-volatile agents with chemical and physical stability. (iv) Compatibility with other ingredients in the formulation. (v) Sufficiently soluble or dispersible in cosmetic oils, emollients or in the water phase. (vi) Absence of any dermato-toxicological effects with minimum skin penetration. (vii) Resistant to removal by perspiration.

Sunscreen agents may be classified into organic light filters of synthetic or natural origin and barrier substances or physical sunscreen agents. Examples of UV-B filters are cinnamates, benzophenones, p-aminobenzoic acid, salicylates, camphor derivatives and phenyl benzimidazosulphonates. Examples of UV-A filters are dibenzoyl methanes, anthranilates and camphor derivatives. Several natural sunscreen agents are available, e.g. camomile or aloe extracts, caffeic acid, unsaturated vegetable or animal oils. However, these natural sunscreen agents are less effective and they are seldom used in practice.

The barrier substances or physical sunscreens are essentially micronized insoluble organic molecules such as guanine or micronized inorganic pigments such as titanium dioxide and zinc oxide. Micropigments act by reflection, diffraction and/or absorption of UV radiation. Maximum reflection occurs when the particle size of the pigment is about half the wavelength of the radiation. Thus, for maximum reflection of UV radiation, the particle radius should be in the region of 140 to 200 nm. Uncoated materials such as titanium and zinc oxide can catalyse the photo-decomposition of cosmetic ingredients such as sunscreens, vitamins, antioxidants and fragrances. These problems can be overcome by special coating or surface treatment of the oxide particles, e.g. using aluminium stearate, lecithins, fatty acids, silicones and other inorganic pigments. Most of these pigments are supplied as dispersions ready to mix in the cosmetic formulation. However, one must avoid any flocculation of the pigment particles or interaction with other ingredients in the formulation which causes severe reduction in their sunscreening effect.

A topical sunscreen product is formulated by the incorporation of one or more sunscreen agent (referred to as UV filters) in an appropriate vehicle, mostly an O/W or W/O emulsion. Several other formulations are also produced, e.g. gels, sticks, mousse (foam), spray formulation or an anhydrous ointment. In addition to the usual requirements for a cosmetic formulation, e.g. ease of application, pleasant aspect, colour or touch, sunscreen formulations should also have the following characteristics: (i) Effective in thin films, strongly absorbing both in UV-B and UV-A. (ii) Non-penetrating and easily spreading on application. (iii) Should possess a moisturizing action and be waterproof and sweat resistant. (iv) Free from any phototoxic and allergic effect. The majority of sunscreens on the market are creams or lotions (milks) and progress has been achieved in recent years to provide high SPF at low levels of sunscreen agents.

14.10 Make-up products

Make-up products include many systems such as lipstick, lip colour, foundations, nail polish, mascara, etc. All these products contain a colouring agent which could be a soluble dye or a pigment (organic or inorganic). Examples of organic pigments are red, yellow, orange and blue lakes. The inorganic pigments comprise titanium dioxide, mica, zinc oxide, talc, iron oxide (red, yellow and black), ultramarines, chromium oxide, etc. Most pigments are modified by surface treatment using amino acids, chitin, lecithin, metal soaps, natural wax, polyacrylates, polyethylene, silicones, etc.

The colour cosmetics comprise foundation, blushers, mascara, eyeliner, eyeshadow, lip colour and nail enamel. Their main function is to improve appearance, impart colour, even out skin tones, hide imperfections and produce some protection.

Several types of formulations are produced ranging from aqueous and nonaqueous suspensions to oil-in-water and water-in-oil emulsions and powders (pressed or loose).

Make-up products have to satisfy a number of criteria for acceptance by the consumer: (i) Improved, wetting spreading and adhesion of the colour components. (ii) Excellent skin feel. (iii) Skin and UV protection and absence of any skin irritation.

For these purposes, the formulation has to be optimized to achieve the desirable property. This is achieved by using surfactants and polymers as well as using modified pigments (by surface treatment). The particle size and shape of the pigments should also be optimized for proper skin feel and adhesion.

Pressed powders require special attention to achieve good skin feel and adhesion. The fillers and pigments have to be surface treated to achieve these objectives. Binders and compression aids are also added to obtain a suitable pressed powder. These binders can be dry powders, liquids or waxes. Other ingredients that may be added are sunscreens and preservatives. These pressed powders are applied in a simple way by simple "pick-up", deposition and even coverage. The appearance of the pressed powder film is very important and great care should be taken to achieve uniformity in an application. A typical pressed powder may contain 40–80 % fillers, 10–40 specialized fillers, 0–5 % binders, 5–10 % colourants, 0–10 % pearls and 3–8 % wet binders.

An alternative to pressed powders, liquid foundations, have attracted special attention in recent years. Most foundation make-ups are made of O/W or W/O emulsions in which the pigments are dispersed either in the aqueous or the oil phase. These are complex systems consisting of a suspension/emulsion (suspoemulsion) formulation. Special attention should be paid to the stability of the emulsion (absence of flocculation or coalescence) and suspension (absence of flocculation). This is achieved by using specialized surfactant systems such silicone polyols, block copolymers of poly(ethylene oxide) and poly(propylene oxide). Some thickeners may be also added to control the consistency (rheology) of the formulation.

The main purpose of a foundation make-up is to provide colour in an even way, even out any skin tones and minimize the appearance of any imperfections. Humectants are also added to provide a moisturizing effect. The oil used should be chosen to be a good emollient. Wetting agents are also added to achieve good spreading and even coverage. The oil phase could be a mineral oil, an ester such as isopropyl myristate or volatile silicone oil (e.g. cyclomethicone). An emulsifier system of fatty acid/nonionic surfactant mixture may be used. The aqueous phase contains a humectant of glycerine, propylene glycol or polyethylene glycol. Wetting agents such as lecithin, low HLB surfactant or phosphate esters may also be added. A high HLB surfactant may also be included in the aqueous phase to provide better stability when combined with the oil emulsifier system. Several suspending agents (thickeners) may be used such as magnesium aluminium silicate, cellulose gum, xanthan gum, hydroxyethyl cellulose or hydrophobically modified polyethylene oxide.

A preservative such as methyl paraben is also included. The surface treated pigments are dispersed either in the oil or aqueous phase. Other additives such as fragrances, vitamins, light diffusers may also be incorporated.

It is clear from the above discussion that liquid foundations represent a challenge to the formulation chemist due to the large number of components used and the interaction between the various components. Particular attention should be paid to the interaction between the emulsion droplets and pigment particles (a phenomenon referred to as heteroflocculation) which may have adverse effects on the final property of the deposited film on the skin. Even coverage is the most desirable property and the optical property of the film, e.g. its light reflection, adsorption and scattering play important roles in the final appearance of the foundation film.

Several anhydrous liquid (or "semi-solid") foundations are also marketed by cosmetic companies. These may be described as cream powders consisting of a high content of pigment/fillers (40–50 %), a low HLB wetting agent (such as polysorbate 85), an emollient such as dimethicone combined with liquid fatty alcohols and some esters (e.g. octyl palmitate). Some waxes such as stearyl dimethicone or microcrystalline or carnauba wax are also included in the formulation.

One of the most important make-up systems are lipsticks which may be simply formulated with a pure fat base having a high gloss and excellent hiding power. However, these simple lipsticks tend to come off the skin too easily. In recent years, there was a great tendency to produce more "permanent" lipsticks which contain hydrophilic solvents such as glycols or tetrahydrofurfuryl alcohol. The raw materials for a lipstick base include: ozocerite (good oil absorbent that also prevents crystallization), microcrystalline ceresin wax (which is also a good oil absorbent), Vaseline (that forms an impermeable film), beeswax (that increases resistance to fracture), myristyl myristate (that improves transfer to the skin), cetyl and myristyl lactate (that form an emulsion with moisture on the lip and are non-sticky), carnauba wax (an oil binder that increases the melting point of the base and gives some surface lustre), lanolin derivatives, olyl alcohol and isopropyl myristate. This shows the complex nature of a lipstick base and several modifications of the base can produce some desirable effects that help good marketing of the product.

Mascara and eyeliners are also complex formulations that need to be carefully applied to the eye lashes and edges. Some of the preferred criteria for mascara are good deposition, ease of separation and lash curling. The appearance of the mascara should be as natural as possible. Lash lengthening and thickening are also desirable. The product should also remain for an adequate time and it should also be easily removable. Three types of formulations may be distinguished: anhydrous solvent based suspension, water-in-oil emulsion and oil-in-water emulsion. Water resistance can be achieved by addition of emulsion polymers, e.g. polyvinyl acetate.

References

[1] de Polo, K. F., "A Short Textbook of Cosmetology", Verlag für Chemische Industrie H. Ziolkowsky, Augsburg, Germany (1998).
[2] Tadros, Th. F., "Applied Surfactants", Wiley-VCH, Germany (2005).
[3] Tadros, Th. F., "Rheology of Dispersions", Wiley-VCH, Germany (2010).

Index

absorbance 232
aerosols 7, 271, 272
adhesion tension 251
adsorbed layer thickness 58, 64, 242
adsorption isotherm 57, 60, 61, 62, 63, 238, 243, 262
after-shave preparations 275
agglomerate 253
alcohol ethoxylate 17
alkyl polyglucoside 20
alkyl sulphate 202
amine ethoxylate 19
amphoteric surfactants 16, 202
anionic surfactants 11, 202
antiperspirants 6
associative thickeners 211

bar soaps 273
bath oils 275
benzalkonium chloride 15
block copolymer 52, 55, 183, 199, 234
bridging flocculation 100
bubble bath 275

carboxylates 12
cationic surfactants 14
cleansing agent 204
cloud point 17
cohesive energy ratio (CER)
colour cosmetics 249
– application of fundamental principles in 265
– competitive interaction in 269
– principles of preparation of 250, 267
constant stress (creep) measurements 143, 187
contact angle 220, 221, 251
cosmetic emulsions 105
– manufacture of 124
– reological properties of 135
counter-ion 111
creaming 108
critical flocculation concentration 91
critical micelle concentration (cmc) 17, 24, 25, 29
critical packing parameter (CPP) 27, 76, 122, 197
cubic phases 29, 74, 78

depletion flocculation 98
disjoining pressure 69, 209
dispersants 262
dispersing agents 260
dispersion structure 255, 268
dispersion process 250
double layer extension 86
double layer interaction 87, 256
double layer structure 87
droplet size distribution 125

elastic interaction 95
electrical double layer 86
electrokinetics 223
electrostatic repulsion 85
elongational flow 135, 158
emulsification 125
– mechanism of 129
– methods of 130
emulsion breakdown processes 106, 107
emulsion formation 105
energy of interaction 95
ether sulphates 202
ethoxylated surfactants 17
eye-liner 284

fatty acid ethoxylates 18
flocculation 109
– free energy of 236
– kinetics of 89
– of sterically stabilised dispersions 97
foam bath 205, 208, 275
foundation 7

gelling agent 183
Gibbs-Marangoni effect 130, 154
graft copolymer 52, 53, 234

hair 217, 278
– surface properties of 220
hair bleaching 280
hair care formulation 277
hair conditioner 217, 279
– role of surfactants and polymers in 224

Hamaker constant 85
hand creams 3
hand soaps 274
hexagonal phase 29, 73, 74
HLB group numbers 114
hydrophilic-lipophilic-balance (HLB) 111
– determination of 116
homopolymer 54
hydrosomes 81

imidazolines 15
incipient flocculation 98
interaction forces 83
interfacial dilational elasticity 127, 129, 152
interfacial tension 126
interfacial tension gradient 128
isethionates 14
isoelectric point 16, 256

lamellar micelle 28
lamellar phase 29, 75
liposomes 3, 193
– manufacture of 194
– stability of 199
liquid crystalline phases 3, 29, 75
– driving force for 76
– formulation of 80
– investigation of 79
– texture of 79
liquid foundation 283
liquid soap 274
lotions 3

make-up products 282
mascara 284
micelles 3, 26
– relaxation time of 30
– thread-like 210, 274
micellisation 23
– driving force of 34
– energy of 33
– enthalpy of 33
– equilibrium aspects of 31
– kinetic aspects of 30
– in surfactant mixtures 35, 38
milling 254
mixing interaction 93
moisturiser 276

multiple emulsions
– breakdown processes of 180
– characterisation of 186
– drop 185
– formulation of 179
– preparation of 181
stability of 190
types of 180
multi-phase systems 3

nail polish 6
nano-emulsions
– based on polymeric surfactants
– formulation of 147
– low energy methods for preparation of 158
– practical examples of 163
– preparation by dilution of microemulsion 162
– preparation of 149
naturally occurring surfactants 21
non-aqueous dispersions 237
non-ionic surfactants 17, 203

oleosomes 81
oscillatory measurements 141, 188
Ostwald ripening 110, 148
– kinetics of 149
– reduction of 149

phase inversion 111
phase inversion composition (PIC) 159
phase inversion temperature (PIT) 117, 160
phosphates 14
photon correlation spectroscopy (PCS) 59
pigment particles 250
poly-quaternary surfactants 10, 225
polymer adsorption 57
– kinetics of 66
polymeric surfactants 21, 72
– adsorption and conformation of 53, 55, 56
– emulsion stabilisation by 67, 68
– in cosmetic formulations 51
polymer-surfactant interaction
powder dispersion 254
powder wetting 250
power density 134, 152, 158
preservative 206
pressed powder 283
– rate of penetration of liquid in 253

radial discharge mixer 132
Reynolds number 131, 151, 156
rheological measurements 263
rheology modifiers 209, 211
Rideal-Washburn equation 253
rod micelles 28
rotor-stator mixer 131

scattering 232
self assembly structure 73, 74
self emulsification 158
shampoos 6, 201, 279
– composition of 204
– hair conditioner in 217
– preparation of 203
– rheology modifier in 211
– role of components in 206
– silicone emulsion in 211
– surfactants for use in 201
shaving cream 273
shaving foam 272
silicone surfactants 22
skin care products 276
skin cross section 2
silicone oil emulsions 211
solubility parameter 235
sorbitan esters 18
spans 18
spherical micelle 28
stability ratio 90
steady state measurements 140, 186
steric repulsion 92, 234
steric stabilisation 256
– criteria of 96
stratum corneum 2
sulphates 12
sun protection factor (SPF) 232
sunscreens 231, 288
– Competitive interactions in 245

– for UV protection 232
surface forces 153
surfactant classes 11
surfactants in cosmetics 11
surfactant-polymer interaction 39, 46
– driving force for 44
– mode of association of 43
– structure of 45
surfactant solution 23

taurates 14
thickening agents 205
thickeners 209, 210
TiO2 232
Turbulent flow 136, 151, 157
Tweens 19

Ultrasonic waves 156
UV-vis attenuation 239, 241, 266

Van der Waals attraction 83, 234
vesicles 193
– driving force for formation of 195
viscous stress 134, 152

Weber number 135, 152
weak flocculation 91, 97
wettability 220
wetting line 221, 251
work of adhesion 222, 251
work of cohesion 252
work of dispersion 253

Young's equation 251

Zeta potential 256
ZnO 232
Zwitterionic surfactants 16